Quantitative methods are bringing a number of changes in the contemporary pattern of university geography. Significant among these is the rediscovery that physical and human geography—often treated as irreconcilable components—have much in common in their basic models, their analytical techniques, and their philosophic methodology.

This volume explores some of the common analytical elements in the study of geographic networks. Part One, *Spatial Structures*, discusses ways of describing networks in both topologic and geometric terms and introduces the role of graph theory in geographic analysis. Part Two, *Evaluation of Structures*, sets the spatial structure of networks against their primal and dual roles—as conductors of, and barriers to, flows. Part Three, *Structural Change*, traces the distinctive regional patterns of network evolution and attempts to build projective growth models through simulation and allied techniques. Throughout the emphasis is on relevant regional case studies drawn from the familiar phenomena of physical geography and human geography.

The text is accompanied by over 200 figures (many of them specially drawn for this volume), and by 38 summary tables. It also contains a detailed cross-referencing system, a research bibliography of over 400 items, and guides for further reading on central topics.

Network analysis is designed as supplementary reading to *Locational analysis in human geography*. It forms the first of a new trilogy of volumes exploring some spatial structures of direct interest to geographers. A second volume on the analysis of regional trends is in preparation.

WITHDRAWN

Network Analysis in Geography

Explorations in Spatial Structure, I

Network Analysis in Geography

PETER HAGGETT

Professor of Urban and Regional Geography in the
University of Bristol

RICHARD J. CHORLEY

Lecturer in Geography in the University of Cambridge;
Fellow of Sidney Sussex College, Cambridge

EDWARD ARNOLD

FOR SARAH AND RICHARD

GF
23
H3

© Peter Haggett & Richard J. Chorley 1969
First published 1969 by
Edward Arnold (Publishers) Ltd
41 Maddox Street, London W.1

SBN : 7131 5459 4

Printed in Great Britain by
Butler & Tanner Ltd, Frome and London

Preface

If books are traceable through misty taxonomic trees to distant intellectual forebears, then this book might claim ancestry in Robert Horton's remarkable paper on the erosional development of streams. This 1945 paper by an American civil engineer laid the basis not only of much analytical work in fluvial geomorphology but more general underpinnings for a process approach to quantitative morphology. Its findings were taught to one of us by Vaughan Lewis at Cambridge and to the other by Arthur Strahler at Columbia; for both of us it represented the start of two trails of work, the one moving away from geomorphology into locational analysis and the other towards the study of drainage basins as the fundamental geomorphic unit.

It was the crossing of these paths at Cambridge in the early sixties that led us to check some of the on-going assumptions on the dichotomy of physical and human geography and to draw together the strands of this book. In it we have tried to explore ways in which the analysis of a topologically distinct class of spatial structures—linear networks—might throw light on common geographic problems of morphometry, origin, growth, balance and design. It builds on an earlier volume by one of the authors (*Locational analysis in human geography*) and in places incorporates relevant sections, but it widens the area of spatial analysis to include physical networks. We are too conscious of the dangers of easy analogy and strained metaphor to claim that, for example, stream systems and transport systems are geographically 'the same'; to do so would force us to ignore aspects of network structure and evolution that are intrinsically important to physical and human geographers respectively. Rather we argue that there are equivalent spatial structures common to both fields and that we can use common mathematical models for both kinds. Hence we lay no claims to completeness or rigour: we have drawn on models from our own areas of experience and the mathematical level is elementary. We hope however that some of the problems raised will attract mathematicians, for their solutions will surely demand an elegance and power beyond our command. For geographers we hope that some familiar material will be cast in a new light and that we may start some discussions on alternative ways of viewing our field.

The length of the acknowledgments section of this book suggests something of the debt we owe to others. A number of workers, notably Christian Werner, Michael Woldenberg and Ove Pedersen, allowed us access to unpublished papers. Northwestern University provided a superb transportation library and stimulating colleagues for writing part of the book. Other parts of the book were tried with varying success on colleagues and students at Cambridge and Bristol. The assistant staff at both departments helped in the preparation of the book and we are particularly grateful to Margaret Reynolds, Mary Norcliffe, Simon Godden, Tony Philpott and Robert Bignall for help at various crucial stages. Our families, by now hardened to the litter of galleys around the house, assisted and retarded the project in characteristic fashion. We dedicate this book to our two eldest children in the faint hope that they may sometimes encourage the others in the former activity.

PETER HAGGETT

RICHARD CHORLEY

Chew Magna, Somerset. *Summer, 1969*

Contents*

* A detailed contents list is given at the beginning of each of the five chapters.

Acknowledgments

The Authors and Publisher wish to express their thanks to the following for permission to reprint or modify material from copyright works:

The Association of American Geographers for the following from *Annals of the Association of American Geographers*: Fig. 17, p. 531, Vol. 53, 1963 (Paper by C. B. Beaty), Fig. 2, p. 91, Vol. 57, 1967 (Paper by P. J. Rimmer), Fig. 1, p. 362, Vol. 51, 1961 (Paper by J. E. Vance), and Fig. 4, p. 322, Vol. 48, 1958 (Paper by W. H. Wallace); the American Association of Petroleum Geologists for Figs. 1–3, p. 2248, *Bulletin*, Vol. 51, 1967 (Paper by A. D. Howard); the American Geophysical Union and the author for Figs. 5 and 6, p. 917, *Transactions*, Vol. 38, 1957 (Paper by A. N. Strahler); the American Society of Civil Engineers for Fig. 4, p. 1434, *Transactions*, Vol. 110, 1945 (Paper by C. O. Clark); the Editors and Yale University Press for the following from *American Journal of Science*: Figs. 5 and 6, pp. 334–5, Vol. 262, 1964 (Paper by J. K. Lubowe), Fig. 8, p. 171, Vol. 255, 1957 (Paper by S. A. Schumm and R. F. Hadley), and Figs. 4, 6, 7 and 8, pp. 874, 876–7, 879, Vol. 263, 1965 (Paper by C. W. Carlston); the Editor for Fig. 2, p. 87, *Australian Geographical Studies*, Vol. 6, 1968 (Paper by E. D. Ongley); Lawrence A. Brown for Figs. 4 and 5, pp. 44–5, 'Models of spatial diffusion research—a review', ONR report, 1965; Ian Burton for Table 12, p. 22, 'Accessibility in northern Ontario', Ontario Department of Highways report, Dec. 1962; the Syndics of the Cambridge University Press for the following: Fig. 8.31, p. 245, *Some lessons in mathematics* by T. J. Fletcher, 1964, and Fig. 106, p. 516, *On growth and form* by W. D'Arcy Thompson, 1942; the author for Figs., pp. 38–9, *Canadian Geographer*, Vol. 12, 1968 (Paper by C. Werner); the Canadian Operational Research Society for Figs. 4 and 5, p. 81, *Journal*, Vol. 2, Dec. 1964 (Paper by J. C. Clapham); E. Casetti for Maps 5 and 13, pp. 71 and 79, 'Classificatory and regional analysis', ONR report, 1964; the Editor and the authors for the following from *University of Chicago, Department of Geography, Research Papers*: Figs. 92–3, pp. 240–1, No. 111, 1966 (Paper by B. J. L. Berry), and Figs. 16 and 17, pp. 32 and 42, No. 84, 1963 (Paper by K. J. Kansky); David Cope for Figs. 1–3, *A network analysis of the growth of the London underground system and its relation to population changes*, unpublished Cambridge University, B. A. Dissertation, 1967; M. F. Dacey for Table 2, p. 18, 'Certain properties of edges on a polygon', MS paper, 1963; the authors and Walker Art Centre, Minneapolis, for the following from *Design Quarterly*: Fig. (upper left), p. 67, 1966–7 (Paper by A. M. Noll), and Fig., p. 38, 1966–7 (Paper by M. L. Manheim); the Elsevier Publishing Company, Amsterdam for Fig. 4, p. 70, *Theory of traffic flow*, edited by R. Herman, 1961 (Paper by J. C. Wardrop); W. H. Freeman & Company, San Francisco and London, for Figs. 5–4, 7–39, 7–43 and 10–10, pp. 140, 293, 303 and 426, *Fluvial processes in geomorphology* by L. B. Leopold, J. P. Miller and M. G. Wolman, 1964; W. L. Garrison and M. Marble for Figs. 19 and 20, p. 70, U.S. Army Transport Command, *Technical Report* 62–11, May 1962; the American Geographical Society for the following from *Geographical Review*: Fig. 1, p. 332, Vol. 47, 1957 (Paper by H. B. Johnson), Fig. 1, p. 504, Vol. 53, 1963 (Paper by E. J. Taaffe), and Fig. 2, p. 557, Vol. 58, 1968 (Paper by M. J. Woldenberg); the Editor, the Authors and the University of Chicago Press for the following from *Journal of Geology*: Fig. 2, p. 330, Vol. 63, 1955 (Paper by S. Judson and G. W. Andrews), Figs. 1, 12 and 13, pp. 454 and 456, Vol. 66, 1958 (Paper by M. A. Melton), Fig. 1, p. 345, Vol. 67, 1959 (Paper by M. A. Melton), and Fig. 8, p. 28, Vol. 74,

1966 (Paper by R. L. Shreve); the Geological Society of America for the following from the *Bulletin of the Geological Society of America*: Fig. 4, p. 771, Vol. 75, 1964 (Paper by K. L. Bowden and J. R. Wallis), Figs. 1 and 25, pp. 282–3, Vol. 56, 1945 (Paper by R. E. Horton), Figs. 19, 20 and 24, pp. 615 and 617, Vol. 67, 1956 (Paper by S. A. Schumm), Fig. 1, p. 1091, Vol. 74, 1963 (Paper by S. A. Schumm), and Fig. 7, p. 297, Vol. 69, 1958 (Paper by A. N. Strahler); the Editor for Fig. 1, p. 5740, *Journal of Geophysical Research*, Vol. 68, 1963 (Paper by H. Schenck); C. W. K. Gleerup, Publishers, Lund, Sweden, and the authors for the following: Fig. 7.10, p. 183, *Theoretical Geography, Lund Studies in Geography, Series C, General and Mathematical Geography*, 1, by W. Bunge, 1962, and for Fig. 10, p. 130, *Migration and the growth of urban settlement, Lund Studies in Geography, Series B, Human Geography*, 26, by R. L. Morrill, 1965; Peter R. Gould for Fig. 10, p. 23, *Space searching procedures*, Report, 1966; Harvard University Press for Fig. 6, p. 26, *Crude oil pipe lines* by L. Cookenboo, 1955; Budd Hebert for Fig. 10, p. 23, 'Use of factor analysis in graph theory to identify an underlying structure of transportation networks', MS report, 1966; Heinemann Educational, Ltd. for Figs. 22, 25B and 46, pp. 314, 328 and 360, *Essays in Geomorphology* edited by G. H. Dury, 1966; the Highway Research Board, Washington D.C., for the following from *Highway Research Board Record*: Figs., pp. 78 and 81, Vol. 64, 1965 (Paper by H. S. Levinson), and Figs., pp. 19–24, Vol. 6, 1963 (Paper by R. B. Smock); Holt, Rinehart and Winston, Inc., New York, for Map 6, p. 545, *Statistics for sociologists* by M. J. Hagood and D. O. Price, 1952; Alan D. Howard for Figs. 12–15, 'Stream capture, bifurcation angle modification, and rate of work in stream angles', report, 1968; the Hutchinson Educational, Ltd. for Fig. 36, p. 192, *United States and Canada* by W. R. Mead, 1962; the Editor for Figs. 2, 3 and 6, pp. 148, 152 and 159, *Journal of Hydrology*, Vol. 2, 1964 (Paper by E. M. Laurenson); the Editor and the Royal Geographical Society for Fig. 14, p. 73, *Transactions of the Institute of British Geographers*, Vol. 42, 1967 (Paper by Peter R. Gould); the Council of Institution of Civil Engineers for Fig. 1, p. 252, *Proceedings*, Vol. 17, 1960 (Paper by J. E. Nash); the American Institute of Electrical and Electronic Engineers for Fig. 5, p. 931, *Proceedings*, Vol. 44, 1956 (Paper by Z. Prihar); the International Association of Scientific Hydrology and the authors for the following: Fig. 1, p. 65, *Bulletin*, Vol. XI(3), 1966 (Paper by C. W. Carlston), Fig. 1, p. 59, *Bulletin*, Vol. XI(3), 1966 (Paper by A. E. Scheidegger), Fig. 3, p. 9, *Bulletin*, Vol. X(3), 1965 (Paper by A. P. Schick), and Fig. 3, p. 90, *General Assembly of Berne*, 1967 (Paper by J. S. Smart, A. S. Surkan and J. P. Considine); the Editor and the Department of Geography, University of Iowa for Fig. 1, *Discussion Papers*, 5, 1967 (Paper by W. R. Black); Macmillan & Co., Ltd., St. Martin's Press, Inc. and the Macmillan Company of Canada for Fig. IXo, p. 384, *Population growth and land use* by C. Clark, 1967; Collier-Macmillan, Ltd. for Fig. 3.4, p. 69, *Decision-making processes in pattern recognition* by G. S. Sebestyen, 1962; the McGraw Hill Book Company, New York, for the following: Figs. 4–II–1, 4–II–4, 4–II–16, 14–12, *Handbook of applied hydrology* edited by V. J. Chow, 1964, Fig. 2.6, p. 31, *Communication nets* by L. Kleinrock, 1964, Figs. 15–3 and 15–14, pp. 392 and 403, *Applied hydrology* by R. K. Linsley, M. A. Kohler and L. H. Paulhus, 1949, and Fig. 5.15, p. 134, *Traffic systems analysis* by M. Wohl and B. V. Martin, 1967; the Editor and the Elsevier Publishing Company for Figs., pp. 371 and 381, *Marine Geology*, Vol. 3, 1965 (Paper by D. R. Stoddart); the Massachusetts Institute of Technology, Department of Civil Engineering for the following from *Systems Laboratory Research Reports*: Fig. 15, p. 25, Vol. R.62–40, 1964 (Report by P. O. Roberts and M. L. Funk), and Figs. 1 and 2, p. 21, Vol. R.64–15, 1964 (Report by M. L. Manheim); the Massachusetts Institute of Technology Press for Fig. 4, p. 272, *Location and space economy* by W. Isard, 1956; Methuen & Co., Ltd. for Figs. 5.6, 5.10 and 15.4, pp. 161, 166 and 616, *Models in geography* edited by R. J. Chorley and Peter Haggett, 1967, and Fig. 11.II.4, p. 531, *Water, earth and man* edited by Richard J. Chorley, 1969; the Michigan Inter-University Community of Mathematical Geographers for Fig. 5, p. 17, *Discussion Papers*, 6, 1965 (Paper by W. Warntz); the Editor, *Midwest Journal of Political Science* for Fig. 1, p. 210, Vol. 10, 1966 (Paper by H. F. Kaiser); Gordon Mills for Fig. 1, p. 2, *Journal of the Canadian Operational Research Society*, Vol. 6, 1968; the Editor and Northwestern University, Department of Geography for the following from *Studies in Geography*: Figs. 1, 6 and 7, pp. 98, 107–16, Vol. 13, 1967 (Paper by M. J. Beckmann), and Fig. 5, p. 285, Vol. 13, 1967 (Paper by M. F. Dacey); Peat, Marwich, Caywood, Schiller & Co. and the Editor for the following from *Operations Research*: Fig. 1, p. 219, Vol. 2, 1954 (Paper by G. Dantzig, O. R. Fulkerson and S. Johnson), Fig. 1, p. 1000, Vol. 13, 1965 (Paper by S. W. Hess *et al.*), Fig. 7, p. 987, Vol. 11, 1963 (Paper by J. D. C. Little *et al.*), and Fig. 9, p. 241, Vol. 6, 1958 (Paper by W. Miehle); the author and VEB Hermann Haach, Gotha, for Fig., p. 250, *Petermanns Mitteilungen*, Vol, 111, 1967 (Paper by F. Topper); P. O. Pedersen for Fig. 4, p. 8, 'On the geometry of administrative areas', MS, 1967; Prentice Hall Inc., Englewood Cliffs, New Jersey, for Fig. 3, p. 73, *Die zentralen orte in Süddeutschland* by W. Christaller, 1933, translated by C. W. Baskin, 1966; Princeton University Press for Fig., pp. 123–7, *Flows in networks* by C. R. Ford Jr. and D. R. Fulkerson, copyright by the RAND Corporation, 1962; the Association of American Geographers for the following from *Professional Geographer*: Figs. 1–3, pp. 8–9, Vol. 15(6), 1963 (Paper by M. Yeates), and Fig. 2, p. 17, Vol. 17(5), 1965 (Paper by F. R. Pitts); the Instytut Geografii, Warsaw for Fig., p. 272, *Przeglad Geograficzny*, Vol. 33, 1961 (Paper by B. J. L. Berry); the Regional Science Association, University of Pennsylvania for Fig. 4, p. 194, *Papers*, 14, 1965 (Paper by R. Lachene); Pergamon Press, Ltd and the author for Fig. 9, p. 101, *Regional Studies*, Vol. 2, 1968 (Paper by N. A. Spence); the Controller, H.M.S.O., the Road Research Laboratory, Crowthorne for the following: Fig. 8, p. 17, *Technical Paper*, 46, 1960 (Paper by T. M. Coburn), Fig. 3, p. 9, *Laboratory Report*, 1968 (Report by D. Owens), and Fig. 3, p. 14, *Technical Paper*, 68, 1967 (Paper by J. C. Tanner) (Crown Copyright Reserved); the author and the American Association for the Advancement of Science for the following from *Science*: Fig. 1, p. 295, Vol. 154, 1966 (Paper by A. Michael Noll, Bell Telephone Laboratories), Fig. 3, p. 276, Vol. 157, 1967 (Paper by D. C. Gazis), and Fig. 1, p. 636, Vol. 156, 1967 (Paper by B. Mandelbrot); Scientific American Inc. for Maps, p. 23, *Scientific American*, Vol. 213(5), 1965 (Paper by R. C. Silva); Pion Ltd and the author for Fig. 2, *Studies in Regional Science* (Paper by A. Scott); Stanford University, Department of Civil Engineering for Fig. 3.10, p. 25, *Technical Report*, 39, 1966 (Report by N. H. Crawford and R. K. Linsley); J. A. Silk for Figs. 8–12, 'Road networks of Monmouthshire', MS report, 1965; the Editor for Fig., p. 333, *Tijdschrift van het Koninklijk Nederlands Aardrijkskundig Genootschap*, Vol. 15, 1965 (Paper by D. Steiner); the Editor and Van Waesberg, Hoogewerff & Richards, N.V. for the following from *Tijdschrift voor Economische en Sociale Geografie*: Fig. 1, p. 232, Vol. 56, 1965 (Paper by K. R. Cox), and Figs. 3–5, pp. 204–5, Vol. 57, 1966 (Paper by Peter R. Gould and T. R. Leinbach); the Editor for Fig. 2, p. 394, *Traffic Engineering and Control*, Vol. 9(8), 1967 (Paper by J. A. Timbers); the Editor for Figs., pp. 5 & 23, *Journal of Transport Economics and Policy*, Vol. 2, 1968 (Paper by R. J. Smeed); the author and the Network Theory Unit for Fig. 1, p. 17, *Transport Networks Theory Unit Report*, Vol. 22, 1965 (Paper by J. D. Murchand); the Editor for Fig. 7, p. 7, *Journal of Tropical Geography*, Vol. 22, 1966 (Paper by R. J. Eyles); the University of Washington Press for Fig., p. 266, *Geographic impact of highway change* by W. L. Garrison *et al.*, 1959 (Paper by R. L. Morrill); the American Geographical Union for Fig. 1, p. 774, *Water Resources Research*, Vol. 3, 1967 (Paper by J. S. Smart); A. Werrity for Figs. 3 and 4, 'Expansion of the railway network in north-east London', MS report, 1967; John Wiley & Sons Inc. for the following: Fig. 6.16, p. 201, *Combinatorial mathematics* by E. F. Beckenbach, 1964, Fig. 2–14 and 2–43, pp. 43 and 72, *Geohydrology* by R. J. M. De Weist, 1965, Fig. 17.4, p. 422, *Spatial analysis in the geological sciences* by R. L. Miller and J. S. Kahn, 1962, and Fig. 19.10, p. 274, *Introduction to Physical Geography* by A. N. Strahler, 1965; M. J. Woldenburg for Figs. I–1 and III–2, *Hierarchical structures*, PhD dissertation, Columbia University, 1968; Yale University Press for Fig. 23, p. 110, *Economics of location* by A. Lösch, 1954; and the Yale Law Journal Company, Fred B. Rothman Company and the authors for Fig. 1, p. 306, *Yale Law Journal*, Vol. 73, 1963 (Paper by J. B. Weaver and S. W. Hess).

The Authors and Publisher further acknowledge the modification of material from United States Government non-copyright publications:

The following from the United States Geological Survey *Professional Papers*: Fig. 21, p. 26, No. 422–D, 1964 (Paper by J. C. Brice); Figs. 102A and 104, pp. 162 and 164, No. 282–F, 1961 (Paper by L. M. Brush); Fig. 3, p. 6, No. 422–C, 1963 (Paper by C. W. Carlston); Fig. 2, p. 3, No. 452–A, 1964 (Paper by G. H. Dury); Fig. 3 and 5, p. 152 and 156, No. 422–G, 1965 (Paper by E. V. Giusti and W. J. Schneider); Figs. 6 and 10, pp. 12 and 17, No. 504–B, 1965 (Paper by J. T. Hack); the following from papers by W. B. Langbein and L. B. Leopold: Figs. 1, 2, 3B, 5 and 6, pp. 3, 6, 9 and 11, No. 422–H, 1966, Figs. 1, 3 and 15, pp. 3, 5 and 17, No. 422–L, 1968, and Figs. 6, 8 and 9, pp. 16, 18 and 19, No. 500–A, 1962; Fig. 4B, p. 7, No. 252, 1963 (Paper by L. B. Leopold and T. Maddock); Figs. 20 and 21, pp. 23–4, No. 282–A, 1956 (Paper by L. B. Leopold and J. P. Miller); Fig. 8, p. 21, No. 352–B, 1960 (Paper by S. A. Schumm); Fig. 16, p. 16, No. 271, 1955 (Paper by M. G. Wolman).

Part One: Spatial Structures

Physical phenomena are spread out in the continuous medium of space and time: in the last two decades, however, discontinuous and combinatorial structures have become of increasing significance. (HERMANN WEYL, Philosophy of mathematics and natural science, 1949, *Appendix D.*)

Chapter one   Topologic Structures

Geographers regularly encounter problems involving flows in their research programmes. Consider the physical geographer studying streaming of water within a drainage basin or the human geographer studying interregional streams of migrants. Both examples represent distinct functional systems but have the fundamental property in common that they involve the flows of some commodity through a channel or a network of channels. The problems in organizing flows into efficient channel patterns and the interpretations of the distinctive channel networks that have emerged—often as major features on the earth's surface—form the major theme of this volume.

The approach followed here is to begin in Part One with the networks as they exist and to analyse their spatial structure in terms of topologic (Chap. 1) and geometric (Chap. 2) components. Having established the main structural characteristics of channel networks the focus is shifted in Part Two to their evaluation. The relation of network structure to flow demands (Chap. 3) and the problem of optimally locating networks (Chap. 4) are the main concern in this section. Part Three analyses the growth and transformation of networks over time (Chap. 5). Throughout, the emphasis is on the general properties and performance of networks and the spatial problems they pose, but these are illustrated with reference to a number of commonly-occurring network systems in physical and human geography.

I. NETWORKS AS GRAPHS

The topologic structure of a network involves the reduction of the channel pattern to its most basic and elemental form. Whether we conceive a network in terms of a standard dictionary definition of 'a meshed fabric of

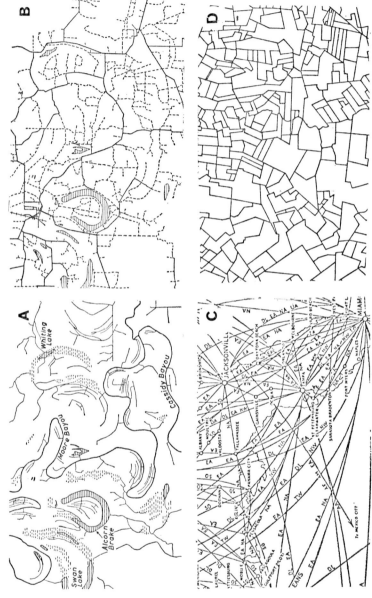

Fig. 1.1. Sample network systems. (*A*) Channel patterns of the Mississippi near Clarkesdale, Miss., U.S.A.; branching networks. (*B*) Road communications system for same area as *A*; circuit networks (planar). (*C*) Airline pattern for Florida, U.S.A.; circuit networks (non-planar). (*D*) Property subdivisions, southern Ohio U.S.A.; barrier networks. Source: Mead and Brown, 1962. p. 191; Thrower, 1966.

intersecting lines and interstices' or in geographers' terms as 'a set of geographic locations interconnected in a system by a number of routes' (Kansky, 1963, p. 1), we are automatically concerned with a very complex bundle of characteristics: viz. how far the locations are from one another, whether the routes joining them are straight or curved, what commodity the network carries, whether the flow is continuous or intermittent, and so on (Fig. 1.1). Clearly all these aspects are highly relevant to individual networks, but they also make comparisons between networks very difficult except at a trivial and self-evident level. In order to get at the basic spatial structure of the network much of this information must be

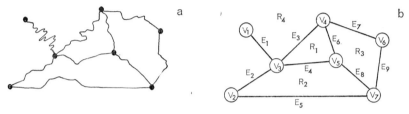

Fig. 1.2. Reduction of a map of a transport network (*A*) to a graph (*B*).

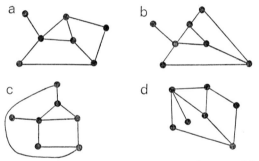

Fig. 1.3. Alternative topologic forms for the graph mapped in Fig. 1.2.

initially discarded, although of course it must be reintroduced later in the analysis.

In this first chapter networks are reduced to the level of *graphs*. Graphs are arrays of points, which are connected or not connected to one another by lines. There is no concern with the length or orientation of the lines nor whether they are straight or curved (although lines may be given positive or negative signs, i.e. *directed graphs*; or assigned numerical values, i.e. *valued graphs*), so that analysis of graphs may reveal common topological structures buried in apparently unlike networks. Fig. 1.2 shows the reduction of a simple road network (with each node representing an origin, intersection, terminal, or settlement on the route) to a system of nodes (*vertices* V_1–V_7), links (*edges* E_1–E_9) and (*regions* R_1–R_4). Since we may disregard distance and direction it is possible to re-draw the graph

(Fig. 1.2–B) in a series of alternative forms which still preserve the basic pattern of interconnections between nodes and links (see Fig. 1.3). Because a number of graph representations of a given network can be formed, it is more useful to store the information about the network in matrix form. Table 1.1 shows a series of binary matrices for nodes, links and regions in which *1* represents a direct connection between the elements and *0* otherwise. These *connectivity* matrices may be supplemented by *incidence* matrices in which connections between nodes and links, nodes and regions and links and regions may be represented in similar form (Garrison and Marble, pp. 17–20).

The study of networks in this topological sense began with Euler's 1736 paper on the seven bridges of the Prussian city of Königsberg and con-

---

Table 1.1. Connectivity Matrices for Graph mapped in Fig. 1.2.

Vertices (V)

	1	2	3	4	5	6	7
1	0	0	1	0	0	0	0
2	0	0	1	0	0	0	1
3	1	1	0	1	1	0	0
4	0	0	1	0	1	1	0
5	0	0	1	1	0	0	1
6	0	0	0	1	0	0	1
7	0	1	0	0	1	1	0

Edges (E)

	1	2	3	4	5	6	7	8	9
1	0	1	1	1	0	0	0	0	0
2	1	0	1	1	1	0	0	0	0
3	1	1	0	1	0	1	1	0	0
4	1	1	1	0	0	1	0	1	0
5	0	1	0	0	0	0	0	1	1
6	0	0	1	1	0	0	1	1	0
7	0	0	1	0	0	1	0	0	1
8	0	0	0	1	1	1	0	0	1
9	0	0	0	0	1	0	1	1	0

Regions (R)

	1	2	3	4
1	0	1	1	1
2	1	0	1	1
3	1	1	0	1
4	1	1	1	0

1 = Connected
0 = Not connected

---

tinued with Cayley's (1879) map-colouring problem; but the first comprehensive treatment of network topology was not published until 1936 in König's *Theorie der endlichen und unendlichen Graphen*. The branch of topology dealing with elementary structure, which came to be called *graph theory*, has developed rapidly in the three decades since König's seminal work and is now summarized in a range of texts (e.g. Berge, 1962; Busacker and Saaty, 1965; Flament, 1963; Harary, Norman and Cartwright, 1965; Ore, 1963).

One of the complications in applying graph theory to the analysis of network structure is the very confused and overlapping terminology. Lines are commonly referred to as 'links', 'edges', 'sides', 'arcs', 'segments', 'branches', 'routes' or 'one-cells'; while points are described as 'nodes', 'vertices', 'junctions', 'intersections', 'terminals' or 'zero-cells'. Interstices within the network or in the surrounding area are termed 'regions' or 'faces'. Terms are often restricted to a particular applied field (e.g., in the medical literature links and nodes become 'neurons' and 'synapses'), for graph theory has found relevance in a variety of problems. Besides the obvious applications to the design of electrical circuits, and to flows through transportation systems (Avondo-Bodino, 1962; Ford and Fulkerson, 1962), stand its uses in the design of artefacts (Alexander, 1964) or in the analysis of linguistic structures (Busacker and Saaty, 1965). In anthropology, sociology, and social psychology there is a tradition of its use in unravelling 'clique structures' and describing patterns of social interaction which goes back to Kurt Lewin's *Principles of topological psychology* (1936). Links between social interaction and models of neural activity have been forged by Rapaport, while uses of directed and valued graphs have extended

Table 1.2. Topological classification of networks

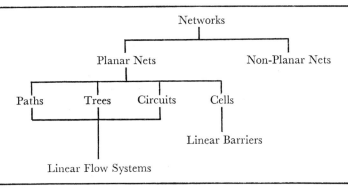

Source: Haggett, in Chorley and Haggett, 1967, p. 10.

from economics out into fields of management (e.g. in the development of 'critical-path analysis' and PERT techniques (Battersby, 1964)).

Treatment of networks in a graph-theoretic context clearly carries certain penalties and benefits. The penalties are the huge losses in relevant information; the gains come in the higher level of abstraction, the relative ease with which large numbers of complex networks can be handled and compared, and in greater flexibility. Flexibility stems not only from analogues between networks in physical and non-physical systems (see Chap. 5.III), but from the ability to switch problems from one 'naturally-occurring' mode to another where solutions may be more easily obtained. Thus we may translate networks into matrices and tap the powerful resources of matrix algebra (see Chap. 1.II(2)), or portray information in input-output matrices as networks (see Chap. 4.II). Similarly we can move rather easily from flow problems to barrier problems, and accommodate routes and boundaries within common network formats (Chap. 1.III). In this chapter the topologic structure of three major classes of networks (Fig. 1.2) is examined. Analysis begins with the relatively simple structure of branching networks (I), and moves on to consider graphs with circuits (II), and finally cellular nets (III). Fig. 1.2 gives familiar map examples of these three topologic classes (Table 1.1).

II. BRANCHING NETWORKS

The first main class of geographical networks considered is distinguished by its tree-like structures (Fig. 1.1–A). These consist of sets of connected lines without any complete loops; in topological terminology they are '. . . connected graphs without circuits' (Ore, 1963, p. 130). Apart from their anastomosing sections (which form closed loops) stream-channel systems have all the properties of the topologic tree: e.g. they follow the rule that trees consist simply of lines successively added to existing lines, so that a tree with n vertices has $n - l$ edges. From the topological viewpoint, stream networks have the simplest possible connectivity and all

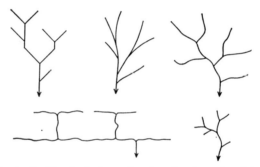

Fig. 1.4. Topologically identical channel networks. Arrowhead indicates outlet. Source: Shreve, 1966, p. 28.

bifurcating networks, such as are illustrated in Fig. 1.4, are topologically similar; indeed many geometrically different stream networks are topologically identical (Shreve, 1966). However, this topological definition of a tree includes forms intuitively recognizable as single paths: e.g. the single line connecting a set of points. This particular kind of topological tree, known as an 'open edge train', poses locational rather than structural problems and is discussed in Chap. 4.I(3). Here discussion is confined to the structure of branching networks as illustrated by the graphs of stream-channel systems.

Stream networks may be viewed as plane graphs (Scheidegger, 1967A) of topological finite rooted trees (Melton, 1959), in which *nodes*, V, (made up of outer tips or sources, t; inner forks or bifurcations, b; and a terminal root or outlet, r) are connected by *links*, E, (exterior, e; interior, i) so that only one link exists between any two nodes, the upstream end of

each link either connecting with two other links or terminating in a
source (Shreve, 1967, p. 178). This statement leads to a problem of
definition in that stream networks involve only flows directed from the
sources to the outlet, and Scheidegger (1967A, p. 104) rejects the term
'tree' in that it implies non-directed arcs (edges). He proposes the directed
term 'arborescence' in which pendant vertices or fingertips emanate from
a root. Whatever the terminology adopted, the following simple relationships exist between the numbers of nodes and links (Melton, 1959;
Shreve, 1967; Woldenberg, 1967) (Fig. 1.5):

$$E = V_{brt} - 1$$
$$E = 2V_t - 1$$
$$V_b + V_r = V_t$$

It is possible to describe the topology of stream networks in a number
of different ways. Shreve (1967, pp. 182–3) has proposed the symbolic
representation of a network by the following procedure: 'Start at the

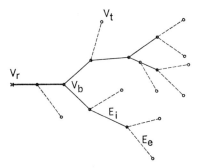

Fig. *1.5*. Ideal channel network showing inner and outer *edges* (E_i, E_e) and root,
branching and terminal vertices (V_r, V_b, V_t). Source: Melton, 1959.

outlet and traverse the network, always turning left at forks and reversing
direction at sources, until the outlet is again reached. During the traverse,
generate a sequence of i's and e's by recording an i the first time an
interior link is traversed and an e the first time a given exterior link is
traversed. Each link will be traversed twice but recorded only once.'
Fig. 1.6-A shows such a symbolic representation, which will ensure that
topologically identical networks will have identical sequences. Because of
the relationship:

$$E_i = E_e - 1$$

the number of e's never exceed the i's (except for the final e), and by
placing i and e equal to $+1$ and -1, respectively, a graph of partial sums
can be generated (Fig. 1.6-B).

Similarly, Scheidegger (1967A, p. 105) would express the arborescence
shown in Fig. 1.6-A as a left to right 'word'. This word is composed of

$2n - 1$ numbers (n 'zeros' and $n - 1$ 'ones'), where 1 is put for each junction and 0 for each pendant vertex, reading from the mouth down and left to right, and ensuring that no more zeros are recorded than there are already ones present to the left (except for the final zero). The network in Fig. 1.6-A is described by the word:

1 1 101 100101001 1 1000010100

Using matrix algebra, Scheidegger (1967A, p. 103) has proposed that a given network can be expressed either as an associated matrix, in which each element (a_{ij}) represents the number of links from x_i to x_j, or as an incidence matrix, where the matrix elements (S_{ij}) are $+1$ if vertex x_i begins link u_j, -1 if vertex x_i terminates link u_j, and 0 if vertex x_i is not on link u_j. Milton and Ollier (1965) uniquely labelled stream segments in a network by assigning multiple code numbers based on their Strahler

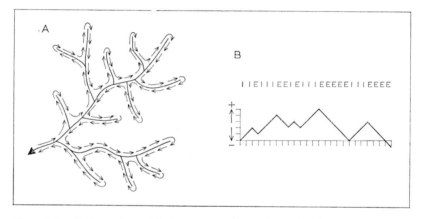

Fig. 1.6. Symbolic and graphical representations of a typical branching channel network. Source: Shreve, 1967.

orders, together with letters describing the sequence of junctions, and other similar locational codes have been devised (Ranalli and Scheidegger, 1968B). Smart (1968B), while agreeing that the Strahler method of ordering provides often too generalized a description of a network, suggests that much more exact descriptions of the topological properties of networks (e.g. by strings of binary digits) are often too detailed. He proposes an 'ambilateral' classification of network topologies which gives the arrangement of topological links, but ignores the character of the junction directions which were considered to be of more minor hydrologic and geomorphic significance (Smart, 1968B).

1. Concepts of hierarchic order

Different segments of a stream system exhibit different morphometric and hydrologic features and relationships, and earth scientists are particularly concerned with their hierarchical organization. This has required the assignment of a level of relative order or magnitude to each segment in a stream network hierarchy, determined by the sequential arrangement of tributaries with respect to the main trunk (Melton, 1957, p. 2; Maxwell, 1960, p. 9; Morisawa, 1959, p. 6). The various methods of ordering to be described 'provided the touchstone by which drainage-net character- istics could be related to each other and to hydrologic and erosional processes' (Bowden and Wallis, 1964, p. 767). As Strahler (1957, p. 914) has pointed out, practical utility is the criterion by which the success of

---

Table 1.3. Measurement levels of some network parameters

1. *Ordinal scale*

System	Unspecified	Horton	Strahler
Symbol for any order	U, V $(U \neq V)$	H	S
No. of streams of first order, etc.	N_{U1}	N_{H1}	N_{S1}
Total no. of streams	ΣN_U	ΣN_H	ΣN_S
Total no. of first order streams tributary to next highest order, etc.	N_{U4}	N_{H1+}	N_{S1+}
Min. possible no. of streams first order, etc.	N_{U1}	N_{H1}	N_{S1}
Highest order in basin	K	K	K

2. *Interval scale*

   Shreve magnitude M

---

3. *Ratio scale*

   Scheidegger order magnitude G (Derived from Index: I; see p. 16)
   Woldenberg order magnitude W

---

stream-ordering techniques must be judged, for 'any usefulness which the stream order system may have depends upon the premise that on average, if a sufficiently large sample is treated, order number is directly pro- portional to relative watershed dimensions, channel size and stream dis- charge at that place in the system'.

   Geomorphic attempts to dissect stream networks can be conveniently viewed within the general context of scales of measurement (Krumbein and Graybill, 1965, pp. 34–8). Some qualitative schemes (e.g. strike stream, but *not* consequent, subsequent, etc.) are purely *nominal*; the Horton and Strahler schemes, in which orders are designated by positive integers, are *ordinal*; Shreve's stream magnitude is an *interval* measure, in that each segment is designated according to how many first-order seg- ments feed it; and the more complex order magnitudes of Scheidegger and Woldenberg are *ratio* scale measures. The symbolism employed in the description which follows is given in Table 1.3.

Apart from early nominal attempts at stream ordering, together with certain qualitative ordinal ones (e.g. Davis, 1899), the first important one was by Gravelius (1914, pp. 1–3) (Horton, 1945, p. 281; Maxwell, 1960, p. 9; Woldenberg, 1967, p. 107). Gravelius first identified the trunk

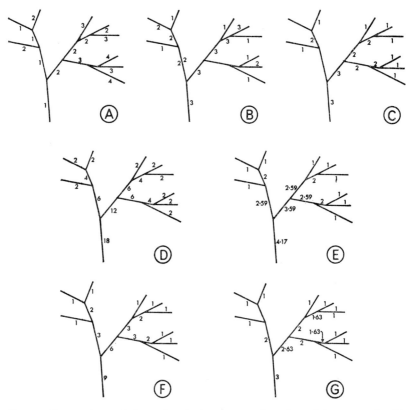

Fig. 1.7. Comparison of channel ordering systems for a branching network with $R_b = 3$. Source: Woldenberg, 1967.

stream (order **1**) by tracing it, explorer-like, from outlet to source, and at every bifurcation following the branch assumed to have the greatest width, discharge, headward branching or junction angle. This process was repeated for each stream directly tributary to order **1**, these being designated order **2** streams, and so on until the most remote fingertip tributaries received the highest order (Fig. 1.7-A). The main disadvantages of this scheme are, firstly, the subjective decisions which have to be taken at each bifurcation, and, secondly, that stream order number is not systematically related to the magnitude of a given segment or link (Woldenberg, 1967, p. 107).

Horton (1945, p. 281) felt that the main stem stream should be of the highest order and 'defined a first-order stream as one receiving no tributaries. A second-order stream is formed by the junction of two first-order streams, and can receive other first-order tributaries. A third-order stream is formed by the junction of two second-order streams, and can receive other second- and first-order tributaries. In short, the junction of two streams of like order forms a stream of next higher order, which can receive tributaries of any order lower than its own. Horton's system further requires that, after all streams have been classified, an investigator start at the mouth of the basin under study and reclassify part of the streams, continuing the higher order headward at a junction, following the tributary that is more nearly in line with the trunk stream' (Broscoe, 1959, p. 1), or along the longer if both enter at the same angle. This reclassification is then repeated for the tributaries to the main stem, and so on, until only fingertip tributaries of first order remain (Fig. 1.7-B). Shreve (1966, p. 21) stated the procedure in terms of four operational rules: '(1) A set of contiguous links of order H extending from a given fork upstream to a source is termed a *provisional Horton stream* of order H. (2) Each link originating at a source is initially considered to be a provisional Horton stream of order 1. (3) If two provisional Horton streams of the same order H join, then the resultant link downstream has order $H + 1$ and that tributary stream is reclassified as order $H + 1$, which either (a) enters the junction in a direction more nearly parallel to the head of the resultant link, or (b) in the case both enter at about the same angle, is longer, or (c) is for some other reason, such as the presence of geologic controls, considered to be the parent stream; and (4) if two provisional Horton streams of different order join, then the resultant link downstream has order equal to that of the tributary of higher order.'

It can be seen that certain difficulties are attendant upon the Horton scheme of ordering, not only in that an element of subjectivity exists in the headward reclassification at bifurcations, but also that some unbranched fingertip tributaries have orders greater than 1, and that not all streams of the same magnitude have the same order. However, Horton was able to show that the number of streams, as well as various other channel and geometrical properties, is consistently related to order. Surprisingly, the subjective ordering decisions in the Horton system have very little influence on the resulting number of streams designated for each order (Shreve, 1966, p. 23).

Strahler (1952, pp. 1120 and 1953) modified Horton's system by allowing his provisional scheme to determine the final ordering (Woldenberg, 1967, p. 110), such that: fingertip channels are designated order 1;

where two first-order channels join, a channel segment of order **2** is formed; where two channel segments of order **2** join, a segment of order **3** is formed; and so on (Fig. 1.7-c) (Strahler; in Chow, 1964, pp. 4–43). In this way the subjective decisions inherent in the Horton system are avoided, all fingertip tributaries are of order **1**, only one stream segment in the basin will have the highest order, and similar orders have more or less similar geometrical magnitudes. The Strahler system maintains the ordinal character of stream ordering (Maxwell, 1960, p. 9), applies to segments and segment orders (although some ordered segments contain more than one topological link), and, in a complete network, yields the same maximum basin order as the Horton system. Shreve (1966, p. 22) has defined the Strahler system of ordering in terms of four rules: '(1) A set of contiguous links of order S extending from a given fork upstream to a fork where two links of order $S - $ **1** join is termed a *provisional Strahler stream* of order S. (2) Each link originating at a source is initially considered to be a provisional Strahler stream of order **1**. (3) If two provisional Strahler streams of the same order S join, the resultant link downstream has order $S + $ **1**. (4) If two provisional Strahler streams of different order join, then the resultant link downstream has order equal to that of the tributary of higher order.' After the order of all the links has been determined the term *provisional* is dropped. Unlike the Horton system, the Strahler system is purely topological, referring only to the interconnections and not to the lengths, shapes, or orientations of the links comprising the network (Shreve, 1966, p. 23) and 'can be derived from concepts of elementary combinatorial analysis without the introduction of arbitrary or non-mathematical geomorphic concepts' (Melton, 1959, p. 345). Another way of describing it is by a progressive pruning sequence such that: (1) when any outer link is cut off at its downstream end, the node at the lower end vanishes; (2) when two outer links are simultaneously cut off at a node, the node remains. When this has been done K times, only the root remains (Melton, 1959, p. 345). As a result of the topological nature of Strahler's ordering system, it is possible to state the following relationships:

$$\underline{N}_s = 2^{K-s}$$

$$\sum_{s=1}^{K} \underline{N}_s = \sum_{s=1}^{K} 2^{K-s}$$

where N_s is the minimum possible number of Strahler streams). Both the Horton system (Leopold and Miller, 1956, p. 16; Brush, 1961, p. 155; Leopold and Langbein, 1962, p. 15) and the Strahler system of ordering (Schumm, 1956, p. 602; Melton, 1957, p. 2; Chorley, 1957; Coates, 1958, p. 4; Smith, 1958, pp. 999 and 1003; Maxwell, 1960, p. 9; Morisawa, 1962, p. 1028) have been widely employed in geomorphic and engineering analysis, and an undesirable confusion has sometimes occurred between them. Bowden and Wallis (1964, p. 767) have pointed to at least one instance of both systems being used without adequate distinction in the same article (Leopold, 1962, pp. 512 and 530). In pursuit of further rationalization, Warntz (1968) and Woldenberg have suggested that if Strahler first-order streams were designated as of zero order the mathematical expression of Horton's laws would be much simplified.

A serious drawback of the Strahler system of ordering, particularly from the practical hydrological standpoint, is its violation of the distributive law (Woldenberg, 1967, p. 111) by not permitting the junction of lower order segments to change the order of the main stream. Shreve (1966; 1967, p. 179) proposed that each exterior link or first-order segment should be given a magnitude (M) of 1, and each successive link a magnitude equal to the sum of all the first-order segments which ultimately feed it (Fig. 1.7-F). Thus a combination (*) of links M_1 and M_2 gives a downstream link magnitude of $M_1 + M_2$:

$$M_1 * M_2 = M_1 + M_2$$

In this way an interval scale of stream ordering is attained, in which each link is described in terms of first-order units, and the basin magnitude equals the magnitude of the greatest link.

Hydrological considerations in stream segment ordering were paramount in the arguments of Rzhanitsyn (1964, pp. 3–5 and 26), who argued that each junction changes the properties of the network and should create a new segment, the more equal the two combining segments the more pronounced the increase of order of the downstream segment. Scheidegger (1965) developed these notions in his 'consistent' law of stream ordering, giving four postulates defining an algebra of combination of stream segments which is commutative as well as associative: (1) When two similar segments (G') are combined the resulting segment has its order increased by one integer:

$$G' * G' = G' + 1$$

(2) A combination of two segments of lower order $(G' - 1)$ with a given order (G') should increase the latter by one integer:

$$G' * (G' - 1) * (G' - 1) = G' + 1$$

(3) To validate the distributive law, it must be postulated that the sequence in which these segments join is immaterial:

$$[G' * (G' - 1)] * (G' - 1) = G' * [(G' - 1) * (G' - 1)]$$

(4) To satisfy the commutative law, it does not matter whether a G' segment joins a G'' one, or *vice versa:*

$$G' * G'' = G'' * G'$$

From the above postulates, a general law for the composition of stream order magnitudes emerges:

$$G' * G'' = \frac{\log (2^{G'} + 2^{G''})}{\log 2}$$

In practical terms, all links are given an index number (I) by giving the fingertip tributaries an index of the exponent unity to the base 2 (i.e. $2^1 = 2$), and indexing all downstream links by adding the index numbers combining upstream at the junction (Fig. 1.7-D) (Woldenberg, 1967). These link indices are then given order magnitude (G) as follows (Fig. 1.7-E):

$$G = \frac{\log I}{\log 2}$$

It is clear that:

$$G = \log_2 2M$$

Scheidegger's order magnitudes give more classes than the Strahler system and a more accurate reflection of segment flow magnitudes, but Scheidegger's system of ordering cannot be adjusted in the case of mis-ordering fingertips (Shreve, 1967, p. 179).

Woldenberg (1967, p. 112) pointed out that Scheidegger's G does not produce the geometrical progression downstream observed in changes of discharge (Q) with order (Leopold and Miller, 1956, p. 18), and returned to Scheidegger's index (Fig. 1.7-D) and its relationship with the basin bifurcation ratio (R_b, the frequency of channel segment branching—see next section) to derive a new order magnitude (W) exhibiting this geometrical progression (Fig. 1.7-G):

$$Q_u = Q_1 (R_b)^{u-1}$$

Substitute I for Q:

$$I = 2(R_b)^{u-1}$$

but:

$$M = \frac{I}{2}$$

therefore:

$$U = W = \frac{\log M}{\log R_b} + 1$$

2. Branching ratios

The implications of segment ordering discussed above arise in terms of the mutual relationships between the different orders and their relationship to other aspects of the transport system (see Chap. 2.II).

(a) Law of path numbers

The relationship between different orders of segments in a simple branching system has been termed by Horton (1945) a 'law of stream numbers' and, more generally, by Haggett (1967, p. 628) a 'law of path numbers'. As stated by Horton (1945, p. 291), 'the numbers of streams of different orders in a given drainage basin tend closely to approximate an inverse geometric series in which the first term is unity and the ratio is the bifurcation ratio'. This can be stated mathematically (where K is the basin order):

$$N_H = R_b^{(K-H)}$$

such that:

$$\sum_{H=1}^{K} N_H = \frac{R_b^K - 1}{R_b - 1}$$

It has been pointed out (Bowden and Wallis, 1964, p. 769; Milton, 1966, pp. 90-1) that this is not a law in the physical sense but an abstract generalization which automatically follows with a high degree of probability from the manner of defining stream orders. Both the Horton and the Strahler systems of ordering allow the calculation of bifurcation ratios (R_{bH} and R_{bS}, respectively), but this can be accomplished in a number of different ways: (1) Individually for each pair of adjacent orders:

$$R_{bS1} = \frac{N_{S1}}{N_{S2}} \text{ etc.} \ldots$$

(2) The arithmetic mean bifurcation ratio (e.g. $R_{\bar{b}S}$) obtained by averaging the $K-1$ individual ratios calculated for the whole basin. (3) The weighted mean bifurcation ratio (e.g. $R_{\bar{b}WS}$), by weighting the ratios $R_{bS1} \ldots R_{bS(K-1)}$ according to the values of $N_{S1} \ldots N_{SK}$, before summing and averaging. This is necessary because the individual bifurcation ratios differ with order within a given basin (Schumm, 1956, p. 603). (4) The geometric mean bifurcation ratio (e.g. $R_{\bar{b}gH}$) is the slope of the line connecting the two end points ($H1$, Log N_{H1}) and (K, 0) on a Horton-type plot (Shreve, 1966, p. 21), such that:

$$R_{\bar{b}gH} = N_{H1}^{1/K-1}$$

(5) The antilog of the best-fit regression coefficient (b) (e.g. $R_{\bar{b}bS}$). This was suggested by Maxwell (1955, p. 520), who fitted the following expression to a plot of stream numbers against order:

$$\log N_U = K \log R_b - (\log R_b)U$$

Maxwell did not make this regression line pass through the point of highest order because, as this order can vary if one or two key first-order channels are added or omitted, it is no more reliable than any other point (Maxwell, 1960, p. 12). As Maxwell (1960, p. 12) himself pointed out, the ordinal scale of U (or H or S) does not strictly permit regressions to be fitted mathematically, and $R_{\bar{b}b}$ might be defined as the antilog of the regression coefficient fitted by eye. In seven basins 98·0 per cent of the variance of stream numbers versus order were 'explained' by the fitted regressions, stressing the highly exponential relationship expressed by Horton's law. Because, as will now be shown, the plot of numbers against Strahler order is more nearly exponential than with Horton ordering, $R_{\bar{b}bS}$ is slightly more accurate than $R_{\bar{b}bH}$.

Because the Strahler system of stream segment ordering is different from the Horton system it produces slightly different stream number versus order relationships; the conversion from one system to the other being effected by (Shreve, 1966, p. 23):

$$N_H = N_S - N_{S-1}$$

However, Strahler's system also produces an inverse geometric series of numbers of stream segments (Fig. 1.8), and Shreve (1964 and 1966, p. 23)

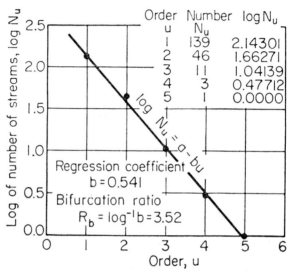

Fig. 1.8. Regression of number of stream segments on order. Source: Strahler, in Chow, 1964, p. 4–44.

found that Strahler's numbers of stream segments were fitted better by the series than Horton's number of streams. In 210 out of 246 published basin data the Strahler system led to smaller root-mean-square deviations from an exponential plot. The two systems also lead to small differences in basin bifurcation ratios. More important, whichever system is employed, there is usually a deviation from a true inverse geometric series—commonly involving an upward concavity, particularly at the higher order end— suggesting to certain workers (e.g. Shreve, 1964; Smart, 1967) that Horton's 'law' represents a statistical relationship. So consistent is this departure from the exponential that Shreve (1967, p. 24) has suggested that the following curve provides a better fit (Fig. 1.9):

$$\log N_S = \frac{K - S}{K - 1}\left[1 - a\,(S - 1)\right] \log N_{S1}$$

In this equation a is a parameter denoting curvature (concave upwards $+$; convex $-$). Values of a calculated for the 246 published basin data for

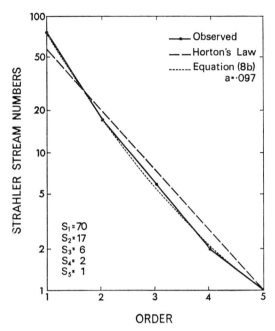

Fig. 1.9. Horton diagram for Home Creek, Tuscarawas County, eastern Ohio (data from Morisawa, 1962, p. 1036), showing Horton's law (long dashes) and Shreves formula (short dashes). Source: Shreve, 1967.

basin orders 3 to 7 illustrate the greater prevalence of concavity, and emphasize the departures from the exponential, particularly in lower order basins.

(b) Regional variations in bifurcation ratios

   The bifurcation ratio is predominantly controlled by the drainage density and, especially, stream entrance angles (Milton, 1966, p. 92), and in regions where the network geometry develops without pronounced lithological or structural controls, bifurcation ratios are highly stable. They show small variation from one region to another (Strahler, 1957, pp. 914–15; Chorley, 1957, pp. 142–3), and fail to confirm Horton's (1945, p. 290) original suggestion that bifurcation ratios vary from about 2·0 for flat or rolling basins to 3·00 to 4·00 for mountainous, highly-dissected basins. 'Because the bifurcation ratio is a dimensionless property, and because drainage systems in homogeneous materials tend to display geometrical similarity, it is not surprising that the ratio shows only a small variation from region to region' (Strahler, in Chow, 1964, pp. 4–45). The minimum possible bifurcation ratio (R_{bs}) of 2·00 (Fig. 1.10) is seldom approached in nature, lying between about 3·00 and 5·00 in basins without dominant differential geological controls (Coates, 1958, pp. 32–3 gives the following mean figures for ten basins in six dominantly sandstone regions having third-order relief of 109–318 feet: $R_{bS1} = 3\cdot98$–$5\cdot07$ and $R_{bS2} = 2\cdot8$–$4\cdot9$), and only reaching higher values where geological controls favour the development of elongate, narrow basins. Recent work, however, is recognizing and attempting to rationalize the small but systematic variations in bifurcation ratios ascribed by Strahler (1964, in Chow, pp. 4–45) to chance variations in the stream numbers of adjacent orders. One such (Giusti and Schneider, 1965) has shown that bifurcation ratios within a given region tend to decrease with increasing order (i.e. $R_{bS1} > R_{bS6}$) because 'as order increases the percentage of streams that coalesce into a higher order tributary also increases and . . . this increase is due to the diminishing amount of area available' (Giusti and Schneider,

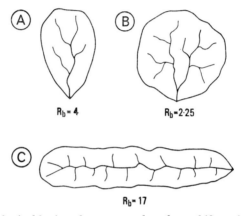

Fig. 1.10. Hypothetical basins of extreme and moderate bifurcation ratios. Source: Strahler, in Chow, 1964, p. 4–44.

1965, p. 3) (Fig. 1.11-A). Using data from the Yellow River basin in the Georgia piedmont, the above authors found that 'basins of equal order but variable areas tend to have the smallest bifurcation ratios in the smallest areas; the ratios increase with increasing areas up to a certain size, beyond which the bifurcation ratios tend to become constant' (Giusti and Schneider, 1965, p. 2) (Fig. 1.11-B). The studies of the probability of occurrence, coalescence and termination of stream segments which are referred to below provide another stochastic approach to the systematic variations in the bifurcation ratio.

Other authors have tried to explain such departures from the so-called law of stream numbers in a more deterministic manner. For example, Gregory (1966) showed that it applies best to active systems of distinct stream channels, mapped in the field, rather than to a valley network which can include fossil elements; and Eyles (1968) suggested that a pronounced upward concavity of the stream number plot may be evidence

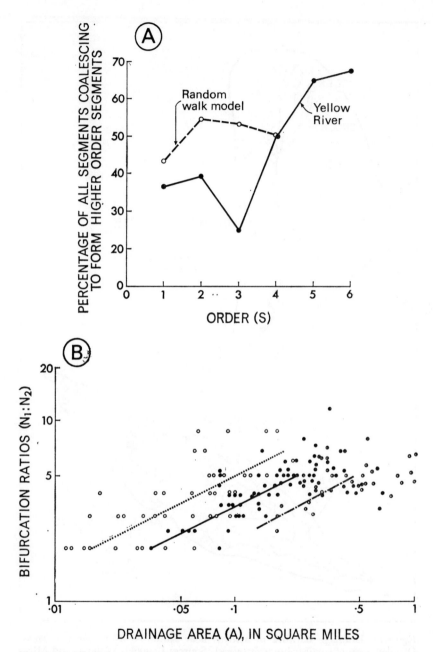

Fig. 1.11. (A) Relationship between percentage of coalescing tributaries and order. (B) Relation between drainage area and bifurcation ratios (N_1 to N_2) for sub-basins with the Yellow River basin. Source: Giusti and Schneider, 1965.

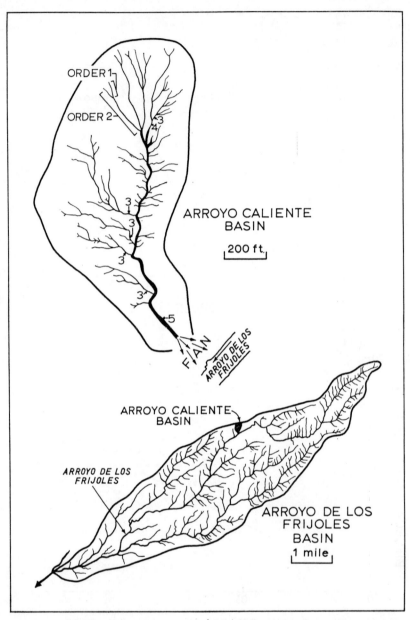

Fig. 1.12. Drainage network of arroyos near Santa Fe, New Mexico (*below*). Detailed map of small tributaries (*above*). Source: Leopold, Wolman and Miller, 1964.

that recent rejuvenation has disproportionately increased the numbers of lower order streams (particularly the first), such that there is a tendency for R_b to be inversely related to order.

3. Data limitations for topologic analysis

The scale of maps plays a large part in determining the accuracy with which channel networks can be inferred from the map. Maps at the scale of about 1 : 24,000 produced from air photographs appear to represent the stream networks of medium-textured terrain with quite a high degree of truth, but this is certainly not the case for areas of high drainage density where, even in the field, the distinction between rills and fingertip channels is difficult to make. Leopold, Wolman and Miller (1964, pp. 138–9) give an extreme instance of the role of map scale in the interpretation of stream channel networks. Fig. 1.12 shows the network inferred for the Arroyo De Los Frijoles from a U.S.G.S. topographic map

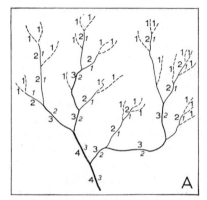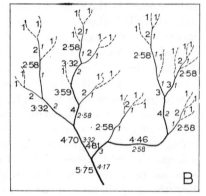

Fig. 1.13. Hypothetical stream-net: (*A*) Strahler orders. (*B*) Consistent orders. Source: Scheidegger, 1965.

at 1 : 31,680, together with a field map of the network in a previously-designated first-order tributary of the Arroyo Caliente basin. The network mapped within this area of 0·027 square miles caused it to be redesignated as fifth order. Such local mapping of sample areas, particularly of first-order basins, allows an estimation to be made from the map regarding some of the true features of the stream network in the field: in this instance, the orders of all the channels in the Arroyo de Los Frijoles could be increased by **4**, and inferences regarding stream numbers, lengths, etc. of orders lower than **5** made from Horton-type plots involving stream order. However, as Scheidegger (1966A) has pointed out, this simple method of increase of order only holds good for the Strahler (1952) system of ordering (Fig. 1.7-C) and for Scheidegger's (1965) 'consistent' system of stream ordering (Fig. 1.7-E) where the original network is extremely regular and the additional mapped channels appear only at the headwaters. In the latter, random addition of low-order channels to the network results in increases of higher order designations by variable amounts. Fig. 1.13 shows the effects on ordering produced by fingertip additions to a network. In the first instance, of the Strahler ordering system, these additions result in a regular increase of all orders, but the second example shows the variable order change effected in the Scheidegger system.

A number of geomorphologists have checked the stream channel networks depicted on topographic maps in the field; their results are instructive but somewhat conflicting, depending on the types of map and field locations. Melton (1957, p. 1), checking basins shown for a number of western States on the U.S. Geological Survey 1 : 24,000 maps, found that a high percentage of first-order channels were represented faithfully by the smallest cusps in the contours, and that in a few cases contour crenulations did not even represent persistent drainage lines but merely surface irregularities. He concluded that about ninety-five per cent of the small basins studied in the field required minor channel network corrections on maps compiled from air photographs. Morisawa (1957; 1959) checked similar 1 : 24,000 maps produced from aerial photographs by multiplex methods at a contour interval of twenty feet for part of the Appalachian Plateaus. She concluded from an analysis of eighteen small basins of less than 2·68 square miles that there was a great discrepancy between the total lengths of blue lines in each basin and the length of channels mapped in the field, but no significant difference between the latter and the channel lengths inferred by means of the contour inference method. Schneider (1961) re-analysed Morisawa's data, but considered the correlation between the field measured lengths and the contour inference lengths to be too poor for stream-channel lengths to be accurately inferred from these topographic maps. Coates (1958, pp. 24–6) tested the channels inferred from the 1 : 24,000 map of the Nashville Quadrangle in Southern Illinois (contour interval of ten feet) with his field maps on a scale of 1 : 600; he found that fingertip tributaries could seldom be inferred from the U.S.G.S. map, and that most mapped first-order channels were in reality third-order. In general he found that total stream channel lengths mapped in the field were three to five times greater than those inferred from the mapped contour crenulations, and that the drainage density obtained from the map was only twenty to sixty-six per cent of that mapped in the field. Maxwell (1960) took similar 1 : 24,000 maps surveyed in 1925–33 by plane table (contour interval twenty-five feet) of the rugged San Dimas Experimental Station in the San Gabriel Mountains of Southern California, photographically enlarged them and checked them against the channel network observed in the field. This strongly suggested that much of the mapping had been done from the ridge crests giving substantial errors in some cul-de-sac tributary canyons in the central and less accessible parts of the watersheds. From such studies as those above it appears that the degree of inaccuracy in channel depiction by modern medium-scale contour maps is very much a function of the scale of the smallest channels (i.e. the drainage density) (Chap. 2.II(1)), and that maps surveyed by ground methods alone are particularly suspect in the inaccessible parts of more rugged terrain.

A rigorous attempt to compare the stream channel networks inferred from the contours on published topographic maps with those visible on aerial photographs has been made by Eyles (1966) for Malayan maps on the scale of 1 : 63,360 with a contour interval of fifty feet. Photographs

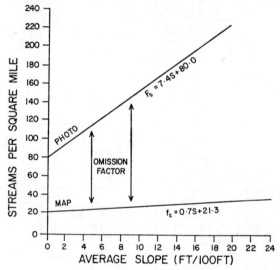

Fig. 1.14. Regression lines showing increase of stream–number omissions per unit area, with increase in average slope, on Malayan map representation. Source: Eyles, 1966.

ranging in scale from 1 : 8,000 to 1 : 25,000 (maximum lateral error of 0·04 inches at a scale of 1 : 15,840) were compared with the topographic maps of fifty third- to fifth-order basins in open country and it was found that only seven basins coincided in order between map and photograph, thirty-three being reduced by one order on the maps and ten by two orders. Fig. 1.14 shows the increase of streams omitted from the maps related to the average slope of the terrain.

4. Stochastic approaches to network topology

The severe data problems outlined in the preceding section have led to theoretical investigations stressing random processes. Shreve (1963, p. 44) has argued that Horton's phrase 'tend closely to approximate' (1945, p. 291) suggests that Horton's 'law' is simply 'a statistical relationship,

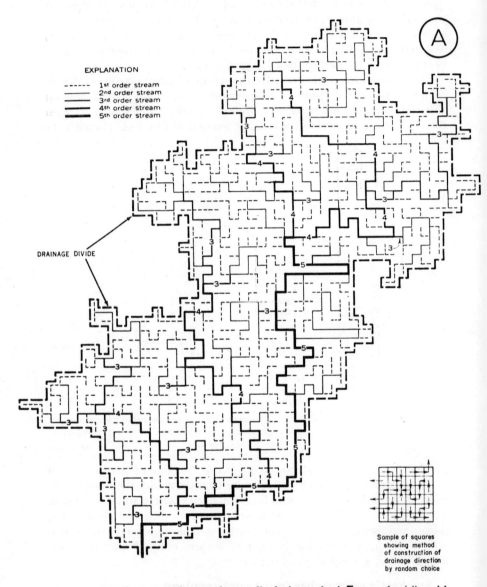

Fig. 1.15. Development of a random-walk drainage basin network. (*A*), with relation of number and average lengths of streams to stream order, (*B, opposite*). Source: Leopold and Langbein, 1962.

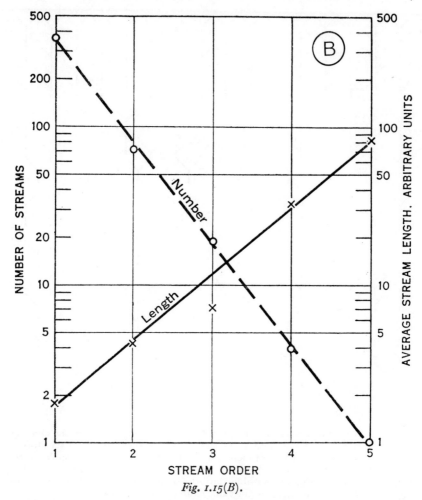

Fig. 1.15(B).

resulting from random development of drainage networks, rather than from orderly evolution as generally assumed'. This implies that among all the possible sets of stream numbers the most probable in nature are those with stream numbers close to an inverse geometric series (Shreve, 1966, p. 17). With the tendency of erosional processes to produce bifurcating networks, the implication is that 'the law of stream numbers arises from the statistics of a large number of randomly merging stream channels in somewhat the same fashion that the law of perfect gases arises from the statistics of a large number of randomly colliding gas molecules' (Shreve, 1966, p. 18). It is interesting, therefore, that the drainage simulated for a series of squares, freely-developing from each in any of the four cardinal directions (with reverse flows prohibited) by a purely

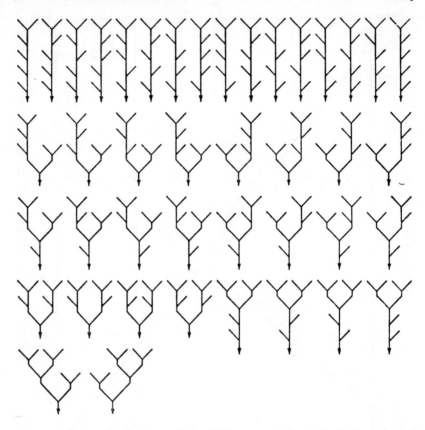

Fig. 1.16. Schematic diagrams of the N(6) = 42 topologically distinct channel networks with 1 = 11 links and n = 6 first-order Strahler streams or six Horton streams. In a topologically random population these networks would all be equally likely. The top row shows the possible second-order networks and the bottom four rows show the possible third-order networks. Arrowhead indicates outlet in each diagram. Source: Shreve, 1966.

random process, generated a synthetic stream-like network (Fig. 1.15-A) which had stream number relationships satisfying Horton's law (Fig. 1.15-B) (Leopold and Langbein, 1962). Milton (1966, p. 92) also reports that five randomly-generated stream networks produced values of R_b ranging from 3·9 to 5·9, suggesting that values for natural stream networks would be lower because external controls bring about more frequent stream unions and discourage the twisting patterns characteristic of random walk nets. More will be said regarding the significance of synthetic random walk nets in a later section on drainage evolution.

An infinitely topologically random channel network, similar to random walk graphs, theoretically produces a bifurcation ratio of 4·00 (Shreve, 1967, p. 184). Smaller topologically random networks form 'a population within which all topologically distinct networks with given number of links are equally likely' (Shreve, 1966, p. 27). Networks with the same number of forks, sources, Horton streams and first-order Strahler streams are of comparable *topological complexity*, and are *topologically distinct* (Shreve, 1966, pp. 27–9). Fig. 1.16 shows forty-two topologically distinct networks, having eleven links, six first-order streams and five forks. Given the number of sources and the network order, the number of topologically distinct networks satisfying this can be calculated, and from this number the probabilities of occurrence of a given series of stream numbers can be

Table 1.4. Probability of occurrence of alternative branching networks

Rank:	Orders				Probability of occurrence (p)
	1	*2*	*3*	*4*	
1	27	7	2	1	0·263
2	27	6	2	1	0·198
3	27	8	2	1	0·169
. . . .					
8	27	9	3	1	0·045*
. . . .					

Source: Shreve, 1963.

* Satisfies Horton's law.

obtained. For example, for a fourth-order basin with twenty-seven finger-tip tributaries only thirty different sets of stream numbers are possible (Shreve, 1963) (Table 1.4).

In practice, the random model has been shown to accord well with data from Western stream systems (Smart, 1968B, p. 25), except that a systematic deviation from randomness is provided by an excess of lower order streams (possibly due to geological controls) giving values of R_b greater than those predicted by theory. From a rather different viewpoint Smart (1967) proposes that even if a large network exhibits Horton's law exactly, the numbers of streams observed in smaller networks within it may show definite and predictable deviations from it. Employing probabilities that streams will be 'excess' (i.e. will terminate by joining a higher order stream, as distinct from combining with another of similar order to generate a higher order stream), Smart (1967, pp. 774–5) demonstrates this deviation by showing the decrease of a theoretical bifurcation ratio with order (Fig. 1.17). Another similar approach by Scheidegger (1966B) assumes that a stream of Strahler order S will either: (1) Combine with another order S to generate order $S + 1$ (probability $= p$), or (2) Terminate by joining one of greater order (probability $= 1 - p$, since this is the only other alternative). Employing a Markov process Scheidegger (1966B,

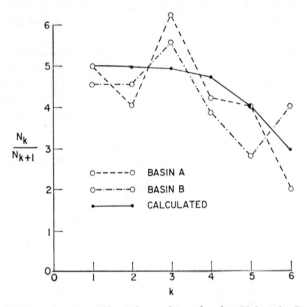

Fig. 1.17. Relations between bifurcation ratios and order. Values for Basins A and B are from data of Giusti and Schneider (1965). Calculated values were obtained using $R_b = 5$ and length exponent $= 1.5$. Source: Smart, 1967.

p. 202) obtains the following relationships between order (S) and number of segments (N):

$$
\begin{array}{ll}
\text{Order } (S) \quad 1 & N_{S1} \\
2 & N_{S1}\,p^1\,2^{-1} \\
3 & N_{S2}\,p^2\,2^{-2} \\
\cdots & \cdots\cdots\cdots\cdots\cdots \\
K & N_{SK}\,p^{K-1}\,2^{-(K-1)}
\end{array}
$$

Thus $N_S = N_{S1}\,p^{S-1}\,2^{-(S-1)}$ (a geometric series). From this the bifurcation ratio can be calculated:

$$
R_{bS} = \frac{N_S}{N_{S+1}} = \frac{N_{S1}\,p^{S-1}\,2^{-(S-1)}}{N_{S1}\,p^{S+1-1}\,2^{-(S+1-1)}} = \frac{2}{p}
$$

By observation R_{bS} is approximately $4\cdot00$ and p must then be $0\cdot5$, and the stochastic branching process predicts that half the stream segments of given order should combine with others of the same order to generate segments of the next highest order, whereas the other half terminate.

The law of stream numbers is capable, therefore, of interpretation in terms of two different basic models (Scheidegger, 1968A; 1968B; Ranalli and Scheidegger, 1968A):

1. *The Horton model* which possesses a hierarchy of Strahler channels arranged to form different 'generations' of streams, within each of which, on average, R_b channels of order S combine to form a channel of order $S + 1$. This gives a constant R_b at different areal scales throughout the system, which can be referred to as a 'Horton net'. However, R_b shows high internal variation within most natural stream networks (with a possible tendency to be inversely related to order) which, while satisfying Horton's law of stream numbers as a whole, cannot be structurally defined as 'Hortonian'.

2. *The random graph model* which, as a whole, statistically approximates Horton's law of stream numbers, but lacks the structural regularity implied by the Horton model. The random manner which characterizes the joining of channels internally implies that the system is topologically random (Shreve, 1966), rather than regularly hierarchical.

It is apparent that whichever model is adopted has a strong influence over one's attitude to network evolution, and we will return to this conflict in a later chapter.

III. CIRCUIT NETWORKS

The second class of regional networks considered are structures with closed loops or circuits (Fig. 1.1-c). The problems of order posed by such networks are very similar to those of trees, but less progress has been made to date in analysing them and reducing them to predictable forms; discussion here centres on the structure of such networks as revealed by the study of graph-theoretic indices. The problem of transforming circuit networks into simpler tree-like forms is postponed to Chap. 5.III where wider problems of network transformation are raised. Since transport systems with their intricate pattern of road, rail, pipeline, air and telecommunications forms a major class of geographically-relevant circuit systems, they are used to illustrate the structural problems posed by this class of graphs. Substantive geographical accounts of these transport systems are given in a number of texts (e.g. Clozier, 1963; Taaffe and Gauthier, 1969; Ullman, in James and Jones, 1954; see also the extensive literature reviews by Leinbach, 1957; Siddall, 1964; Wolfe, 1961). The approach followed here is to look first at the simple indices that can be developed from an initial appraisal of graph structure, and subsequently to consider the extra structural information that can be derived via matrix analysis.

1. Elementary graph-theoretic measures

One of the simplest applications of graph theory has been its use in comparing sets of networks. By reducing the network to a topological graph, by selecting three basic measures and by manipulating the three measures in a series of ingenious indices (Table 1.5), we can form some useful

---

Table 1.5. Elementary topological measures of network structure

A. *Measures based on gross characteristics:*

Cyclomatic number $= E - V + G$ $E =$ number of links (edges) in the network

$V =$ number of nodes (vertices) in the network

$G =$ number of sub-graphs

Beta index $= E/V$

Alpha index $= \left(\dfrac{E - V + G}{2V - 5}\right)100$

Gamma index $= \left(\dfrac{E}{\dfrac{V - (V-1)}{2}}\right)100$ $\Big\}$ For Planar Graphs

Alpha index $= \left(\dfrac{E - V + G}{\dfrac{V(V-1)}{2} - (V-1)}\right)100$

Gamma index $= \left(\dfrac{E}{3(V-2)}\right)100$ $\Big\}$ For Non-Planar Graphs

B. *Measures based on shortest-path characteristics:*

Diameter $=$ maximum d_{ij} $d_{ij} =$ shortest paths (in links) between ith and jth node

Accessibility index $= \displaystyle\sum_{i=1}^{v} d_{ij}$

Dispersion index $= \displaystyle\sum_{i=1}^{v} \sum_{j=1}^{v} d_{ij}$

Source: Garrison and Marble, 1962; Kanstry, 1963.

---

common yardsticks for network comparison. The three basic parameters on which the indices are based are (i) the number of separate (i.e. non-connecting) *subgraphs* in the network (G), (ii) the number of *edges* (links) in the network (E), and (iii) the number of *vertices* (nodes) in the network (V).

     Fig. 1.18 shows a series of networks in which the number of subgraphs $(G = 1)$ and the number of vertices $(V = 10)$ is held constant, but in which the number of edges is successively increased. Visual comparison of the four graphs shows that the network is becomingly successively more connected through the *ABCD* sequence. The simplest

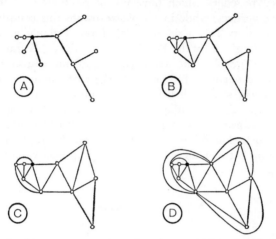

Fig. 1.18. Graph with constant number of vertices and increasing number of edges.

description of increasing complexity is to relate the number of edges and vertices in a ratio E/V. This *Beta* index differentiates simple topological structures (with low Beta values) from complicated structures (with high Beta values). The lower portion of the Beta scale (from zero to one) differentiates between different types of branching networks (Chap. 1.II) and

Table 1.6. Changes in topological indices for graphs mapped in Fig. 1.18 with increasing connectivity

Graphs:	Planar				Non-Planar*
	A	*B*	*C*	*D*	*E*
Cyclomatic number	0	5	10	15	36
Beta index	0·90	1·40	1·90	2·40	4·50
Alpha index*	0	33%	67%	100%	100%
Gamma index*	37%	58%	79%	100%	100%
Mean local degree	1·8	2·8	3·8	4·8	9·0
Diameter	5	5	4	3	1
Mean dispersion index	2·58	2·45	1·90	1·60	1·00
Mean shape	1·78	1·64	1·68	1·56	1·00

* Direct one-link connections between all ten vertices. Not mapped in Fig. 1.18.

disconnected networks. Values of one and above differentiate circuit networks. For non-planar graphs values may range to ∞, but for planar graphs the maximum value is 3·00 (Kansky, 1963). *Planar* graphs are characterized by edges which have no intersections or common points except at the vertices, while in *non-planar* graphs this condition does not hold (Ore, 1963, p. 12): a rigorous definition of planar and non-planar graphs has been given by the Polish mathematician, Kuratowski. A city map is a planar graph if the streets (i.e. edges) meet in squares or street intersections (i.e. vertices), but need no longer be planar if two roads cross at different levels (so that their intersection on the map does not coincide with a vertex). Air-route maps represent a major case of non-planar graphs. The four graphs shown in Fig. 1.18 are clearly all planar networks, but Table 1.6 includes Beta values for a fourth graph with complete communication between all ten vertices. Typical Beta values for transport

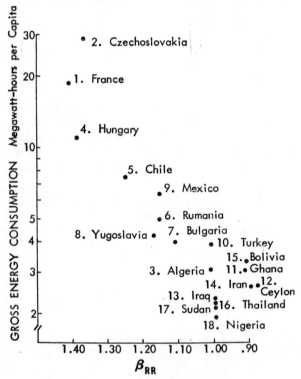

Fig. 1.19. Relationship of Beta index to development levels for sample national railroad networks. Source: Kansky, 1963, p. 42.

systems for a cross-section of countries range from around 0·9 for poorly developed railway systems (e.g. Nigeria) to 1·4 for more complex networks (e.g. France) (Kansky, 1963) (Fig. 1.19).

A more useful index of the connectivity of networks is the Alpha index or 'redundancy index'. This consists of the ratio between the observed number of fundamental circuits to the maximum number of circuits that may exist in a network (Garrison and Marble, 1962, p. 24). The observed number of fundamental circuits is given by the cyclomatic number (μ), while the maximum number of circuits, the divisor, is given by $2V - 5$ for planar graphs and by $\dfrac{V(V-1)}{2} - (V-1)$ for non-planar graphs. The cyclomatic number (or first Betti number) is itself a fundamental measure of network complexity ($\mu = 0$ for branching networks and disconnected graphs), but combined into the Alpha index it gives a sensitive measure of connectivity. Values of zero indicate a branching network or 'minimal-spanning tree' in which the removal of any one edge would break the graph into two unconnected sub-graphs; conversely values of one indicate a completely connected graph in which no further edges may be added without duplicating existing links. Although the number of edges in the planar (Fig. 1.18-D) and non-planar versions of a completely connected graph will differ (see values in Table 1.6), both carry values of $a = 1\cdot00$. By multiplying the ratio by one hundred the Alpha index is given a range of zero to one hundred allowing an interpretation of 'per cent' redundancy. For the sample countries studied by Garrison and Marble (1962) the mean value of this index was $7\cdot039$ with a standard deviation of $6\cdot234$.

2. Binary connectivity matrices

More efficient descriptions of the topologic structure of circuit networks may be obtained by treating the graph as a connectivity matrix (Table 1.1).

(a) *Original connectivity matrices*

The connectivity matrix (C) of a graph shows direct links between vertices by binary coding: *one* if direct link exists, *zero* if no direct link

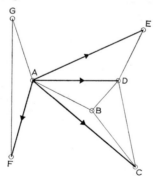

Fig. 1.20. Sample graph with directed arcs: transmitter system. Source: Prihar, 1956, p. 931.

Table 1.7. Connection and shortest-path matrices for the network mapped in Fig. 1.20.

Connection Matrices

$C^1 =$ (First-order)

	A	B	C	D	E	F	G
A	0	1	0	1	0	1	1
B	1	0	1	0	1	0	
C	1	1	0	1	0		
D	1	1	0	1			
E	0	1	1	0			
F	0	1	0				
G	1	0					

$C^2 =$ (Second-order)

	A	B	C	D	E	F	G
A	2	2	2	3	1	1	1
B		3	2	1	1	1	1
C		3	0	1	1	1	2
D		1	1	3	0	1	2
E	1	1	1	1			1
F	1	1	1	1			1
G	2	2	2	2			2

$C^3 =$ (Third-order)

	A	B	C	D	E	F	G
A	3	7	7	5	3	3	
B	4	5	5	2	1	1	
C	4	3	5	1	1	1	
D	5	5	3	4	1	1	
E	1	3	1	0			2
F	1	1	1	1			1
G	3	2	2	3			1

Fourth-order ($= k^{th}$ order):

	A	B	C	D	E	F	G
A	10	17	17	22	10	6	6
B	5	**19**	15	15	11	5	5
C	5	9	**10**	10	6	2	2
D	6	9	9	**15**	4	2	2
E	1	5	5	3	4	1	1
F	3	2	2	3	1	**2**	**4**
G	3	8	8	8	6	**4**	**5**

$\sum C^4_{ij} =$ (Row Sums): 88, 75, 44, 47, 20, 14, 42

$\sum C'^{4}_{ij} =$ (Column Sums): 33, 69, 66, 76, 42, 22, 22

$\sum\sum C^4_{ji} = 330$ (Matrix Sum)

$C^1 + C^2 + C^3 + C^4 =$

	A	B	C	D	E	F	G
A	15	27	27	33	17	11	11
B	10	**26**	23	25	15	7	7
C	7	15	**15**	17	9	3	3
D	8	16	16	**21**	9	3	3
E	4	3	3	4	5	1	1
F	7	7	7	5	2	**4**	**8**
G	7	11	11	12	8	**8**	**8**

$\sum C^n_{ij} =$ (Row Sums): 141, 113, 69, 76, 30, 24, 66

$\sum C^n_{ji} =$ (Column Sums): 53, 105, 102, 119, 65, 37, 37

$\sum\sum C^n_{ij} = 519$ (Matrix Sum)

Shortest Path Matrices

$D_1 =$ (First-order)

	A	B	C	D	E	F	G
A	0	1	0	1	0	1	1
B	1	0	1	0	1	0	
C		1	1	0	1	0	
D		1	1	0	1		
E	0	1	0				
F	1	0					
G		1	0				

$D_2 =$ (Second-order)

	A	B	C	D	E	F	G
A	0	1	1	1	1	1	1
B	1	0	1	1	1	2	2
C	2	1	0	1	2	2	2
D	2	2	1	0	2	2	2
E	2	2	2	2	0	1	
F				1	0		
G	2	2	2	2	1	0	

$D_3 =$ (Third-order)

	A	B	C	D	E	F	G
A	0	1	1	1	1	1	1
B	1	0	1	2	2	2	2
C	2	1	0	1	3	3	3
D	2	2	1	0	3	3	3
E	3	0	1	0			
F	2	1	0				
G	2	2	2	2			

$D_4 =$ (Fourth-order $= k^{th}$ order)

	A	B	C	D	E	F	G
A	0	1	1	1	1	1	1
B	1	0	1	2	2	3	3
C	2	1	0	1	3	3	3
D	2	2	1	0	3	3	3
E	3	0	4	1			
F	3	0	1	0			
G	2	2	2	1	0		

$\sum d_{ij} =$ (Row Sums): 6, 9, 12, 11, 16, 15, 10

$\sum d_{ji} =$ (Column Sums): 11, 10, 10, 9, 11, 14, 14

$\sum\sum d_{ij} = 79$ (Matrix Sum)

Source: Prihar, 1956, pp. 931–2.

exists. Table 1.7 shows the connectivity matrix for the simple graph of
a radio-communication system (Prihar, 1956, p. 931) mapped in Fig. 1.20.
Station A is a powerful transmitting station capable of contacting the other
six stations in the system. While some of the other stations $(B \ldots G)$ can
respond directly to station A, others transmit only over shorter distances
and are only able to reach it *indirectly* through relaying stations. This
network contains therefore two-way contacts (*symmetries*) and one-way
contacts (*antimetries*): the connections matrix (C^1) is therefore assymetric

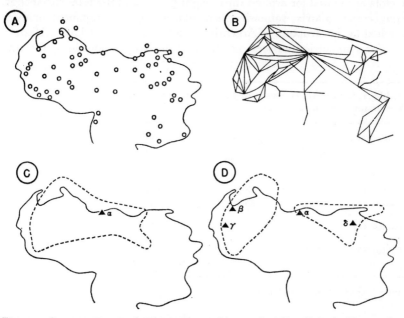

Fig. 1.21. Latent structure in the pattern of internal airline links in Venezuela:
(A) vertices, (B) links between vertices, (C) Factor II fields, (D) Factor III fields:
Source: Garrison and Marble, 1962.

about the principal diagonal. Compare the symmetric matrices for two-
way links shown in Table 1.1.

Direct analysis of binary matrices through principal component
analysis has been attempted by Garrison and Marble (1962, pp. 65–71;
(1964). They studied the airline system of Venezuela (essentially a non-
planar network) which connects fifty-nine cities (Fig. 1.21-A) via a system
of 104 routes (Fig. 1.21-B). This was reduced topologically to a system of
vertices and edges displayed in the form of a connection matrix, which
was analysed through principal component analysis to produce a set of
four basic patterns of variation (Table 1.8). Factor I scaled the vertices
in terms of their total number of direct connections with other vertices
and was a nearly linear scaling of the cities by size. This 'basic' factor

accounted for nineteen per cent of the total observed variation. Factor II showed up a field effect centring on the leading vertex, the city of Caracas (α) (Fig. 1.21-c), and raised the level of explanation by a further seven per cent. Somewhat similar in importance was factor III (Fig. 1.21-D), which showed up a major regionalization effect: a western system centring on Maracaibo (β) and Santa Barbara (γ) stands out from the 'eastern' system centring on Caracas and Maturin (δ). The fourth factor showed a weak but detectable minor regionalization effect. Together all four factors accounted for around thirty-eight per cent of the total variance in connectivity matrix, somewhat less than a similar component analysis applied to the structure of Argentine airline system (*Cf.* Chap. 4.II.2).

For symmetric matrices (i.e. without one-way connections) the row

---

Table 1.8. Factor Analysis of Latent Structure in Venezuelan Airline Routes

| Components: | Variation explained | | Interpretation |
	Per cent	Cumulative per cent	
Factor I	18·7	18·7	Size axis
Factor II	7·5	26·1	Caracas 'field' axis
Factor III	6·6	32·7	East–West 'field' axis
Factor IV	5·2	38·0	Minor regionalizing axis

Source: Garrison and Marble, 1962, p. 69.

---

or column-wise addition of the ones in the binary matrix give the number of links incident at each node. The average for the nodes in the network is termed the *mean local degree*. Values for this parameter have not been widely studied, but Webb (1955) made use of the measure in comparing the topology of the Martian 'canal' system with a range of natural and man-made earth patterns. The use of similar measures in the study of the contact patterns of cellular networks is discussed in Chap. 1.IV.

(b) Powered connectivity matrices

The original binary matrix (C^1) can be powered to give a series of matrices $C^2, C^3, \ldots C^n$ (see Table 1.7). Following the rules of matrix algebra:

$$C^2 = C^1 C^1 = \begin{pmatrix} a & b \\ c & d \end{pmatrix}\begin{pmatrix} a & b \\ c & d \end{pmatrix} = \begin{pmatrix} aa + bc & ab + bd \\ ca + dc & cb + dd \end{pmatrix}$$

Matrix C^2 contains the following information about the network: (1)

diagonal cells c_{ii} show the total number of two-step connections for each station; (2) off-diagonal cells c_{ij} show the total number of two-step connections for the pairs indicated. For example, there are three ways of contacting station D from station A (via routes ABD, ACD and AED), and only one two-step route for station E (i.e. EDE). By indicating the number of alternative ways of linking stations, the matrices C^2, C^3, . . . C^n indicate 'safety' factors in the system. The utility of this measure must however be modified by the fact that large numbers of 'redundant' paths are included in the matrices. Higher-order matrices are derived by continuing the powering process, and this process is terminated when all

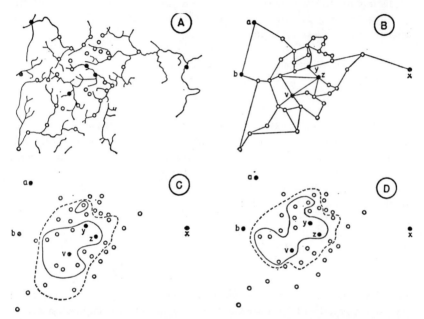

Fig. 1.22. The use of graph-theoretic measures to compute the relative accessibility of Moscow (y) in twelfth- and thirteenth-century Russia. In both (C) and (D) the heavy line encloses the ten 'most connected' places and the broken line the twenty 'most connected' places. Source: Pitts, 1965.

off-diagonal elements are filled. For the graph shown in Fig. 1.20 this is achieved when $n = 4$, when n is the diameter (or solution time) of the graph. It is clear by inspection that messages can be transmitted throughout the entire network in at most four steps. With the assymmetric graph mapped in Fig. 1.20 it is important to distinguish between the direction of accessibility. Inspection of Table 1.7 shows that since elements $c_{ij} \neq c_{ji}$, the row totals and column totals may be different. Thus while node A occupies the most flexible location within the network in terms of originating four-step paths ($\Sigma c^4_{ij} = 88$), node D occupies the most flexible location for four step terminating paths ($\Sigma c_{ji}^4 = 76$). The advantages to be derived from the two sets of information are discussed in Chap. 1.III. (4b).

Pitts (1965) has provided a very useful geographical application of powered matrices in the derivation of measures describing the relative locational advantages of thirty-nine settlements in twelfth- and thirteenth-century Russia. Fig. 1.22-A shows the location of the settlements in relation to the rivers and trade routes and Fig. 1.22-B gives a simplified representation of this map as a planar graph. The information contained in this graph may be reduced to a one-zero array (a connection matrix) by listing all the settlements along both the rows and columns of the 39×39 matrix. The first four settlements in the matrix (C^1) would appear in the following form:

$$C^1 = \begin{pmatrix} 0 & 1 & 0 & 0 & . \\ 1 & 0 & 1 & 0 & . \\ 0 & 1 & 0 & 1 & . \\ 0 & 0 & 1 & 0 & . \\ . & . & . & . & . \end{pmatrix}$$

Thus the direct link between the first settlement ($a = $ Novgorod) and the second settlement ($b = $ Vitebsk) is indicated by a 1 in the second cell of the top row. Absence of direct one-step connections is shown by a zero. By raising the matrix to the power of the graphs diameter δ (eight in Pitt's example) it is possible to derive a matrix, C^8; which is shown for the first four settlements as:

$$C^8 = \begin{pmatrix} 110 & 15 & 143 & 16 & . \\ 15 & 155 & 21 & 167 & . \\ 143 & 21 & 580 & 32 & . \\ 16 & 167 & 32 & 257 & . \\ . & . & . & . & . \end{pmatrix}$$

The above matrix contains the following information about the network: (1) the diagonal entries indicate the total number of eight-step routes usable in going out and back from a settlement (thus Novgorod (a) has a value of 110 routes, Vitebsk (b) has 155 routes, and so on); (2) the off-diagonal entries indicate the total number of eight-step routes going between pairs of places. To obtain the total number of alternative routes we add across the rows of all the powered matrices up to and including C^8, to give a total connectivity measure for each vertex, gross vertex connectivity. Fig. 1.22-C shows the location of the 'most connected' vertex ($v = $ Kozelsz) and the 'least connected' ($x = $ Bolgar). Moscow (y) is the fifth most-connected settlement.

(c) Weighted connectivity matrices

A procedure for deriving a modified accessibility index based on both direct and indirect connections between nodes was developed by Shimbel and Katz (1953). The original connection matrix C is converted into a powered matrix T^n where:

$$T^n = a^1 C^1 + a^2 C^2 \ldots + a^n C^n = \sum_{i=1}^{n} a^i C^i$$

The original connection matrix C is binary ($1 =$ direct connection; $0 =$ otherwise) and a is a scalar varying between 0 and 1. The powered matrix provides accessibility indices for any node in the network or for the network as a whole. An element of the matrix T^n, t_{ij}^n measures accessibility of node i to node j. Similarly the sum of the elements in a row (or column):

$$\sum_{j=1}^{n} t^n$$

denotes the accessibility of node i to all the other nodes in the network. The sum of all the elements:

$$\sum_{i=1}^{n} \sum_{j=1}^{n} t_{ij}^n$$

measures the accessibility of the network as a whole.

Two important areas for experiment with the Katz index concern the power of the matrix n and the values of the scalar a:

(i) *Power level of the matrix:* By continued powering of the matrix it will clearly be possible to eliminate all zero elements when n- is equal to the diameter of the graph (i.e. when the solution time of the graph is reached), and this represents an effective limit to the value of n (see Table 1.7). Migayi (1966) argues that interesting results may be derived by intermediate power levels of the matrix. Linkage analysis was conducted on three power levels, T^3, T^5 and T^{11} for the interstate highway systems of the eastern United States. All intersections, as well as cities with populations of 100,000 or more, are regarded as nodes, and the final network selected consisted of sixty nodes and one hundred edges. Although the boundaries of the network on the Gulf Coast and the Atlantic are realistic, the region was arbitrarily truncated. Nodes on these western and northern frontiers have, therefore, values less than those on the actual continent-wide network. Regionalization was revealed by a linkage analysis of the T^n network (Nystuen and Dacey, 1961) conducted on three levels of the original T matrix: $T^3 T^5$ and T^{11}. For the latter, two main regions were identified, a northern region centring on Indianapolis and a southern region centring on Atlanta. Results of the analysis clearly demonstrate that with successively higher values of the exponent n, higher orders of nodal regionalization are produced. Accessibility increases most rapidly for nodes with a high number of incident links.

Gauthier (1967) has used a powering procedure in a study of the São

Paulo (Brazil) highway network. Binary values were replaced by sets of cost values related to length of haul, and these were scaled so that direct connection between any two nodes was proportional to the highest cost linkage in the network. As the upper right part of Table 1.9 shows, values in the matrix are positive decimal loadings for each of the arcs; only the first six of the 120 Paulista nodes are shown. Power expansion of the matrix was carried out in the normal way and the summed values of elements used to give a matrix T, which may be interpreted as a numerical representation of the 'accessibility surface' of the matrix. Values for the first five nodes are given in the lower half of Table 1.9. Each element t_{ij} reflects 'the number of direct linkages incident on the node, the number of open and closed path sequences, and the length and constructional quality of both direct and indirect node-linkage assemblages' (Gauthier, 1967, p. 6). Higher values indicate higher accessibilities between the nodes

---

Table 1.9. Part of connection and accessibility matrices for regional highway sub-graph*

Nodes:	S.P.	M.C.	S.J.C.	A.	P.	B.P.
São Paulo	—	0·81	0·63	0·75	0·57	x
Mogi das Cruzes	480·0	—	0·69	x	x	x
São Jose dos Campos	363·0	135·0	—	x	x	x
Atibaia	754·0	266·0	203·0	—	0·88	0·92
Piracaia	484·0	174·0	133·0	283·0	—	0·78
Braganca Paulista	473·0	159·0	121·0	274·0	182·0	—

Source: Gauthier, 1967, pp. 5–10.

* São Paulo, 1960.

---

and vice versa. Thus in Table 1.9 the greatest accessibility is between São Paulo and Atibaia, and the lowest between São Jose dos Campos and Braganca Paulista.

(ii) *Scalar:* The scalar a^k introduced by Katz has received some criticism. The objective of the scalar is to add an arbitrary weight to indirect links: as the powering of the matrix increases to allow k-link connections, so the value of a becomes progressively less. Although the value of $a = 0·30$ is suggested by Katz, there seem no *a priori* grounds for retaining it at this level, and Hebert (1966) has investigated the effect of changing values for various T matrices. Values of a equal to 0·30, 0·40, 0·50 and 0·60 were used in an analysis of the resulting T^n matrices, which showed that variations in the scalar a caused significant variations in regional interpretation; but results were not systematic enough over the range of values studied to suggest general conclusions. Migayi (1966, p. 20) has suggested that values of a might be varied heuristically to accentuate or dampen the accessibility of central places on the network in any investigation. Larger values of a would clearly tend to emphasize indirect links as the expense of local direct links and thus the accessibility scores of nodes near the centre of the graph. Analysis of the changing frequency distributions of the elements within the powered connection matrices (C^1, C^2, ... C^n in Table 1.7) may suggest an internal basis for the weighting decision, i.e. values of a might be individually determined for each graph by its internal structure to provide a locally-consistent standard.

3. Shortest-path matrices

The shortest path between two nodes in a network is measured topologically by the number of intervening links to be traversed. It may be derived by inspection, by shortest-path algorithms (see the 'cascade' algorithm discussed in Chap. 4.I), and as a by-product of the matrix-powering procedures discussed in the preceding section. Inspection of the matrices for the radio-transmitting network shown in Fig. 1.20 demonstrate that the first-order shortest-path matrix (D_1) and the first-order connectivity matrix (C^1) are identical (Table 1.7). Clearly diagonal elements must be zero while ones indicate one-step links between nodes; the unfilled off-diagonal elements in the matrix indicate unknown shortest-paths with values greater than one. These 'missing' elements are filled by reference to the connection matrices. The infilling of non-zero elements in matrix C^2 (as compared to matrix C^1) is recorded by values of two in the matrix D^2; each new element in the shortest-path matrix indicates a two-step link. This process is continued with matrices C^3 and is completed with matrix C^4 when all off-diagonal elements are filled with non-zero numbers. The final stage of the shortest-path matrix is always derived from matrix C^n, where n is the solution time of the matrix. Values of elements in the shortest-path matrix may be used for both the comparison of networks and for the evaluation of nodes within the same network.

(a) Comparison of graphs

Two gross measures of the size and connectivity of a graph are given by the *diameter*, $\delta(G)$, and its dispersion, $D(N)$. The diameter is the maximum number of edges in the shortest path between each pair of vertices (Kansky, 1963, p. 12), and is shown by the largest value (i.e. maximum d_{ij}) in the shortest-path matrix. It is equivalent to the 'solution time' of the graph. Thus the diameter of the radio-transmitting network shown in Fig. 1.20 is four, and the values for the four graphs of increasing connectivity shown in Fig. 1.18 is five, five, four, and three respectively. As shorter pathways are opened up by new links, the diameter decreases. The diameter is then a crude 'range' measure which is somewhat weak for differentiating transportation networks, since it is directly affected by the size of the graph, but inversely affected by its connectivity.

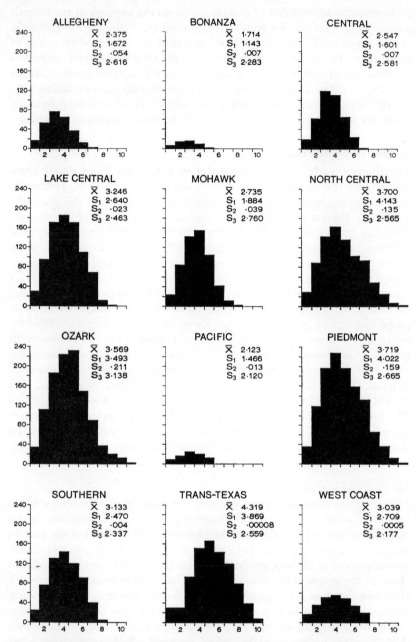

Fig. 1.23. Histograms of shortest-path routes for graphs of twelve United States domestic airline networks, Summer 1966. X = Arithmetic Mean, S_1 = Standard Deviation, S_2 = Skewness, S_3 = Kurtosis.

Similar criticisms may be levelled at the dispersions index. The dispersion of a network (Shimbel, 1953; Shimbel and Katz, 1953) is the sum of all the elements in the shortest-path matrix; viz.:

$$D(N) = \sum_{i=1}^{n} \sum_{j=1}^{n} d(i,j)$$

where d_{ij} is the shortest path between two nodes in the graph. The value for the radio-transmitting station is seventy-nine (see Table 1.7) and the corresponding values for the four graphs in Fig. 1.18 are 232, 210, 170, and 144 respectively.

The instability of the diameter and dispersion indices in comparing

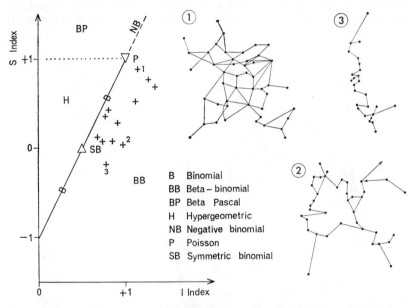

Fig. 1.24. Allocation of networks to appropriate discrete frequency distributions. I, S diagram for real roots distributions (*left*). Sample graphs (*right*).

groups of networks lead to investigation of a series of standard statistical parameters for the array of shortest-path elements. The sample population studied was the scheduled airline networks for the eleven local airline operators in the United States; the array of shortest-path elements for each network are shown as histograms in Fig. 1.23. Statistical analysis demonstrates the wide variation in the average path lengths (from only 1·71 links for the Bonanza system to 4·32 links for Trans-Texas) and less obvious variations in the form of the shortest-path frequency distributions. Frequency distributions for the airlines studied were analysed in terms of Ord's *S–I* diagram (Ord, 1967) in which systems of discrete distributions are mapped in terms of their first three moments. All the airlines studied showed Beta-binomial characteristics (Fig. 1.24) but with some strong variations within the class. Research on large numbers of simulated and real networks currently under way (Haggett, James, Cliff and Ord, 1969) suggests that rather distinct pattern groupings within S–I space may be distinguished. Similar work on the frequency distributions shown by König numbers and Shimbel indices is reported by Werner (in Horton, 1968).

(b) Comparison of nodes and links

The shortest-path matrix may also be used to compare the relative accessibility of nodes within the network. The rows in the shortest-path matrix in Table 1.7 indicate the number of steps needed to connect the node indexed on the left of the row to all other nodes indexed at the heads

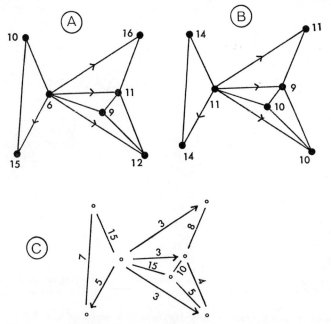

Fig. 1.25. Node accessibility on basis of (*A*) originating and (*B*) terminating paths for network mapped in Fig. 1.20. (*C*) Link usage index.

of the columns. One gross measure of accessibility is the largest value in the row (the *König number*), which ranges from only one for node *A* to four for node *E*. A more refined measure is given by the sum of the row values (the *vertex accessibility number*), which ranges from only six for node *A* to sixteen for node *E*. The second index uses more of the information in the matrix and can distinguish the relative accessibility of nodes with the same König numbers (e.g. nodes *C*, *D*, and *F*). It should be noted, however, that these values refer to rows and therefore to originating paths only. Similar procedures applied to the columns reveal different information for terminating paths (i.e. $\Sigma d_{ij} \neq \Sigma d_{ji}$). Figs. 1.25-A and -B show the importance of this distinction. Thus while node A might be regarded as the optimal location within the network for a facility which demanded rapid outward movement (e.g. fire stations, police stations within an urban area), node D might be regarded as optimal for facilities with static terminal facilities (e.g. schools, surgeries, etc.).

Vertex accessibility was also studied by Pitts (1965) for his system of Muscovite settlements. Since the matrix was assumed to be symmetric a single vertex accessibility number may be derived. Use of this index (mapped in Fig. 1.22-D) shows Moscow (y) as the second most accessible place on this more refined index, with Kolomna (z) the most connected and Bolgar (x) the least connected. These results should be compared with the gross vertex accessibility index (discussed in Chap. 1.II(2b)) which placed Moscow fifth (Fig. 1.22-C). The use of vertex accessibility numbers as a measure of the impact of alternative highway schemes is discussed by Burton (1963) for the Ontario road system (Fig. 4.2). (Chap. 4.I.1(a).)

Although the relative importance of links within a network has attracted less attention than vertices, it may be approached in directly similar ways. Connectivity matrices for links may be established (see Table 1.1) and sets of comparable indices derived. Fig. 1.25-C shows the relative importance of the links in Prihar's system (Fig. 1.20) in terms of the number of shortest paths between all nodes that use a given link. The critical importance of AB and GA is indicated. More advanced methods for evaluating critical links within a transport network are discussed by Wollmer (1963).

IV. BARRIER NETWORKS

Barrier networks are clearly different in function from the two types of networks previously examined. For while branching and circuit networks consist of channels conducting flow, barrier networks consist of links which either block flow or resist flow. A number of geographers have analysed the empirical characteristics of barrier systems in terms of political geography (Minghi, 1963) or land settlement systems (Thrower, 1963), while Yuill (1965) has proposed more general classifications of such barriers in terms of reflecting and absorbing functions. Such barrier networks have tensional analogues in physical geography. However, from the structural viewpoint, the fundamental feature of such barrier systems is their closure: i.e. they consist of closed loops which may be isolated (e.g. island boundaries) (Fig. 1.1-C) or contiguous (e.g. the loops surrounding sets of administrative areas and forming the mutual boundaries between them) (Fig. 1.1-D). Barrier networks may therefore be regarded as *duals* of flow networks in both a functional sense and in a topological sense. As Fig. 1.26 shows networks occurring in one mode give rise to corresponding duals in the second (Warntz, 1966).

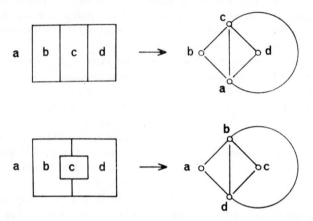

Fig. 1.26. Examples of dual structure of regions and networks.

1. Structural regularities

Cellular nets form a distinct topological class within the general group of planar graphs. They are polygonal graphs (Ore, 1963, p. 98), defined as a planar graph in which the edges form a set of adjoining polygons in the plane, polygons being used in a general sense to indicate figures with either straight or curved boundaries. Thus a map showing the division of the United States into states would be such a polygonal graph, including as it does both straight meridional lines (the Nevada–Utah interface), curved parallel arcs (the Wyoming–Montana interface) and irregular natural lines (the Kentucky–Indiana interface).

For polygonal graphs we may apply formula first derived by Euler which states that:

$$V - E + C = 2$$

where V, E and C are the number of vertices, edges and cells respectively (Ore, 1963, p. 99). We must recall in applying this formula, known as Euler's polyhedral formula, to the conterminous United States that we must include the infinite set of all points lying outside the United States (the 'outside' cell) as well as the internal cells: C is forty-nine, not forty-eight. Euler's formula has an interesting extension to a problem that worried Lösch (1954) in his work on the modal cell for central-place hierarchies, i.e. 'what regular polygons of equal area may be used to completely fill a plane without overlapping or unused spaces?' Euler's formula may be extended to show that conditions for the establishment of these regular tesselations are only fulfilled when:

$$I + \frac{p}{p^*} = \frac{p}{2}$$

both the number of edges at each vertex (p) and the number of edges bounding each cell (p^*) are the same. Only regular triangles ($p = 6$,

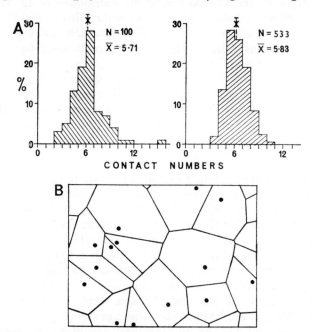

Fig. 1.27. Histograms for contact numbers for sample of (A) Brazilian counties and (B) Danish communes. (C) Randomly generated polygons. Source: Haggett, 1965, p. 52; Pedersen, 1967; Dacey, 1963.

$p^* = 3$), quadrangles ($p = p^* = 4$) and hexagons ($p = 3$, $p^* = 6$) satisfy this equation (Ore, 1963, p. 108).

In a regular hexagonal system the contact number would clearly be six, as one area would be contiguous with its six neighbours, each of which would in turn have six neighbours. Haggett (1965) studied a random sample of one hundred counties from the 2,800 counties (*municipios*) for Brazil. Since coastal counties and those on the international frontier had their fields truncated, these were eliminated from the sample; the remaining eighty-four were examined and their contact numbers recorded. The frequency curve of the results is shown in Fig. 1.27-A. It demonstrates that although the number of contacts varied from two to fourteen nearly one in three counties had exactly six neighbours. The mean contact number for the sample was 5·71.

Pedersen (1967) has extended Haggett's observations with counts of contact numbers in the administrative communes in Jutland and Fyn in Denmark. For the 553 communes studied the average number of neighbours was found to be 5·83. It was shown, however, that this number should not be interpreted to indicate the existence of a hexagonal net structure, for 'blurring' of basic triangular or quadratic nets could give very similar averages. Comparison of the distribution of all communes, together with an analysis of their shape characteristics, lead Pedersen to conclude that the administrative net in Denmark is basically triangular (about sixty per cent) with a smaller quadratic component (forty per cent). Hexagonal elements are largely absent.

Experimental evidence on contact numbers is available from work by Dacey and Tung (1962) who constructed a series of synthetic point patterns each randomly dislocated a mean distance ρ from positions on a hexagonal lattice. The quantity of random disturbance was measured by an index, k, where:

$$k = \rho/(1\cdot075\lambda^{-1/2})$$

and λ is the expectation that a unit area contains a nucleus. Points were used as the nuclei for the construction of Dirichlet regions (see Chap. 4.II.1), so that sets of pseudo-random polygons were produced (Fig. 1.27-c) with the properties that (i) any two polygons have at most one edge in common; (ii) the tesselation of polygons is uniquely determined, and (iii) each vertex is connected to three or four other edges (Dacey, 1963, p. 3). Contact numbers were studied for patterns with slight disturbance ($k = 0\cdot05$) to patterns in which the degree of random disturbance was so large that all trace of the underlying hexagonal lattice was lost ($k = 3\cdot00$). For each of the patterns studied the mean number of edges was found to be less than six, and for values of k greater than 0·30 the mean approaches values between 5·7 and 5·8. On theoretical grounds Dacey (1963) derives an expected mean value of 5·7888 for random polygons rather than the value of six assumed in random crystal aggregates (Meijering, 1953; Johnson and Mehl, 1939).

2. Equilibrium models

The empirical regularities shown by barrier networks has naturally led to a search for explanatory models. The two groups of models reviewed here deal with space partition under equilibrium conditions for two distinct geographical situations; the evolution of barrier networks over time is treated in Chap. 5.I.3.

(a) Unbounded nets: the 'mid-continental' problem

With unbounded networks the basic problem of network structure may be stated in terms of the efficient partitioning of areas between competing centres. We may define efficiency in two ways: efficiency of movement as measured by the distance from the centre to outlying parts within the territory, and efficiency of boundaries as measured by the length of the territory's perimeter. Three geometrical principles are important in applying these minimum energy criteria to the division of an area (Coxeter, 1961):

(i) Regular polygons are more economical shapes than irregular polygons. If we take the familiar four-sided polygon, we can illustrate that for the regular square shape with an area of one square kilometre, the furthest movement (i.e. from the centre to the furthest point within the square) is 0·707 kilometres and the perimeter is four kilometres. If we convert the regular square form to a rectangle of similar area but with two of its sides twice as long as the others, the furthest movement goes up to 1·031 kilometres and the perimeter to five kilometres. Experimentation demonstrates how the greater the contrast in the sides of the rectangle, the less economical it becomes in terms of both accessibility from the centre and length of perimeter.

(ii) Circles are the most economical of the regular polygons. If we imagine a continuum of regular polygons running from the triangle (3-gon), square (4-gon), pentagon (5-gon), and hexagon (6-gon) upwards, then at each stage we are increasing the number of sides and vertices by one. The limiting case is clearly the circle which we may regard as a regular polygon with an infinite number of sides and vertices. If we examine this sequence we see that, if the area remains constant, the accessibility from the centre as measured by the maximum radial distance improves and the perimeter becomes shorter (Haggett, 1965, p. 49). Although the improvement in economy is consistent, the gains are not regular: the square is about half as efficient as the circle and the 10-gon is about ninety per cent as efficient as the circle.

(iii) Hexagons are the regular polygons which allow the greatest amount of packing into an area, consistent with minimizing movement and boundary costs. The problem of packing circular fields into a hypothetical area is illustrated in Fig. 1.28 which shows with shading its inefficiency as measured by the unused areas lying between the circles. The problem of filling a plane with equal-area regular polygons was first investigated by Kepler in the early seventeenth century, who suggested there were three solutions: the regular triangle, the regular square, and the regular hexagon. Of these three regular tessellations (Coxeter, 1961, pp. 61–4), the hexagon retains most of the advantages of the circle: it is about four-fifths as efficient as the circle in terms of maximum radial distance and perimeter.

Hexagons have held a fascination for natural scientists and mathematicians since the Greeks; concepts of hexagonal symmetry played a key role in the growth of crystallography and Thompson (1942, pp. 102–25) has shown its importance throughout the biological sciences. It is not surprising therefore that the two main theoretical works on settlements and their support fields, Christaller's *Die zentralen Orte in Suddeutschland* (1933) and Lösch's *Die raumliche Ordnung der Wirtschaft* (1940; 1954) should have used the hexagon as the modular unit in their models of settlement structure. Isard (1956) has shown that the regular (i.e. equal-area) pattern of hexagons suggested by Christaller and Lösch are unlikely to occur in practice. Because of the high density of population at the central core postulated by Lösch, the size of the market area here is likely to be

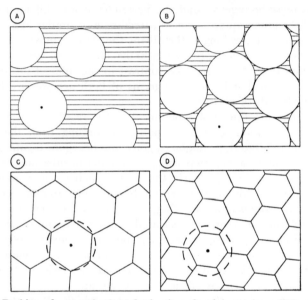

Fig. 1.28. Packing of centres in the colonization of a plain to give cellular hexagonal territories. Source: Lösch, 1954, p. 110.

smaller, while away from the market it is likely to be larger. Isard has produced a figure (Fig. 3.11-D) which retains as many of the assumptions of the Löschian system as possible, but introduces this concept of more closely packed centres near the overall nodal point. Extreme difficulty was found in working with the hexagonal form and, as the figure shows, it was impossible to retain both the hexagon and urbanization economies. As Isard points out, the hexagon is a pure concept much as perfect competition is a pure concept to the economist. It loses its significance as a spatial form once the inevitable agglomeration forces—which are themselves inherent in the Löschian system—are allowed to operate. We should not expect regular hexagonal territories to be generally visible on the earth's surface, since they are probably related not to geographical space but to population or income space. Hexagons may therefore be thought to be latent in most human organization, but only through appropriate transformations of geographical space (see Getis, 1963) is their form likely to become **apparent**.

(b) Bounded nets: the 'island' problem

To extend the equilibrium solution from the semi-finite plane to small geographical areas requires some modification. If we retain the concept of the co-equal 120° junctions of the hexagon but introduce the concept

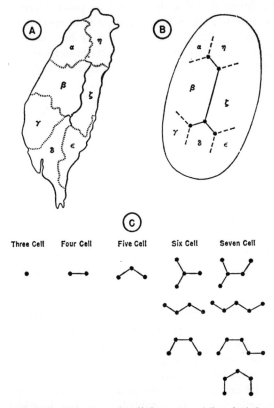

Fig. 1.29. Topological character of cellular nets: *(A)* administrative areas of Formosa, showing 'polar furrow'; *(B)* generalized form of *(A)*; *(C)* range of polar furrow arrangements for fixed cell numbers. Source: Thompson, 1942, p. 516.

of a fixed boundary then we have a problem in space partitioning not unlike that of dividing an island into *N* counties, all of which have a maritime boundary (i.e. are not encircled by the rest). The Island of Formosa (Fig. 1.29-A) with its seven maritime provinces forms such a case. Topologically we may simplify the Formosa case by generalizing the coastline of the island as an elipse (the boundary), regarding the seven

NAG—C

provinces as cells and replacing the province boundaries by cell 'walls' set at co-equal 120° angles (Fig. 1.29-B).

If within our island we have only three cells meeting together, then, in terms of our steady state theory, we may assume they would meet at a point at equal angles. If we have four cells they need an 'internal' wall (i.e. one that does not have one end touching the coastal boundary). With five cells we need two such internal walls, with six cells three, and so on, following the rule that we require $N - 3$ internal walls to separate N cells. These internal walls, designated as polar furrows by the cell biologist (Thompson, 1942 edn., p. 516), are of special interest to our model in that (as Fig. 1.29-C shows) they may be arranged in numbers of different ways. With the four- and five-cell island there is no ambiguity, but clearly with six cells there are three different arrangements of the

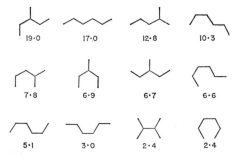

Fig. 1.30. Twelve possible forms of the polar furrow for eight-cell divisions of an island, arranged in a stability sequence. Source: Thompson, 1942, p. 607.

three-link polar furrow. Seven cells give four alternatives, eight cells give twelve alternatives, nine cells give twenty-seven alternatives, and beyond that the number increases rapidly with estimates by Brückner of 50,000 alternative arrangements for thirteen cells and around 30,000,000 alternatives for sixteen cells (Thompson, 1942 edn., p. 598). These higher numbers are largely of academic interest since with real geographical situations (e.g. islands divided into counties) it seems likely that internal non-maritime cells would be established in the interior; Brückner's estimates refer to cases where all the cells retain a maritime boundary.

The realization that the number of cell patterns is, from the topological viewpoint, strictly limited leads on to a consideration of (i) how many of the possible patterns actually occur and (ii) in what proportions they occur. Although the strictly geographical answers to these questions have yet to be determined, there is some evidence from the biological sciences that there may be a continuum of cell forms ranging from rather stable (and therefore common) to highly unstable (rare) arrangements. Fig. 1.30 shows the twelve possible arrangements of polar furrows (each five edges in length) for eight-cell islands arranged in a stability sequence. The frequencies are based on laboratory study of more than a thousand epiblastic cells (Thompson, 1942 edn., pp. 606–7) and shows ranges from forms as common as 1 : 5 to as rare as 1 : 50. Thompson's findings suggest a rich field of research in the stability of geopolitical patterns over a wide time-space range, for really equable divisions were not in each case a regular criterion of stability.

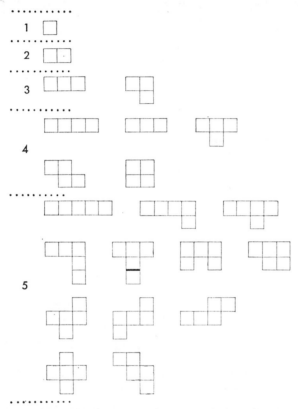

Fig. 1.31. Alternative topologic arrangements for networks with up to five cells. Source: Harary, in Beckenbach, 1964, p. 201.

Thompson's problem of cell division has a parallel in the *ising* problem (Beckenbach, 1966, p. 199). The simplest case is the growth of a one-celled pattern that has a square shape and which can only grow in the plane by adding one square cell of the same size to any of its sides. If the patterns are simply connected (i.e. they contain no holes) then Fig. 1.31 shows the form of all such patterns with up to five cells. Beyond five the number of combinations increases very rapidly: e.g. Reed (1962) has shown that for seven cells there are over one hundred

combinations, and for ten cells, over 4,000 alternative combinations. The
four-cell system may be of particular interest to settlement geographers
through its link to the quarter-section system. Johnson (1957), in a study
of land settlement in the Whitewater watershed in Minnesota, United
States, shows that the quarter-sections were taken up by settlers in a
number of ways (Fig. 1.32). Square arrangements of forty-acre units into
a single quarter-section (160 acres) were common only in the open prairie
areas, and in the broken timberland along the bluffs and the alluvial

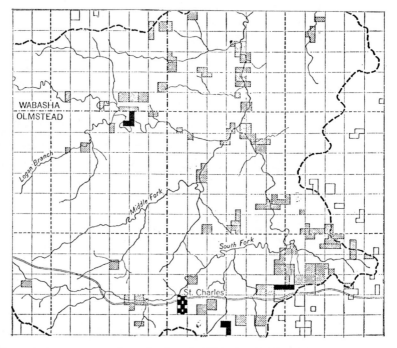

Fig. 1.32. Arrangement of regular farm holdings in 1855 in the Whitewater basin,
Missouri, U.S.A. Source: Johnson, 1957, p. 332.

land along the river bottoms, various combinations of four contiguous
squares were taken up. If one takes geographical orientation into account
then the five mathematical possibilities of combining four squares may
be increased to nineteen. Frequently squares were combined to allow
more rational use of available land resources, even where this meant farms
straddling township boundaries. Such topological ingenuity by the settlers
in southeast Minnesota in the 1850s represents what Johnson (1957,
p. 348) describes as 'a first, if short, step to adjust a rational land system
to geographical reality'.

Chapter two Geometric Structures

Whatever the gains that accrue from viewing a network in terms of its basic topology, it is clear that large and significant parts of its spatial structure are missing. In this chapter some of these missing elements are re-introduced: the fundamental relationship to area is brought back through the analysis of density characteristics, while the more elusive concepts of shape, orientation and spacing are explored. There remain, however, large areas of pattern analysis in which the treatment of networks remains primitive, and, with the surge of interest in pattern recognition that has accompanied satellite photography, we anticipate rather rapid advances in structural differentiation of network geometry in this area. Again the emphasis in the chapter is on the common structural characteristics of networks rather than their empirical functions, but examples continue to be drawn from a range of familiar geographic networks.

I. NETWORK SHAPE

Shape has proved one of the most elusive of geometric characteristics to capture in any exact quantitative fashion (Bunge, 1962, pp. 73–88). Many of the terms in common usage—circular, ox-bow, shoestring—turn out to be arbitrary so that misclassification is common, while some of the more mathematical definitions fail to do justice to our intuitive notions of what constitutes shape. In this section we look at the problems posed by the shape of the simplest network elements and go on to examine extensions to more complex network forms.

1. Open links : the shape of meanders

We may usefully distinguish the problems of describing the shapes of lines which represent deviations from a straight path between a given origin and destination (*open* links) from one in which either the origin and destination are in common and the line forms a loop or the boundaries of the network

57

are met (*closed* links). The first set of problems are characterized by the problems of describing meander trains and the second by parallel problems in cell description and overall network shape.

(a) Search for parameters

Smart and Surkan (1967, pp. 965–6) distinguish between two main types of deviations from the inter-node shortest path: *wandering* ('all sorts of unsystematic deviations from a straight-line path') and *meandering* ('curves of considerable symmetry whose dimensions are proportional to

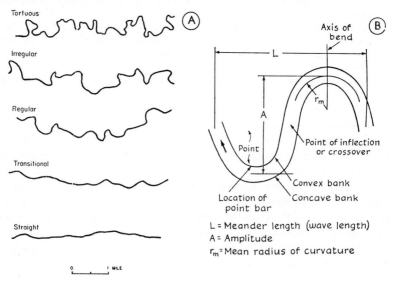

Fig. 2.1. (*A*) Examples of channel patterns. White River near Whitney, Nebraska (*P* = 2·1); Solomon River near Niles, Kansas (*P* = 1·7); South Loup River near St Michael, Nebraska (*P* = 1·5); North Fork Republican River near Benkleman, Nebraska (*P* = 1·2); Niobrara River near Hay Springs, Nebraska (*P* = 1·0). (*B*) Definition sketch for meanders. Source: Schumm, 1963, Leopold, Wolman and Miller, 1964.

the size of the channel') (Fig. 2.1-A). Meandering paths pose much greater problems in measurement in that (i) the paths may be many times longer than the direct *O–D* distance, and (ii) deviations are not random but rather highly symmetric.

One of the simplest measures proposed to describe path wandering is to relate the length of the observed path (O_L) to the length of the expected path (E_L), where this is measured as the direct 'desire-line' distance between the two ends of the path. Thus Schumm (1963, p. 1089) has proposed descriptive categories of channel sinuosity which range from 'straight' courses (O_L/E_L = 1·00) through three intermediate classes—'transitional', 'regular', and 'irregular'—to 'tortuous' courses (O_L/E_L > 2·00). (Fig.

) There is, in practice, some difficulty in applying this index to river channels since it is not always clear where the terminating points for the observation should be located. For transport lines the terminal points are usually clearly defined, and Kansky (1963, pp. 31–2) has proposed a general formula for a connected network of lines, viz.

$$\left\{ \sum_{i=1}^{n} (O_L - E_L)^2 \right\} \Big/ V$$

where V is the number of vertices connected. Fig. 2.2 shows the desire line

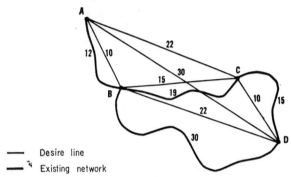

Fig. 2.2. Sample graph for calculation of degree of circuity in a network. Source: Kansky, 1963, p. 32.

and route maps for a hypothetical four-node network, and Table 2.1 the matrix from which the *degree of circuity* or *route factor* can be measured.

Nordbeck (1964) examined the possibility of a short-cut approximation to real-distance paths in road systems using a constant q in the relation: $O_L = qE_L$ where O_L is the real distance and E_L the straight-line distance. The real distance is assumed to lie within limits: $E_L < O_L < \sqrt{2E_L}$. If

Table 2.1. Calculation of circuity index

Links:	AB	AC	AD	BC	BD	CD
Desire-line distance (E_D)	10	22	30	15	22	10
Network distance (O_D)	12	31	40	19	30	15
Squared difference $(O_D - E_D)^2$	4	81	100	16	64	25

Nodes:	A	B	C	D	Mean
Sum of squared differences	185	84	122	189	—
Mean squared difference	46·25	21·00	30·50	47·25	36·25

Source: Kansky, 1963, p. 32.

the distance in a square street pattern is considered, and the distance from the inner part of the block to the nearest street corner is neglected, then at the lower limit the two points will be situated in the same street and in the second case the points will be situated diagonally (Nordbeck, 1964, p. 208). Studies within the town of Trollhattan, Sweden, showed values for the extension constant varied from 1·10 to 1·43 with the extreme values over the shorter distances. The general straight-line relationship yielded values of 1·21. Timbers (1967) investigated variations in this ratio of the road distance to direct distance (termed 'route factor') for journeys within the United Kingdom. On the whole the average route factor for trips between 780 pairs of towns was found to be 1·17 (see Fig. 2.3), with a small but statistically insignificant difference between pairs of large

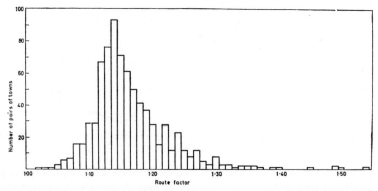

Fig. 2.3. Histogram of route factors between 780 pairs of British towns. Source: Timbers, 1967, p. 392.

towns (1·12 for towns with populations greater than 500,000) as compared with pairs of small towns (1·14 for towns less than 100,000). Regional differences within Britain were small: Wales (1·28) with its peripheral location and rugged terrain, has a higher ratio than the Midlands (1·12), a central location where routes are unbroken either by estuarine indentations or mountain masses. Journeys within towns were somewhat higher (see Table 2.2), and there were two noticeable trends—firstly for route factors to decrease with increasing distance between the trip terminals, and secondly for journeys to and from the centre to be less deflected than cross-town journeys. The relationship of Timber's findings to theoretical values for triangular, square and hexagonal networks is taken up in Chap. 4.I.2b.

    Mueller (1968) has drawn attention to the deficiency of the sinuosity index as applied to streams because it fails to distinguish between hydraulic sinuosity (*HSI:* i.e. that freely developed by the channel uninfluenced by valley-wall alignment) and topographic sinuosity (*TSI:* i.e.

Table 2.2. Route factors for journeys between points in towns

Town:	Random journeys		Journeys to or from the town centre		
	1 mile	½ mile	1 mile	½ mile	¼ mile
Bracknell	1·21	1·33	1·21	1·29	1·35
Cardiff	1·26	1·41	1·19	1·30	1·31
Eastbourne	1·30	1·36	1·14	1·16	1·22
London	1·20	1·22	1·13	1·13	

Source: Timbers, 1967, p. 394.

that imparted by the geometry of the valley). These two components of sinuosity, totalling hundred per cent, can be expressed as:

$$HSI = \frac{C/L_s - V/L_s}{C/L_s - 1} \times 100\%$$

$$TSI = \frac{V/L_s - 1}{C/L_s - 1} \times 100\%$$

Where C is length of channel; V is length of valley thalweg, and L_s is length of the straight line connecting the two ends of the channel. A comparison of the upper and lower courses of the Mancos River, Colorado,

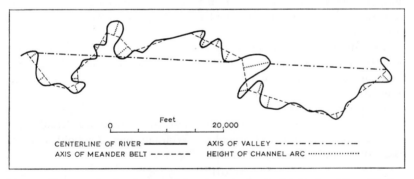

Fig. 2.4. Construction for characterizing sinuosity for the Maiandros River, Turkey. Source: Brice, 1964.

showed that, although their sinuosity indexes are similar (1·07 and 1·23), the components of sinuosity are very different (*HSI* is 44 and 71, respectively). Brice (1964) has further pointed to the difficulty of using some arbitrary value of sinuosity (i.e. ⩾ 1·5; see Leopold and Wolman, 1957) to specify meandering, because it ignores the properties both of arc symmetry and of the repetitive symmetry of a sequence of arcs. The first property can be assessed for a reach by comparing the average arc length, height and form ratio (i.e. arc length/arc height) with that of an 'ideal reach' in which the average arc is repeated along a straight line of the same length as the meander-belt axis, viz.:

$$\text{Symmetry of: arc length} = 100 - \frac{100 \; (\text{mean deviation of arc length})}{\text{mean arc length}} \%$$

$$\text{arc height} = 100 - \frac{100 \; (\text{mean deviation of arc height})}{\text{mean arc height}} \%$$

$$\text{form ratio} = 100 - \frac{100 \; (\text{mean deviation of form ratio})}{\text{mean form ratio}} \%$$

The property of repetition of symmetry can be measured by expressing as a percentage the number of arcs in the reach whose length does not deviate more than twenty-five per cent from the average length, divided by the number of arcs in the ideal reach. Fig. 2.4 shows a stretch of the Maiandros River, Turkey (the original River Meander) with a sinuosity index of 1·56 and the following symmetries—arc length seventy per cent, arc height sixty-six per cent, form ratio sixty-seven per cent and repetition thirty per cent. Brice (1964, p. 26) concludes that this river has a low degree of symmetry and suggests that the term meandering should apply 'to a sinuous stream that has, from place to place along its course, one or a series of symmetrical arcs, the length of which is related to the width of the stream'.

Deviations in meandering paths pose far greater problems of measurement than those of sinuosity. Attention has been focused on the simple geometrical relationships of such meander properties as wavelength (L), amplitude (A), channel width (W), meander belt width (W_b) and mean radius of curvature (r_m) (Fig. 2.1-B). The most consistent relationships have been found between wavelength, on the one hand, and channel width and radius of curvature, on the other (Leopold and Wolman, 1960; Leopold, Wolman and Miller, 1964, p. 296), such that (Fig. 2.5):

$$L = 10 \cdot 9 W^{1 \cdot 01} = 4 \cdot 7 \, r_m^{0 \cdot 98}$$

These correlations exist, of course, because the mutual relationships are controlled by such external factors as discharge and roughness. Leopold and Wolman (1960, p. 774) assumed the exponents in the equation to be unity, giving the ratio $\frac{r_m}{W} = 2 \cdot 3$; comparing this with the equivalent ratio measured for fifty streams they found the median to be 2·7 and that two thirds of the measured values fell between 1·5 and 4·3. It is interesting that this ratio similarity makes the patterns of most meandering streams look alike on planimetric maps despite wide difference in their absolute size. Other geometrical aspects of meanders correlate less well, however; amplitude, for example, correlates poorly with wavelength. This suggests that the latter may be more influenced by hydraulic factors, whereas the former is determined more by local factors such as bank erosion characteristics. An extreme example of this has been given by Strahler (1946) where the pronounced slaty cleavage in the Martinsburg Shale has caused the elongation of the meanders of Conodoguinet Creek, Pennsylvania. Attempts to fit geometrical curves to sets of meanders have not proved satisfactory, partly because of the combination of meandering and wandering previously referred to, and partly because it may be that a

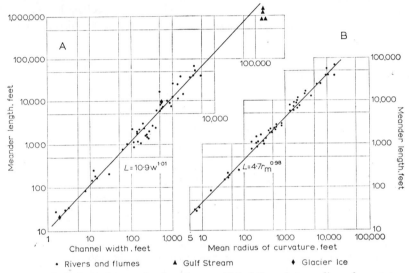

Fig. 2.5. Relation of meander length to width (*A*) and to radius of curvature (*B*) in channels. Source: Leopold, Wolman and Miller, 1964.

single stream channel may exhibit the effects of a number of superimposed meander wavelengths.

    In practice meanders are notoriously difficult to define and slight differences in operational definition yield significant differences in the results obtained. These difficulties led Speight (1965) to investigate the application of power-spectra mathematics to the continuously curving form of the river. Here changes in the angle of direction of the channel measured at standard distances along the talweg are substituted for measurements on individual 'waves'.

    Spectral analysis avoids the implication of earlier meander studies that there is a single dominant wavelength of meandering discernible (even though it may be obscured by irregularities of a quasi-random nature), and substitutes the idea of a 'spectra' of oscillations measured against distance downstream. Fig. 2.6-A shows a characteristic talweg trace analysed into a 'spectral envelope' (Fig. 2.6-B) containing a number of coherent oscillatory peaks. Variations in meander intensity $X(k)$ were calculated from the formula:

$$X(k) = \frac{1}{n}\left[C_x(0) + \sum_{l=1}^{m-1}\left\{ C_x(l)\left(1 + \cos\frac{\pi l}{m}\right)\cos\frac{\pi x l}{m}\right\}\right]$$

where n is the number of direction angles measured (between regularly spaced points on the talweg trace), C_x is the auto-correlation between successive angles (X), and m is the number of frequency bands (Speight, 1965, p. 5). Values of X were plotted against the frequency in cycles per 10^5 foot (this frequency being inversely related to wavelength). We can interpret the meander spectra shown in Fig. 2.6-B as consisting of two

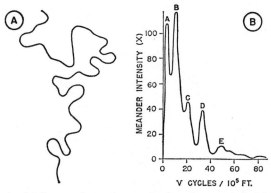

Fig. 2.6. Analysis of (A) meander trains on the Angabunga river, New Guinea, in terms of (B) a power spectrum. Source: Speight, 1965, p. 4.

intense peaks with a wavelength of about four miles (peak A) and two miles (peak B), with a series of smaller peaks at about $\frac{4}{5}$ mile (C), $\frac{3}{5}$ mile (D) and $\frac{2}{5}$ mile. Although the reliability and resolving power of spectral analysis hinge on the number of direction angles measured, the spacing of these measurements, and the number of frequency bands used in producing the spectra (Blackman and Tukey, 1959), it is clear that, with the availability of digital computers, power-spectra analysis is likely to be used increasingly in the future analysis of wandering paths.

A further method for describing meander geometry has been given by Langbein and Leopold (1966) and Scheidegger (1967A), and involves the statistical probability (p) of a stream direction deviating by an angle $(d\phi)$ from its previous direction in progressing an incremental distance (dS). If p is normally distributed, then:

$$dp = C \exp \tfrac{1}{2}\left(\frac{d\phi/dS}{\sigma}\right)^2$$

where σ is the standard deviation and C is defined by $\int d p = 1$. This formulation is identical to a class of random walk problems formulated by Schelling (1951). From the above, the total arc length (S) can be expressed:

$$S = \frac{1}{\sigma} \int \frac{d\phi}{\sqrt{2\,(\cos \omega - \cos \phi)}}$$

where ϕ is the direction angle measured with respect to the datum line

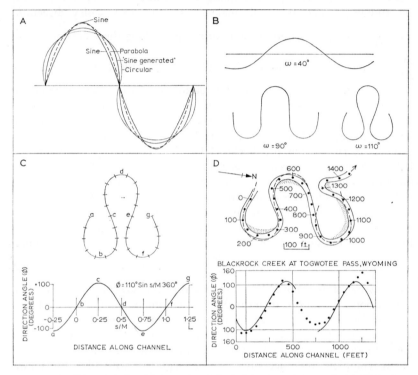

Fig. 2.7. (A) Comparison of four symmetrical sinuous curves having equal wavelength and sinuosity. (B) Examples of most frequent random walk paths of given lengths between two specified points on a plane. (C) Theoretical meander in plan view (above) and a plot of its direction angle as a function of distance along the channel path (below). (D) Map of channel of Blackrock creek, Wyoming, U.S.A., compared with sine-generated curve (dashed) (above) plot of actual channel direction against distance (dots) and a sine-curve (full-line) (below). Source: Langbein and Leopold, 1966; Schelling, 1951.

connecting the ends of the path and ω is the maximum value of ϕ. ω is quite independent of the total path (stream) distance (M) along the meander, and is a unique function of sinuosity, varying from $0°$ (straight) to a maximum of $125°$ (2.2 radians) at the incipient cut-off state, when the limbs intersect (Fig. 2.7-B). Approximating the above equation one obtains a definition of the planimetric geometry of stream meanders (Langbein and Leopold, 1966, p. 3):

$$\phi = \omega \sin \frac{\sigma\sqrt{2(1 - \cos \omega)}}{\omega} S$$

or
$$\phi = \omega \sin \frac{S}{M}2\pi$$

where M = total path distance along the meander; ω = maximum angle (in radians) which the path makes with the down-valley direction; and $\frac{S}{M}$ = the proportion of the total distance along the meander (i.e. at $\frac{S}{M} = \frac{1}{2}$ and 1, $\phi = 0$ and flow is directed down-valley; at $\frac{S}{M} = \frac{1}{4}$ and $\frac{3}{4}$, $\phi = \omega$). Fig. 2.7-c gives a plot of equation for a theoretical meander—which, it should be noted, is not a sine curve but a sine-generated curve—showing that changes in channel direction are a sinusoidal function of distance along the channel. Fig. 2.7-D shows a sine-generated curve fitted to an actual river reach, the Blackrock Creek at Togwotee Pass, Wyoming, U.S.A. A similar approach has been applied by Ongley (1968A) to the pattern of Serpentine Cave, New South Wales. He first took twenty-eight directional bearings along the channel; referring these to 179 degree-intervals, he found that a Chi-square test showed no significant difference between the observed deviations and the random ones predicted by the Poisson distribution, where the probability of a single bearing falling within any one-degree intervals is $\frac{28}{179} = 0.156$. However, recognizing that the drawing of directional lines tangential to the flow involves subjective assumptions, the author then divided the channel into segments of chords equal to four feet and specified the direction of each segment with reference to the angle between it and the preceding segment—plus for a change to the right and minus for a change to the left. One oscillation is then defined as two complete and successive series of segments with plus and minus angles. There were ten complete oscillations for each of which the wavelength (L_w) and the channel length (L_c) could be measured. A measure of sinuosity is then given by L_c/L_w; the wavelength/channel width ratio averaged 5.5, and the average channel length/width ratio was 7.1. A check of the results using a 2-foot channel segmentation gave results differing little from the above. Smart and Surkan (1967, p. 965) similarly define the sinuosity of a meandering-style reach as equal to the channel length for one wavelength divided by the wavelength.

(b) The Steinhaus paradox and its implications

In the attempt to devise operational methods of describing complex paths, the assumption has been made that accurate measurement is possible. Nystuen (1966) has drawn attention to a critical theorem by

Courant (1937, p. 277): 'Every curve $y = f(x)$ for which the derivative $f'(x)$ is continuous in a rectifiable curve, and its length between $x = a$ and $x = b$ $(b > a)$ is given by the formula:

$$S(a, b) = \int_a^b (1 - y'^2)^{\frac{1}{2}}\, dx$$

where
$$y'^2 = \left(\frac{dy}{dx}\right)^2$$

But most routes have many points of discontinuity on them so that 'the first derivative does not exist at all points on the curve and on deductive grounds it is not at all clear that the length of a boundary line can be measured with theoretical rigour' (Nystuen, 1966, p. 7). This problem has been stated by Steinhaus (1954) as a paradox: the more accurately an empirical line is measured the longer it gets.

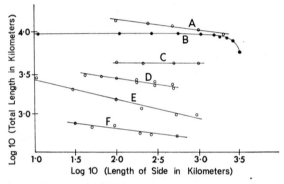

Fig. 2.8. Apparent boundary length as a function of the length of side of measuring polygons with equal sides. *A* Australian coastline. *B* Circle. *C* South African coast. *D* German land frontier, 1900. *E* West coast of Britain. *F* Land frontier of Portugal. Source: Mandelbrot, 1967, p. 636.

Clearly the length of a particular path may be approximated by summing a series of straight-line segments between points located on the line. As the segments become shorter the sums become larger; at the molecular level the length approaches infinity. Richardson (1960) has drawn attention to the changing lengths of political boundaries measured at different scales (Fig. 2.8). However, Nystuen (1966, p. 8) has pointed out that '. . . the paradox is not to be confused with the fact that all physical quantities are subject to errors of measure. The problem remains regardless of the level of accuracy. A finer measure will always be longer.' The paradox casts considerable doubt on the utility of such measures as length of coastline or political boundaries; Richardson (1960, p. 169) records major differences in the common land frontiers claimed by Spain (987 kilometers) with Portugal (1,214 kilometres) or the Netherlands (380 kilometres) with Belgium (449 kilometres).

Perkal (1966) has suggested that definition of the length of a line must involve a scale consideration, and has proposed a standard measure, *epsilon length*, for this purpose. The epsilon length of an empirical line is given by $L_\varepsilon(x)$ where:

$$L_\varepsilon(x) = \frac{a_\varepsilon(x) - \pi\varepsilon^2}{2\varepsilon}$$

(Nystuen, 1966, p. 11). Here $a_\varepsilon(x)$ is the area within the epsilon distance of the line, and $\pi\varepsilon^2$ is the area of the two half-circles at the end of the line. This last term is ignored for closed circles. We can imagine ε as the diameter of a circle which is 'rolled' along both sides of the line. The

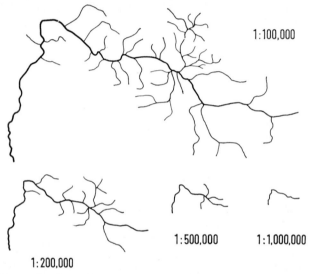

1:100,000

1:500,000   1:1,000,000

1:200,000

Fig. 2.9. Alternative form of stream system at various map scales. Source: Topfer, 1967, p. 250.

smaller the value of ε the more the irregularities of the empirical line are reflected in the area a_ε and the larger the estimate. Clearly the procedure is analogous to finding the length of a rectangle by dividing its area by its width. The advantages of Perkal's method are that: firstly, it ties the estimated length of the line to a standard, but, secondly, allows that standard to be selected by the researcher. Nystuen (1966) illustrated the use of the method in an analysis of the boundary length of Lake Minnetonka, U.S.A., as defined on a 1/126,720 map. Using an epsilon value of five millimetres (or 0·39 miles) the length was computed as 48·0 miles; when ε was reduced to four millimetres (or 0·32 miles) the estimated length increased to 54·0 miles.

Clearly the scale of the map is an important variable in determining the measured length of network segments, and its critical role in the

analysis of stream networks is discussed in Chap. 1.I(3). Fig. 2.9 shows
clear variations in channel plan for a stream system mapped at four scales
from 1/100,000 to 1/1,000,000. Such systematic scale variations are
paralleled in maps of road systems. Other non-systematic sources of bias
have been reported by Board (in Chorley and Haggett, 1967, p. 678),
who shows how map-producing agencies '... often adopt different

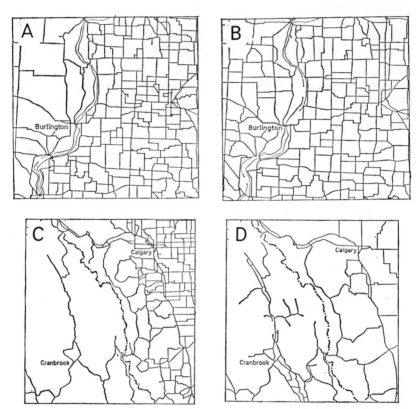

Fig. 2.10. Selectivity in the representation of road patterns on maps. (*A*), (*B*)
Illinois, U.S.A., as shown on State and oil company maps. (*C*), (*D*) Part of western
Canada as shown on British Columbian and Albertan maps. Source: Board, in
Chorley and Haggett, 1967, p. 678.

standards for the inclusion of detail as between their territory and surrounding areas.' Fig. 2.10 shows two examples of network distortion.
The official highway map of Illinois, U.S.A. (Fig. 2.10-A), gives poor
coverage of the network outside its own boundaries (compare the Standard
Oil Company map of the same area in Fig. 2.10-B); likewise maps of a
common area in Western Canada produced by the highway authorities of
British Columbia (Fig. 2.10-C) and Alberta (Fig. 2.10-D) show very

different patterns. Board cites a number of other examples—including the insertion of 'serpentine' bends on rivers to render them more 'river-like'—which suggest the need for some caution in direct map analysis of either shape or other geometrical characteristics.

2. Closed links

(a) Shape of cells

The shape of closed lines leads to the consideration of the shape of individual polygons (e.g. islands) and the average characteristics of polygons within networks (e.g. parish shape within a network of boundary

Table 2.3. Alternative ratios for the comparison of the shape of closed figures

Index:	Formula	Title	Origin
1	A/L^2	Form ratio	Horton (1932)
2	$A/\left\{\pi\left(\dfrac{P}{2\pi}\right)^2\right\}$	Circularity ratio	Miller (1953)
3	$\left\{2\sqrt{\dfrac{A}{\pi}}\right\}/L$	Elongation ratio	Schumm (1956)
4	$L/2\left\{A/\left[\pi\left(\dfrac{L}{2}\right)\right]\right\}$	Ellipticity ratio	Stoddart (1965)
5	$\displaystyle\sum_{i-1}^{n}\left\{100.\dfrac{R_i}{\displaystyle\sum_{i-1}^{n}R_i} - \dfrac{100}{n}\right\}$	Radial-line ratio	Boyce & Clark (1964)
6	$\dfrac{A}{\sqrt{2\pi\displaystyle\int_{a} R^2 dx\, dy}}$	Compactness index	Blair and Biss (1967)

Variables:

A Area (square km)
P Perimeter (km)
L Major axis (km)
B Minor axis (km)
R_i Radial axes from gravity centre to perimeter (km)
aR Radial axes from gravity centre to small area, 2 (km)

Source: Stoddart, 1965, Blair and Biss, 1967.

lines). Shape characteristics of the former have led to a considerable literature within the field of sediment analysis and a wide range of form ratios has been derived (see Table 2.3). Most involve manipulation of basic values for the area, perimeter, and axes of the unit and their comparison

with basic geometric forms like the circle or eclipse. More complex measures involve the determination of the gravity centre of the area and the
length of radial lines from this centre, or, in the case of Bunge's shape index
(Bunge, 1962), the comparison of the shape with sets of equal-sided
polygons. Stoddart (1965) used a battery of the five ratios shown in
Table 2.4 plus the Bunge index to describe the shape characteristics of a
sample of ninety-nine coral atolls drawn from the four major island groups
of the south-west Pacific (Table 2.4). Each index ranked the atolls in
different ways, though there was sufficient general homogeneity in the
results to suggest that Satawan atolls (appearing in the median group of

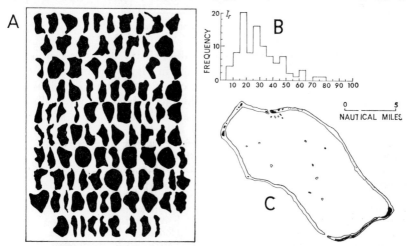

Fig. 2.11. (*A*) Shapes of 99 sample atolls with (*B*) their frequency distribution on
the radial-line ratio. (*C*) 'Typical' shape (Satawan Atoll). Source: Stoddart, 1965,
pp. 371, 381.

three of the indices) may be regarded as having the 'most typical atoll
shape' (Fig. 2.11). Stoddart's study was concerned with building general
models of atoll form, but throws considerable light on the general problem
of cell description within networks. Blair and Biss (1967) critically review
the attempts made to develop significant shape indices with special
reference to the problems of 'fragmented' and 'punctured' units. They
propose a new shape measure (Table 2.3) to overcome some of these
problems; the application of the measure to a series of political units is
shown in Table 2.5.

    Empirical work on the shape of administrative area has been attempted
by Haggett (1965, p. 50) using a simple shape index, S, given by
$S = (1\cdot27A)/l^2$ where A is the area of the county in square kilometres, and
l the long-axis of the county drawn as a straight line connecting the two
most distant points within the perimeter. The multiplier ($1\cdot27$) adjusted
the index so that a circle would have an index of $1\cdot00$ with values ranging

Table 2.4. Summary of atoll-shape data by island groups*

Atoll group:	Number in sample	(1) (mean)	(2) (mean)	(3) (mean)	(4) (mean)	(5) (mean)
Caroline	27	0·348	0·554	0·616	2·750	30·38
Marshall	16	0·395	0·524	0·593	3·656	33·56
Maldive	17	0·375	0·644	0·677	2·429	22·13
Gilbert and Ellice	9	0·324	0·478	0·641	2·284	30·34
Total sample Means	99	0·349	0·569	0·638	2·867	29·78
Standard deviation		0·133	0·186	0·142	—	14·73

Source: Stoddart, 1965.   * For indices refer to Table 2.3.

down towards zero. The actual shape values recorded by this method for a sample of Brazilian counties are shown in Fig. 2.12, and range from values as low as 0·06 for very elongated counties to values as high as 0·93 for compact near-circular counties. In this measuring system the values for the three lattices are 0·42 for triangular, 0·64 for square and 0·83 for

Table 2.5. Shape of major administrative areas*

Rhodesia	0·974	Uganda	0·890
Rumania	0·969	Venezuela	0·890
Uruguay	0·968	Saudi Arabia	0·880
Kenya	0·964	S.W. Africa	0·880
Algeria	0·963	Congo	0·871
Bolivia	0·957	Persia	0·864
Egypt	0·957	Luxembourg	0·860
Nigeria	0·955	China	0·853
Sudan	0·951	Turkey	0·847
Ecuador	0·950	Syria	0·845
Spain	0·947	Costa Rica	0·832
Angola	0·947	Mongolia	0·810
Iraq	0·942	Greenland	0·806
Nicaragua	0·937	Paraguay	0·769
Tanganyika	0·935	U.S.S.R.	0·769
Colombia	0·931	Argentina	0·733
France	0·930 (with Corsica)	Jordan	0·730
Ethiopia	0·925	Mozambique	0·726
South Africa	0·925	Somaliland	0·716
Ghana	0·915	Czechoslovakia	0·711
Libya	0·914	Norway	0·444
Afghanistan	0·907	Pakistan	0·432
Tunisia	0·902	Philippines	0·406
Eire	0·893	Chile	0·304
Burundi	0·893		

Source: Blair and Biss, 1967, p. 11.
      * Compactness measured on Blair—Biss index.

hexagonal; boundary lines have been interpolated on Fig. 2.12 to divide
the sequence into three zones about these values. The results strongly
suggest the generally elongated nature of the counties. However, the
possible correspondence of the lattice boundaries with gaps in the frequency distribution suggested that shapes tend to cluster about the three

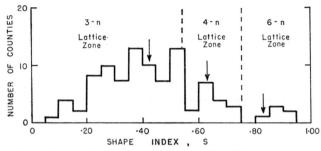

Fig. 2.12. Shape characteristics of a sample of 100 Brazilian counties in relation to
lattice zones. Source: Haggett, 1965, p. 51.

alternative tessellations proposed by Kepler. This possibility was taken up
by Pedersen (1967) for a sample of Danish communes. The frequency
distribution of shapes was broadly comparable to that of Fig. 1.12 with a
preponderance of triangular elements, but the gaps between the three
major lattice classes were not confirmed.

(b) Overall network shape

Shapes of entire network have been studied by Kansky (1963, pp. 21–3)
using a combination of topological and geometrical measures. He begins
with a rather weak topological measure, diameter (δ) (Chap. 1.II(3)),
which gives an index of the topological length or extent of the graph by
counting the number of edges in the shortest path between the most
distant vertices. Diameters rise in value as the 'extent' of the graph increases, but fall as improved connections are made between vertices. From
the primitive concept of diameter Kansky moves on to relate this to the
actual dimensions of the network, using an index π given by C/d, where C
is the total mileage of the transportation network and d is the total mileage
of the network's diameter. In Fig. 2.13-A we have a network where the
total mileage (C) is 150. Inspection of the graph's diameter in terms of the
preceding diagram shows that the value of δ is four. There is, however, no
unique diameter, but six alternative paths that fulfil the minimum diameter criterion; the location and length of these paths are shown in
Fig. 2.13-B. The mean length of the six diameter paths is fifty-five miles;
so that the shape index π is 150 divided by fifty-five, i.e. 2·73. In practice
the shape index varies considerably over the range of transportation networks studied by Kansky. For railway networks, developed countries like
France may have indices approaching thirty, while underdeveloped

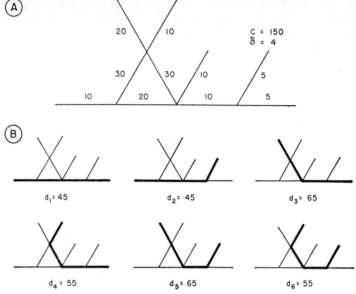

Fig. 2.13. Derivation of a shape index for a hypothetical route network. Source: Kansky, 1963, p. 23.

countries like Bolivia may have values of one. Changes in the value of π over time suggest that the index increases with the general level of regional economic development (Haggett, 1965, pp. 239–40).

II. NETWORK DENSITY

The critical relationships between the magnitude of the network and its areal environment are expressed in terms of density measures of one kind and another. In this section we look at some of the research findings on this interaction for the three major topological classes of network.

1. Branching Networks

(a) *Drainage density*

The most important areal measure of stream network geometry is the *drainage density* (D), which expresses the fineness of erosional texture of the landscape, and is a valuable indicator of the scale of planimetric units (e.g. stream spacing) (Melton, 1957, p. 4). Drainage density (D) was defined by Horton (1932, p. 357; 1945, p. 283) as the average length of stream channel within a basin per unit of area (D is dimensionally equal to L^{-1}, and commonly expressed in miles per square mile):

$$D = \frac{\Sigma L}{A}$$

Drainage density exhibits a very wide range of values in nature and is generally believed to reflect the operation of the complex factors controlling surface runoff. Common values for D are 3·0 to 4·0 for the sandstones of the Appalachian Plateaus (Smith, 1950, pp. 658–9; Morisawa, 1959, pp. 84–6); 8·0 to 16·0 for rocks of moderate resistance in areas of the humid central and eastern United States covered in deciduous forest (Strahler, 1952, p. 1135; Coates, 1958, pp. 19 and 59); 20·0 to 30·0 for the chaparral-covered metamorphic Coast Ranges of California (Smith, 1950, p. 659; Strahler, 1952, p. 1135; Maxwell, 1960); 50·0 to 100·0 for the dryer areas of the West (Melton, 1957); 200·0 to 400·0 for the shales of the Dakota badlands (Smith, 1958, p. 999); and 1,100·0 to 1,300·0 for the Perth Amboy clay badlands (Schumm, 1956, p. 616).

Obviously the values of drainage density obtained are very dependent upon the scale of the investigation. Attempts to circumvent the tedium of measuring the necessary lengths and areas and to predict drainage density from line sampling studies have not proved very promising, although techniques will doubtless improve. For example, Stoddart (in press) attempted to characterize D for fifty 1 : 25,000 map sheets of lowland England by sixteen intercept line sampling methods and six quadrat methods. The best of the former (a set of five lines, one kilometre apart, five kilometres long and rotated twice) gave thirty-seven per cent prediction, and the best quadrat method much less (twenty-five one square kilometre quadrats for each one-hundred square kilometre map sheet).

Preliminary work seems to show that drainage density within a basin is independent of order, and that the drainage density of large basins is not significantly different from that of the included first-order basins (Hack, 1957, p. 66; Morisawa, 1959, p. 9 and 1962, p. 1035). Drainage density is largely controlled by the lengths of the first-order streams (i.e. by the position of the streamheads), and Kirkby and Chorley (1967) have questioned Horton's model in which channel development is merely a question of space-filling with reference to distance from the divides—at least in areas of deep soil cover. Horton (1932; 1945, p. 285) deduced that the *length of overland flow* (L_g) could be expressed as:

$$L_g = \frac{1}{2D \sqrt{1 - \left(\dfrac{S_c}{S_g}\right)^2}}$$

where $\qquad \dfrac{S_c}{S_g} \backsimeq \dfrac{\text{average channel slope}}{\text{average ground slope}}$

and suggested that if S_c is less than $\frac{1}{3}S_g$, then the term $\sqrt{1 - \left(\dfrac{S_c}{S_g}\right)^2}$ may be neglected, giving the more usual form:

$$L_g = \frac{1}{2D}.$$

A measure similar to the length of overland flow is the *constant of channel maintenance* (*C*) which is the area, in square feet, required to maintain one foot of drainage channel (Schumm, 1956, p. 607):

$$C = \frac{1}{D} \times 5280$$

Stated another way, 'the relationship between mean drainage-basin areas of each order and mean channel lengths of each order of any drainage network is a linear function whose . . . regression coefficient is equivalent to the area in square feet necessary on the average for the maintenance of one foot of drainage channel' (Schumm, 1956, p. 607). Strahler (1957,

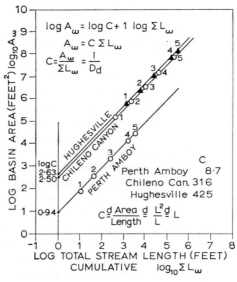

Fig. 2.14. Constant of channel maintenance, *C*. (*ω* = Strahler order). Source: Schumm, 1956, p. 606.

p. 915) expressed the same concept by means of a logarithmic relationship between basin area and cumulative stream length, in which the constant of channel maintenance is the antilog of that area when cumulative length equals unity (Fig. 2.14). The parallelism of the plots for the three basins shown foreshadows the dimensional similarity to be treated shortly.

Another measure of drainage texture is the stream frequency (*F*) (Horton, 1932, p. 357; 1945, p. 285):

$$F = \frac{\sum_{r=1}^{K} N_u}{A_K}$$

This is a measure of the number of stream segments per unit area (Strahler, in Chow 1964), and is a different concept from drainage density of which it may be independent (Melton, 1957, p. 4) (Fig. 2.15). Melton (1958A, p. 36) showed a very high correlation between F and D (for 156 basins, $r = +0.97$) (Fig. 5.15). The same author demonstrated that the ratio F/D^2 is a dimensionless measure of the completeness with which a channel system fills a basin outline, for a given number of channels (Melton, 1958A, p. 35), and that it may be an important evolutionary index of drainage systems. The F/D^2 ratio holds over wide variations in scale.

A less important measure is the *texture ratio* (T) (Smith, 1950), which

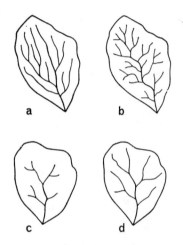

Fig. 2.15. Basins (a) and (b) have the same drainage density, but different stream frequencies; basins (c) and (d) the same stream frequency but different drainage densities. Source: Strahler, in Chow 1964.

is the number of crenulations (N) in the basin contour having the maximum number of crenulations, divided by the perimeter (P) of the basin. This can be weighted (T_m) per unit area (Smith, 1950, p. 660) giving characteristic values for coarse (<4), medium $(4\text{--}10)$ and fine (>10) textured topography. The texture ratio has been shown to bear a fair relationship to measured values of the drainage density (Strahler, 1957) (Fig. 2.16).

Shreve (1967, p. 184) has derived certain of the above relationships from mathematical assumptions. Assuming that all links in the drainage system have the same length, one could state crudely that mean stream length increases with order as a geometric series with ratio two, and total stream length as a geometric series with ratio four (a good agreement with reality). If it is further assumed that link lengths (l) have a logarithmic-normal distribution with a mean inversely proportional to the drainage density (D), then:

$$L = l(2\mu - 1)$$
$$A = Kl^2(2\mu - 1)$$
$$D = \frac{L}{A} = \frac{1}{Kl}$$
$$K = \frac{(2\mu - 1)}{LD}$$

where L is the total channel length in a basin of magnitude μ and area A, and K is a dimensionless constant. Taking the value of K to be 0·96 (the

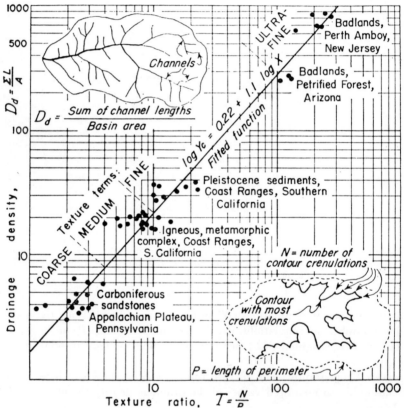

Fig. 2.16. Definitions of drainage density and texture ratio. Source: Strahler, 1957.

mean of Melton's eleven basins with no differential geological control (Melton, 1957, Table 2)), the mean basin area drained by each link is a square of approximate area l^2, and the constant of channel maintenance is approximately l. Assuming that the bifurcation ratio is $\frac{1}{4}$, then: (1) $\mu = 4^{w-1}$ and $A \backsimeq (2/D^2)4^{w-1}$ (where w = basin order). This agrees well with nature. (2) Number of Strahler streams (S_s) in networks of magnitude $\mu = 4/3$, which agrees with Melton's (1957) mean ratio of 1·34. (3) If $F = \dfrac{S_s}{A}$ and $K \backsimeq 1$ then $\dfrac{F}{D^2} \backsimeq 0·694$ for a large basin. Melton's mean value was 0·71.

(b) Length—order relationships

Horton (1945, p. 291) stated a law of stream lengths: 'The average length of streams of each of the different orders in a drainage basin tend closely to approximate a direct geometric series in which the first term is the average length of streams of the first order'

i.e.
$$L_H = L_{H1}\, R_{LH}^{H-}$$

where R_{LH} is the length ratio. Horton also gave a short method for approximating the total length of streams of each order (H):

$$\sum_{i=1}^{N_H} L_H = L_{H1} . R_{bH}^{K-H} . R_{LH}^{H-1}$$

where K is the order of the whole basin, N_H is the number of streams of each order, and R_{bH} is the bifurcation ratio. As Schumm (1956, p. 604) has pointed out, the greatest deviations from this rule occur when the entire length of the highest order stream has not been identified.

The application of the Strahler method of stream ordering showed histograms of individual orders to be strongly right-skewed (Miller, 1953; Schumm, 1956, p. 607; Strahler, 1964, in Chow pp. 4–45) and of logarithmic normal character (Fig. 2.17). The use of the mean lengths of Strahler streams

$$L_s = \frac{\sum\limits_{i=1}^{N} L_S}{N_S}$$

—was also believed to conform to Horton's law of stream lengths (e.g. by Schumm, 1956; Leopold and Miller, 1956; Chorley, 1957; Morisawa, 1959 and 1962, p. 1029) (Fig. 2.18-A), but some workers have questioned the validity of the law as applied to Strahler streams (Melton, 1957; Maxwell, 1960, p. 23). It was clear that Horton's plot of mean length versus order was more nearly exponential than that of Strahler's versus order (Woldenberg, 1967, p. 116), and Strahler (1957, p. 915) suggested a modification involving the application of a power function relationship between his total stream lengths and order. Morisawa (1962, p. 1030) showed that its regression coefficient (b) could be expressed by:

$$\log^{-1}b = R_{LS} - R_{bS}$$

However, this modification is not a great deal more satisfactory than the original Strahler mean length relationship, because total length for each order is simply a function of mean length times number of streams, and, as the latter is a strong inverse geometric series, this largely controls Strahler's plot of total stream lengths (Bowden and Wallis, 1964, pp. 769–770). Consequently Broscoe (1959, pp. 4–5) proposed an improvement of

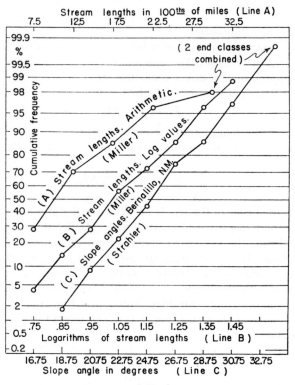

Fig. 2.17. Cumulative frequencies plotted on probability paper of (*A*) arithmetic and (*B*) logarithmic values of sample stream lengths. Slope angles from Bernalillo, New Mexico, are plotted on (*C*). Source: Strahler, 1954.

Strahler's original plot of mean stream lengths by *cumulating* the mean segment lengths (*L′*) for each Strahler order, showing that these are equivalent to Horton's mean stream lengths (Bowden and Wallis, 1964):

$$L' = \sum_{l=1}^{N} \bar{L}_S$$

Broscoe's modification fits Horton's inverse geometric series better than Strahler's uncumulated mean lengths, with a length ratio approximating that of Horton (Fig. 2.18-B) (Bowden and Wallis, 1964), and the culmination is hydrodynamically sensible in that conditions downstream are dependent upon some measure of cumulated length upstream (Woldenberg, 1967, p. 116).

Scheidegger (1966B, p. 202) has shown that the ratio of stream lengths can be deduced as forming a direct geometric series from stochastic branching principles. The distance (L_i, which is also equal to the mean length of stream segments of given order) which a stream has to flow in an area (A) to encounter another of the same order (from a population of n_i) varies as the square of the mean distance between streams of that order, i.e.

$$L_i \sim [(A/n_i)^i]^2 \sim A/n^i$$

and

$$\frac{L_{i+1}}{L_i} = \frac{n_i}{n_{i+1}} = \frac{2}{p}$$

where p is the probability of joining another stream of the same order. The same author (Scheidegger, 1968c) compared actual values of R_L obtained for the Appalachian Plateaus (Morisawa, 1962) with synthetic

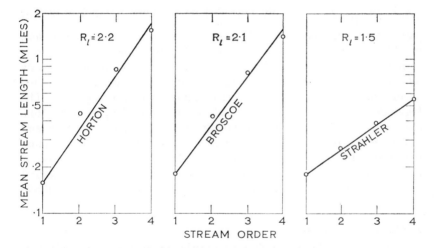

Fig. 2.18. Comparison of the law of lengths and resulting length ratios for Castle Creek near Soda Springs, California. Length is defined after Horton (*A*), after Broscoe (*B*), and after Strahler (*C*). Source: Bowden and Wallis, 1964.

values from randomly-generated networks in which each edge was assumed to have the same length (L) and to drain the same area (L^2). The length ratios fitted far less well than the area ratios, however, and Scheidegger (1968c, p. 1020) concluded that the assumption of a constant edge (or link) length was at fault. Pursuing the same line of reasoning, Smart (1968A), noting that the Horton approach to basin morphometry was essentially an attempt to describe its planimetry in terms of five parameters (R_b, R_L, R_A, $\bar{L}_1$ and $\bar{A}_1$), assumed:

1. That all topologically distinct networks with a given number of sources are equally likely (Shreve, 1966).
2. That the lengths of interior links are independent random variables drawn from the same population.

The results of this simulation seem to agree with field measurements better than does Horton's law of stream lengths, and to account for about sixty-five per cent of the variance in mean stream length data for third- and fourth-order basins. The full development and testing of this model is hampered by lack of field data on the distribution of interior link lengths.

Other linear measures which have been used to describe stream networks include: (1) The *length of overland flow* (L_g) (Horton, 1945, p. 284), is the horizontal length of the flow path from a point on the divide to the appropriate orthogonal point on the adjacent stream channel (Strahler; in Chow 1964). Where this is related to first-order drainage basins it is approximately equal to one-half the reciprocal of the drainage density (i.e. $L_g \simeq \dfrac{1}{2D}$). L_g is usually averaged by sampling a number of flow paths emanating from uniformly-spaced points around first-order basins. (2) The *interbasin length* is the horizontal length of the triangular areas without channels between adjacent tributary basins and draining directly into channels of higher than first order without themselves being included in any lower order basin (Strahler; in Chow 1964, pp. 4–47). (3) The *longest dimension* is measured from the basin mouth to the perimeter (Horton, 1932). This has been modified by Smart and Surkan (1967) to extend to the highest point on the perimeter of the first order basin containing the source of the master stream (L_0). (4) The *length of the main stream*, or master stream (L'). (5) The *mesh length* of the master stream (L'_μ), measured from the basin outlet along the mainstream and continued to the divide (Smart and Surkan, 1967).

2. Circuit networks

Variations in the density of circuit networks are examined in this section in relation to a wide range of environmental variables. The networks chosen for study are road and rail transport systems arranged in an ascending scale-order from local urban networks up to national transport systems (Haggett, 1965, pp. 73–9).

(a) Regional variations in density

The variations in the density of transport networks at the *local* level are an integral part of central-place theory. The proportion of space occupied by transport channels increases as centres of activity are approached; as Fig. 2.19-B shows, nearly one half of the land use of central

Detroit is taken up by roads, but this proportion falls rapidly with distance from the *CBD*. Similar distance-decline rates have been confirmed for American and British cities. As city size increases the needs for central interaction increase and the proportion of land-use devoted to transport needs rises in a linear fashion (Fig. 1.19-c). Information on density lapse-rates for transport channels has been sharpened by a detailed study by Borchert (1961) of the road pattern in the 'twin-cities' area of Minneapolis –St Paul in the north central United States. Instead of measuring road

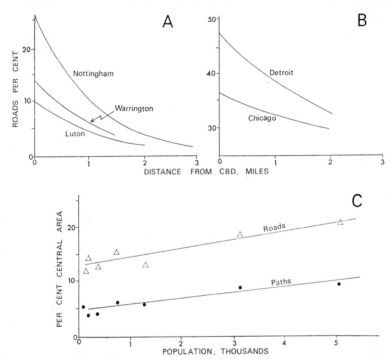

Fig. 2.19. Proportion of land devoted to transport channels with distance from town centre (*A, B*) and size of town (*C*). Source: Owens, 1968; Clark, 1967.

density by length of road per unit, he evolved a simple measure of counting all the road junctions on the map. The density of junctions was found to be so highly correlated with road length (coefficient of correlation of +0·99) that it could usefully be substituted for conventional and slower length measure. Borchert's findings show for the Minneapolis–St Paul area the very strong association between population density, as measured by number of single family dwellings, and the network density, as meas-ured by intersections. The concentric pattern of the network density zones about the two city centres was found to be strongly marked.

At the *regional* level a very thorough investigation of the distribution of road densities in Ghana and Nigeria has been made by Taaffe (Taaffe, Morrill, and Gould, 1963). Route density for each of Ghana's thirty and Nigeria's fifty sub-regions was measured for first- and second-class roads and related in the first instance to the population and the area of each unit. Using regression analysis, population was found to explain about fifty per cent of the variation in road density in both Ghana and Nigeria. When area was included with population in multiple regression, the level of explanation rose to seventy-five and eighty-one per cent respectively.

Taaffe went on to suggest four other less important variables that might help to solve the 'unexplained' differences between the actual densities and the expected densities on the regression analysis. These four variables were identified as: (i) hostile environment; (ii) rail-road competition; (iii) intermediate location; and (iv) commercialization and relation to the development sequence.

Hostile environment, a familiar and basic geographical theme, was illustrated in Ghana by the very low route densities in the swampy lands of the Volta river district and where the Mampong escarpment sharply restricts the development of feeder routes. Railroad competition was found to be a more complex factor in that one could argue either that railroads would curtail the need for roads by providing an alternative form of transport, or that railroads would stimulate road building by its gingering effects on production for inter-regional trade. The second argument appeared the stronger in Ghana and Nigeria. Units with an intermediate location, between two important high-population areas, were found to have densities well above those expected on the basis of population and area alone. Road density was positively associated with the degree of commercial activity, the more productive areas having a heavier road pattern than more backward areas. One anomaly noted from this pattern were the mining areas which, relying largely on rail movements, did not follow the resource development-road density relationship.

Difficulties in comparing network densities at the *world* level between countries raise acute problems of differences in operational definition of routes. Not only are definition problems multiplied (i.e. differences between single- and multiple-track railways or farm roads and eight-lane freeways), but similar information is recorded and classified in very different ways. Ginsburg has attempted to standardize these conflicting figures in his *Atlas of economic development* (1961) and his findings will be used here as a basis for argument. Two critical maps in the Ginsburg atlas are of railway density (Map XXIV) and road density (Map XXIX). Both maps show density as length of road per hundred square kilometres, though it is emphasized that there are a number of other and equally valid ways of showing density (e.g. in relation to population or population and distance). For our purposes the density per unit area provides the more basic parameter in that it describes the actual existence on the ground of specialized routes, whether those routes be intensively or lightly used.

The basic characteristics of the world distribution pattern are shown in summary form in Table 2.6. Road density, an index compiled from a a variety of sources and with rather unstandardized figures, gives a world average of around ten kilometres/one hundred square kilometres or about ten times as great a density as that for railways. The gap between the maximum values and the means is however considerably greater for roads; Belgium, reported with the highest road density, was about thirty times as dense a network as the world mean, while Luxembourg with the highest railway network was only about twenty times as dense as the world mean. At the other end of the distribution, one country (Greenland) is reported with zero road density, and twenty-seven countries have no railways. The distribution is then a very skewed one, with a few countries

Table 2.6. Distribution of route density

Route media:	Roads	Railways
Number of countries compared	126	134
World mean density, km/100 km²	10·3	0·95
Maximum density, km/100 km²	302·0	17·90
Minimum density, km/100 km²	0·0	0·00
Countries below world mean, per cent	64%	67%

Source: Ginsburg, 1961, pp. 60, 70.

with very dense networks and many countries with very sparse networks. Nearly two-thirds of the countries have densities below the world mean.

Transport networks are demonstrably part of the development 'infrastructure' and the distribution of countries with high and low densities may be reasonably linked to their general level of economic development. This hypothesis may usefully be explored by adopting the economic-demographic development scale developed by Berry (1960-B), based on the values in the Ginsburg atlas (Ginsburg, 1961, pp. 110–19). The scale is derived from forty-three separate indices of economic development which plots countries on a demographic scale along the shorter x-axis and a technological scale along the longer y-axis. Some ninety-five countries are distributed along this scale (Fig. 2.20), with highly developed countries on the upper left of the scale and poorer countries on the lower right.

On this continuum countries with high and low road and rail densities have been superimposed: the first ten countries with high densities are shown with large solid circles and the last ten countries with the lowest densities are shown by large open circles. The location of the United States on this continuum has been marked with an asterisk (*) for

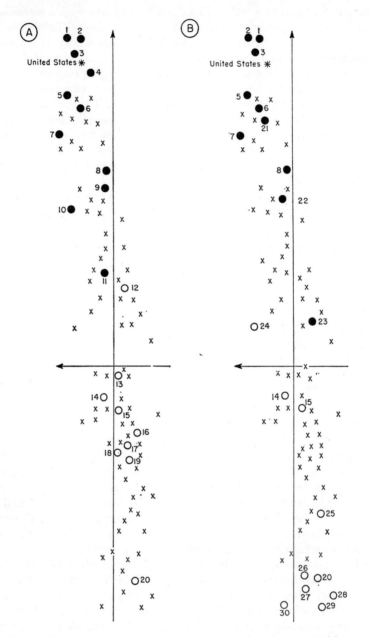

Fig. 2.20. Relationship of countries with high and low road (*A*) and railway (*B*) density to Berry's development spectrum. Source: Berry, 1960, p. 91.

reference. The road density pattern (Fig. 2.20-A) shows a cluster of high-ranking countries at the developed end of the spectrum. Nine of these ten are European countries; United Kingdom (1), West Germany (2), Belgium (3), France (4), Switzerland (5), Netherlands (6), Denmark (7), Poland (8), and Ireland (10). Only one other country, Hong Kong (11), lies outside this pattern. Indeed the only highly developed countries with relatively modest road densities are the United States, Canada, and Sweden. The pattern for railway density (Fig. 2.20-B) in general follows the same pattern, with European countries in leading positions. Seven of the countries re-occur with two eastern European countries, Czechoslovakia (21) and Hungary (22), and the far eastern state of Taiwan (23) coming into the picture. The apparently anomalous position of Taiwan reflects the relatively high-ranking position on the railway density map of a number of south-east Asian countries like India and Burma which were developed under the railroad-building British colonial administration, or Japan and Taiwan (a former Japanese colony) in which transport was deliberately developed about the railroad net.

At the other end of the extreme the position of low-ranking countries is complicated by the absence from Berry's economic-demographic development scale of most of the very underdeveloped countries. Not enough data was available to place them accurately on the scale and the 'ten lowest density countries' are drawn from the more restricted population of the ninety-five countries on the continuum. Nevertheless the pattern shown is an interesting one. In terms of railway density (Fig. 2.20-B) the low-ranking countries cluster strikingly at the bottom of the development ladder. Six of the seven countries at this base are African states, Sudan (25), Ethiopia (28), Libya (30), Liberia (20), Gambia (27), and French Equatorial Africa (26), together with Afghanistan (29). More developed colonial countries with low-density railroad nets were Surinam (14) and British Guiana (15), both with excellent river transport, and the only major anomaly, Iceland (24).

For road density (Fig. 2.20-B) the pattern of the low-ranking countries was not so clear. Relatively developed countries with very large land areas, U.S.S.R. (9) and Brazil (12), stand out as major anomalies, while at the lower extreme only Liberia represents the African cluster noted on the railway density map. Surinam and British Guiana here form the centre for a cluster of non-African tropical states in the lower-middle range of development with Costa Rica (13), Ecuador (16), Bolivia (18) from the Americas, and Iran (19) and British Borneo (17) from Asia. In general the road density pattern is less easy to interpret and reflects in part the wide range in definition of 'roads'. The lack of correspondence between the lows on the two media suggests that railways have served as substitutes for roads, and in other cases, like British Guiana, river and coastal shipping has served as a substitute for both.

(*b*) *Structural interrelationships*

The relationship of density to other structural parameters for transport systems was explored in studies carried out at Northwestern University by Garrison and Marble (1962, pp. 27–53) and Kansky (1963, pp. 37–80). A series of measures was examined of the road and railway network of a group of fifteen countries, varying in development level from Sweden down to Angola. Fig. 2.21 is an attempt to summarize the findings and relate them to other work by Ginsburg (1961). The diagram consists of two main systems, a *transport system* and an *environmental system*. These major systems are sub-divided in turn into road and rail sub-systems and socio-economic and physical sub-systems respectively. Within each sub-system 'box' are a series of parameters which describe either the form of the transport network (density, load, etc.), or the characteristics of the environment within which that network is located (urbanization, terrain ruggedness, etc.). The parameters are largely self-explanatory: 'Berry I' refers to an index of Technical Development developed by Berry (1960) from factor analysis of forty-three indices of economic development for a sample of ninety-five countries: 'Berry II' refers to an index of Demographic Status also determined from the same study.

Links within the system are shown by vectors. The internal links (the broken arrows) show the interlocking of the various structural parameters within the transport system: only one vector is shown for each parameter, its direction being determined by the highest correlation level (*rho* value), achieved with the other parameters within the system. The important links between the two density measures stand out with vectors clustering around the railroad-density 'box'. In general the rail sub-system was much more closely interlocked than the road system, which showed more links with the rail system than with itself.

Internal links within the environmental system were not of direct interest, and therefore vectors were drawn only between the environmental parameters and the transport box with which it was most closely correlated. The strength of these inter-system links are generally less than those within the transport system itself. Again the highest links are with railroad density (closely associated with the Berry I Index of general economic development). However, the most striking feature of the diagram is the convergence of vectors from the environmental system on the railroad sub-system. Roads appear to be much less closely linked with either the economy or the physical environment of the countries in which they are set.

In interpreting the diagram it should be remembered that the standard and comparability of rail data is generally better than road data, that relationships are based on a sample of only fifteen countries, and that only gross differences between countries are shown. Analysis is also hindered

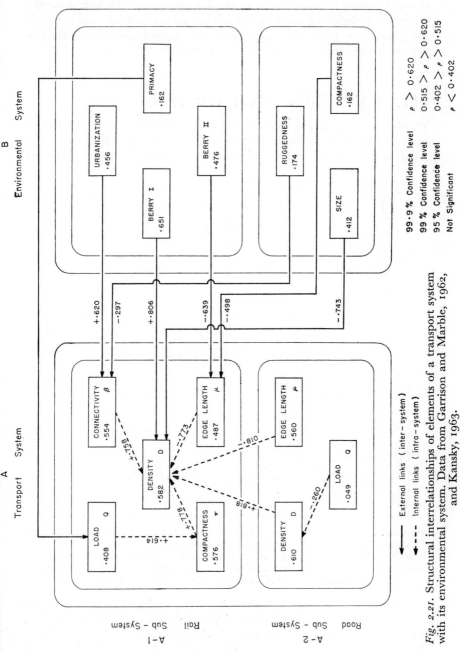

Fig. 2.21. Structural interrelationships of elements of a transport system with its environmental system. Data from Garrison and Marble, 1962, and Kansky, 1963.

by the need, for the sake of clarity, to show only the dominant links in the system. However, values given in the left corner of each box show the general correlation co-efficient of each parameter with all others within the transport system; these suggestions confirm the importance of the two density parameters as general guides to network structure.

III. PATTERN AND ORDER

Within the geographical literature there have been a number of attempts to classify networks in simple pattern terms. For example Howard (1967) recognizes four basic stream patterns, each broadly related to regional

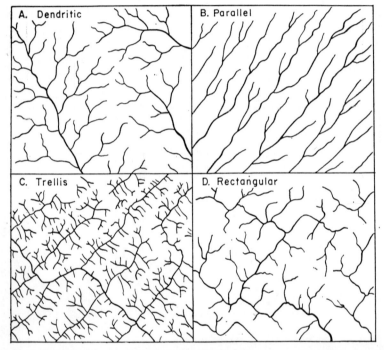

Fig. 2.22. Basic drainage patterns. Each pattern occurs in a wide range of scales. Examples shown may be regarded as types. Dendritic pattern resembles spreading branches of oak or chestnut tree with tributaries entering at wide angles. In trellis pattern, small tributaries to long parallel subsequent streams are about same length on both sides of subsequent streams. Source: Howard, 1967.

geological structure, and each occurring over a variety of scales (Fig. 2.22). *Dendritic* networks are associated with a lack of structural control (Fig. 2.22-A); *parallel* networks with steeply-dipping rocks or elongated landforms (Fig. 2.22-B); *trellis* networks are formed on dipping or folded sedimentary rocks, or on parallel fractures (Fig. 2.22-c); and *rectangular* networks on joints and faults at right angles, but they lack the orderly trellis pattern (Fig. 2.22-d). These basic patterns may be modified by local influences, leading to a further subdivision. Thus trellis networks (Fig. 2.23) may be subdivided into 'directional' trellis, with greater tributary development on one side of master streams characteristic of gently-dipping homoclinal structures; 'recurved' trellis, developed on plunging folds and

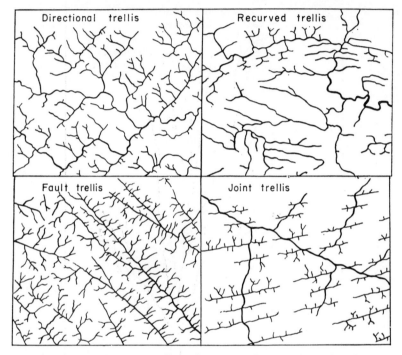

Fig. 2.23. Modifications of 2 basic pattern. Source: Howard, 1967.

on contorted metamorphic terrain; 'fault' trellis, developed on a right-angular set of faults; and 'joint' trellis, developed on joint sets, as in Zion Park, Utah. It is apparent that such pattern classification as shown above includes, and often confuses, a number of the distinct geometrical characteristics already discussed (i.e. shape and density) with other less easily defined parameters. In the following discussion an attempt is made to identify some of these other more elusive geometrical elements.

1. Orientation of networks

(a) Parameters of stream-network orientation

One of the most obvious methods of denoting flow direction of a network is with reference to some basin base-line. Shykind (1956; Miller and

Kahn, 1962, pp. 422–4) described dendritic drainage patterns with reference to a line connecting the head and mouth of the highest order segment, denoted as 180° in full azimuth (Fig. 2.24). The angles between flow directions of segments of all orders—similarly measured by joining head and mouth—and the base line were measured, showing that: (1) Orientations ordered according to length (i.e. a Strahler-type ordering) fluctuate

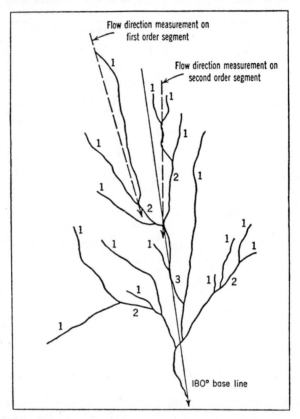

Fig. 2.24. Flow direction measurements on a youthful dendritic basin. Source: Shykind, 1956.

randomly around the flow direction of the master stream base line. (2) Channel lengths ordered according to orientation of flow direction fluctuate randomly around the median channel length. Strahler (1954) measured the azimuths of stream segments of different order with respect to the axis of basin symmetry (Fig. 2.25), and found that, for a third-order dendritic system developed on flat-lying rocks, first-order directions show a platykurtic distribution of wide dispersion, whereas second-order directions are bimodal with smaller dispersion. More recently,

Ongley (1968B) has defined the basin vectorial axis as the resultant of the lengths and directions of vectors linking the individual heads and mouths of all Strahler streams of order K (the maximum basin order) and $K-1$ (Fig. 2.26).

Another manner of treating stream flow directions is to do so stochastically. Morisawa (1963) proposed that on a homogeneous flat surface the probability (p) of flow in any of the four cardinal directions is twenty-five

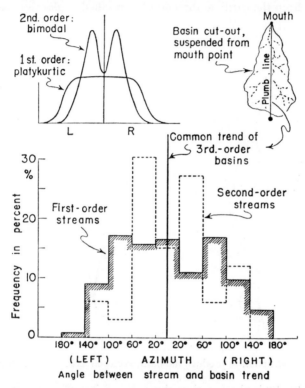

Fig. 2.25. Azimuths of first- and second-order stream segments plotted relative to the trend of the main third order basin. Combined data of ten third-order basins developed on cherts and cherty limestones in the Wolf Lake, Illinois, quadrangle of the Ozark Plateau province, central U.S.A. Source: Strahler, 1954.

per cent, and measured the percentage flow directions for first-order segments in four dendritic networks in the Appalachian Plateaus (Table 2.7). This seems to fit the random theory quite well, even though basins A and D have been developed on rocks with a slight south-easterly dip. The fingertip tributaries were thought to be more sensitive to rock structure than those of higher orders and least likely to reflect any original topographic slope of the basin, this being likely to become more and more masked by the random dendritic development of tributaries with the passage of time. Morisawa (1963, p. 529) also gave first- and second-order directional data for streams flowing on the Wissahickon Schist near Philadelphia which has pronounced NE–SW foliation (Table 2.7). Similarly, Schick (1965) analysed the probabilities of local streamflow directions deviating to the right (+) or left (−) of the cardinal direction of drainage of the reach. Taking third- and fourth-order basins of low slope on homogeneous lithology, the probabilities of the flow direction (δ) deviating from the cardinal direction by specified angles were calculated

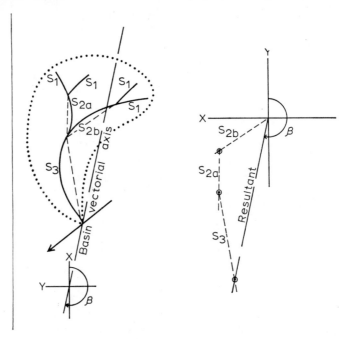

Fig. 2.26. Construction to produce the basin vectorial axis. Source: Ongley, 1968.

(Table 2.8). This showed that there was a less than twenty per cent probability of deviation smaller than 10° from the cardinal direction, a more than fifty per cent probability for deviation greater than 30°, and a sixteen per cent probability for deviation greater than 50° (i.e. one can expect that the local flow direction of a stream will deviate more than 50° from the cardinal flow direction for sixteen per cent of its distance). Of course these figures depend very much on the scale of the source of information, and for steeper basins the variance of the distribution of deviations will be smaller.

This stochastic model has been used by Schick (1965) to measure the effect of non-homogeneous rock structures on local stream directions (Fig. 2.27). This compares the observations of channel direction made from maps on scales of 1 : 5,000 to 1 : 62,500 for eight streams, with those of the model, and measures the deviation of a stream from this model in terms of the percentage of observations which fail to accord (i.e. too few or too many) with the model. Schick suggests that if this percentage is below five per cent there are no important lineative controls, whereas a percentage of more than fifty per cent is characteristic of meandering. In Fig. 2.27, for example, stream 4 has its cardinal direction parallel to jointing,

Table 2.7. Probabilities of flow in four cardinal directions for sample appalachian stream networks

Directions:	N–S	E–W	NE–SW	NW–SE
(A) Basins:				
A	21·74	27·74	23·19	27·54
B	25·27	25·27	27·47	21·99
C	25·83	23·17	26·49	24·51
D	25·83	24·17	22·92	27·08
(B) Orders:				
First	13·50	10·43	52·35	23·72
Second	10·78	10·78	50·98	24·75

Source: Morisawa, 1963.

Table 2.8. Probabilities of deviation of local streamflow directions from overall direction of segment

Flow range	$-90°$ to $-70°$	$-70°$ to $-50°$	$-50°$ to $-30°$	$-30°$ to $-10°$	$-10°$ to $+10°$	$+10°$ to $+30°$	$+30°$ to $+50°$	$+50°$ to $+70°$	$+70°$ to $+90°$
Frequency %	1	7	16	17	18	17	16	7	1

Source: Schick (1965).

and a percentage deviation of thirteen per cent; stream 5 is the most nearly in accord with the homogeneous model (deviation four per cent); and stream 8 has the greatest deviation of sixty-one per cent.

A further method of specifying stream directions is by reference to the entrance angles (Z_e) of tributaries developed on a valley-side slope (θ) at junctions with larger streams of lower slope (γ) (Horton, 1932, p. 360, and 1945, pp. 349–50; Schumm, 1956, p. 617; Melton, 1957, pp. 3–4). Hence a tributary, developing normal to the contour lines of the valley side, would tend to enter at an angle:

$$\cos Z_e = \frac{\tan \gamma}{\tan \theta}$$

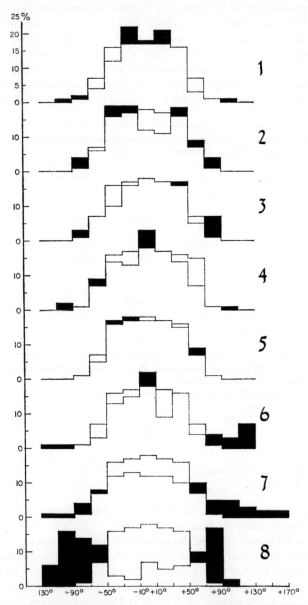

Fig. 2.27. Distribution of local stream directions from eight sample stream net-works in relation to a 'normal' distribution. Black indicates excess of observed over normally-distributed directions, the upper white sections a deficiency.

Thus, where the valley-side slope is steep in relation to the gradient of the master stream, a tributary tends to join at almost right angles; but where the master-stream gradient and valley-side slope are almost the same, the tributary virtually parallels the main channel, joining it at a small angle (Schumm, 1956, p. 617). It might also be expected that, with the passage of time, entrance angles would decrease as θ approaches γ; and Schumm (1956, p. 618) showed that entrance angles in mature basins are significantly less than those in youthful basins. The most specific work on the topic of stream entrance angles has been done by Lubowe (1964), who defined entrance or junction angles as 'the angle, projected to the horizontal, between average flow directions, determined by the ends of the stream segments extending from the junction point to an upstream point at a distance equal to 0·2 the average length of the second-order streams' (p. 331). Lubowe defined the channel networks from the Monterey Quadrangle, Kentucky and three other areas (two in the Appalachian Plateaus and one in the San Gabriels) on the basis of U.S.G.S. 1 : 24,000 topographic maps. From these data Horton's equation was tested and found to accord quite closely with the observed angles, except for junctions of streams with the same order, which were smaller than predicted. As expected, the mean junction angle increases as the order of the receiving stream increases, though (for given orders of junction) it was also found to decrease with increasing relief between the four areas (Fig. 2.28), probably due to the high gradients of the receiving streams.

(b) Interpretation of orientations

Few attempts have been made to estimate the influence of surface fracture patterns on the orientation of drainage lines, and these have been largely qualitative. Judson and Andrews (1955) showed that there is a striking correlation between one of two dominant joint directions and that of first-order streams west of Shullsburg, Wisconsin, but that the influence varied inversely with order (Fig. 2.29). They concluded that lower order streams are most susceptible to influence by joint directions, providing the orientation of the higher order streams allows this. Another visual correlation between joint and stream directions was made by Milton (1965), who measured the length and orientation of each channel segment which had an approximately straight course for more than the arbitrary distance of a quarter of a mile—measured on maps of Victoria, Australia on a scale of 1 : 63,360. He plotted the sums of these lengths in 10° classes based on flow direction, and in directional classes showing symmetrical orientation without sense of flow direction. These were found to bear a good relationship with the local orientation of joints. Little work has yet been carried out on the quantitative statistical comparison of stream and fracture orientations, partly because of difficulties in dealing with the skewed and polymodal circular distributions which would be involved. Potter and

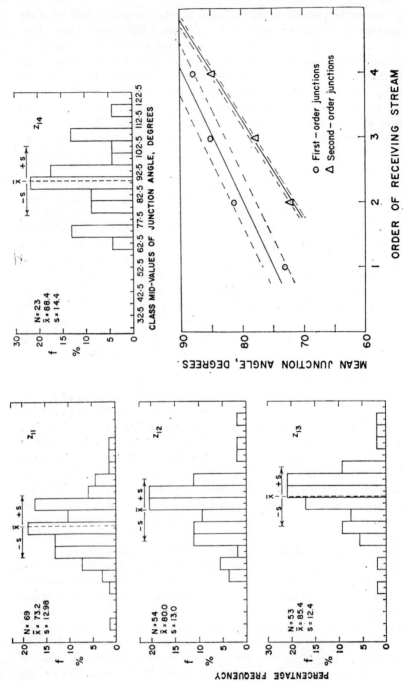

Fig. 2.28. Histograms of first-order stream junction angles with first (Z_{11}), second (Z_{12}), third (Z_{13}) and fourth (Z_{14}) order streams, and relation of mean junction angle to order. Monterey Quadrangle, Kentucky. Source: Lubowe, 1964.

Pettijohn (1963, pp. 263–4) demonstrate that directional data can be graphically depicted both as histograms and current roses, and they recommend that comparisons of different orientations can be made by the calculation of the azimuth $(\bar{x})$, magnitude (R) and variance of the

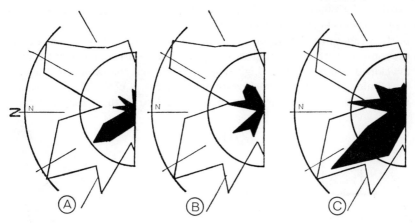

Fig. 2.29. Relation of jointing to stream flow west of Shullsburg, U.S.A. White represents joints, black represents total stream flow. (*A*) First-order streams. (*B*) Greater than first-order streams. (*C*) All streams. Each circle represents $2\frac{1}{2}$ miles of stream flow or five joint localities. Source: Judson and Andrews, 1955.

resultant vector means for each of the sets of directional data grouped into classes, where:

$$x_i = \text{midpoint azimuth of } i\text{th class interval}$$

$$v = \sum_{i=1}^{N} n_i \cos x_i$$

$$w = \sum_{i=1}^{N} n_i \sin x_i$$

$$\bar{x} = \text{arc tan } \frac{w}{v}$$

$$R = \sqrt{v^2 + w^2}$$

The effect of structures upon the orientation of drainage networks occurs in a very direct manner, as streams develop along surface lines of weakness. It has also been suggested by Melton (1959), however, that drainage lines may become at least partially adjusted to subsurface structures *before* being superimposed upon them, as the result of repeated minor uplifts of buried structures, of differential compaction of overlying sediments, and of subsurface influence over ground water (and thereby surface water) flow directions.

(c) Orientation of transport networks

Rather little work has been attempted on the general question of the orientation of transport networks where flows may move in either or both directions along a given link. Routes in areas of strong terrain often show extremely strong directional components, and we might expect levels of economic development to be reflected in weaker but discernible differences in the strength of orientation: e.g. colonial areas with a dominant export economy might be expected to show a stronger directional focus than a developed country with a high degree of internal spatial interchange.

Gould (1967, pp. 71–3) investigated the use of eigenvalue ratios and tensor fields as measures of the orientational strength of road networks

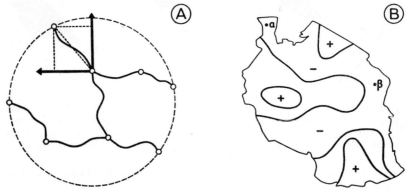

Fig. 2.30. Approach to road network orientation through eigenvalue ratios. (A) Sample quadrat (B) Ratio map for Tanzania. Source: Gould, 1967, pp. 72–3.

within sampling quadrats. For Tanzania, a systematic sample of twenty-seven circular quadrats were drawn and the road network within each quadrat reduced to a series of vectors; these joined either (i) nodes within the network, or (ii) nodes and the intersection with the quadrat perimeter (Fig. 2.30-A). Each vector n_i was reduced to components n_{ix} and n_{iy} in an orthogonal system of coordinates. Ratios between the eigenvalues are mapped in Fig. 2.30-B. These show areas with high values (over 3·20) in the centre and south, which reflect the strong directional focus of feeder links along the railway and in areas such as the Kilombera valley. Conversely, areas with low values (below 1·60) reflecting weak 'directionality' are concentrated in more developed areas like the Bukoba area (α) south of Lake Victoria and the Tanga area (β) on the coastal strip. Gould (1967, p. 72) found broadly similar results for Sierra Leone, and concludes that 'the ratios of eigenvalues seem to reflect the degree to which areas have entered into the internal exchange economy of the country'. Further work incorporating a weighting component for the length or flow along each vector and allowing scale variations in sampling is suggested.

2. Network spacing

Although spacing is inherent in the earlier discussion of network density
(Chap. 2.II), it has also proved useful to set this within a broader probabilistic framework. One of the few attempts to push analysis beyond the

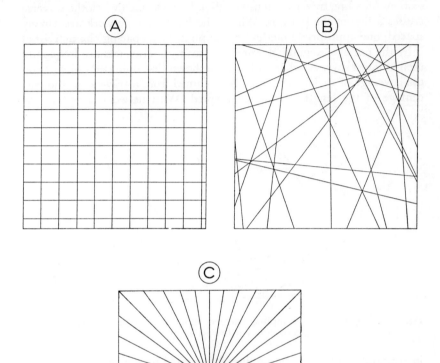

Fig. 2.31. Examples of linear patterns. (*A*) Divergent. (*B*) Random. (*C*) Convergent.
Source: Dacey, 1967, pp. 278–9.

recognition of typed patterns has been put forward by Dacey (1967, pp.
277–9), who proposed discrimination between linear patterns on the basis
of their spacing. Fig. 2.31-A shows a random pattern generated by joining
pairs of points each representing random pairs of plane coordinates. From
these random patterns lines may 'diverge' from each other until in the

limiting case they are parallel and orthogonal (a Manhattan grid) (Fig. 2.31-B), or may converge towards each other until, in the limiting case, they form rays converging to a single point (a radial grid) (Fig. 2.31-C). Average line-length and frequency failed to differentiate between the random and non-random patterns, and Dacey proposed the use of intersections on a line traverse, an intersection being formed when the traverse crosses a line of the pattern. Where the lines on a network are closely spaced, one appropriate method of sampling their spacing characteristics is by the use of systematic or random traverses. Traverses are frequently used in forest surveys, where it has been shown that unbiased estimates on the coverage of various uses may be obtained from a traverse as the percentage of the length of line intercepted. With network sampling the

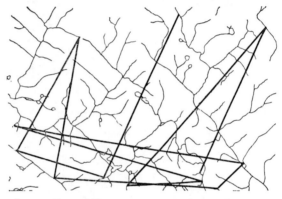

Fig. 2.32. Random-walk sampling traverse superimposed on dendritic channel pattern. Source: Dacey, 1967, p. 285.

interest shifts to the number of lines intercepted by the sampling traverse. Here, Dacey (1967) has suggested that the traverse method may not appear suitable when the number of intersections is small, i.e. the pattern is a widely spaced one, and has proposed instead an alternative approach by combining traverse segments into a 'random walk'. A random walk is constructed by locating an originating point from a pair of random coordinates. From this point a vector is constructed by selecting a real number from the range zero to two. The vector terminates where it intersects the most distant line of the pattern under investigation. From this point a second randomly-oriented vector originates; it terminates at the point at which it intersects the most distant line under investigation. Further vectors are constructed from the terminating points where vectors do not form at least one intersection with the pattern which is being inspected. Each segment of the walk originates at the last intersection formed by the preceding sector, and the walk is continued until the number of intersections reaches the desired sampling size. The segments forming the random walk can then be *'unhinged'* to form a single straight line segment (Fig. 2.32).

In a random spacing of points along a long traverse line, Clark (1956) has shown that two-thirds of the points have a reflexive (first) nearest neighbour, and that the proportion of points which have a reflexive nearest neighbour of order n is $(2/3)^n$. When the proportion of points

Table 2.9. Spacing characteristics of a dendritic channel pattern

Neighbour:	First	Second	Third	Fourth	Fifth	Sixth	Total
Observed frequency	48	31	23	26	21	16	165
Expected frequency	47·36	31·52	21·02	14·06	9·37	6·25	129·58
Difference	+0·64	−0·52	+1·98	+11·94	+11·63	+9·75	+35·42
Chi-square	0·009	0·009	0·187	10·140	14·435	15·210	9·682

Source: Dacey, 1967, p. 287.

having n^{th} order nearest neighbours is significantly greater than $(2/3)^n$ the lines are positively grouped as in the Manhattan grid. Conversely, when the proportion of points having n^{th} order nearest neighbours is significantly less than $(2/3)^n$ the lines are negatively grouped as in the radial grid.

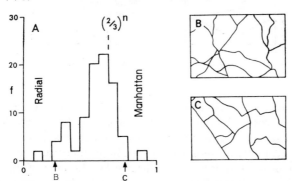

Fig. 2.33. Sample survey of rural road networks in southwest England. (*A*) Histogram of proportion of first-order reflexive pairs. (*B*), (*C*) Sample quadrats.

Tests by Dacey (1967, pp. 281–3) on the patterns shown in Fig. 2.31 showed clear discrimination between the three patterns for the first three-orders of grouping, even though data was gained from a small number of line traverses. The river pattern in Fig. 2.32 is analysed in Table 2.9. With this dendritic river pattern the observed and expected frequencies of reflexive nearest neighbours differ only for the higher-order neighbours.

Dacey (1967, p. 286) suggests that a possible interpretation of this result is 'overall regularity in the distribution of rivers but randomness in the spacing of river junctions along any of the major drainage basins'. Fig. 2.33 shows results obtained by Haggett from a sample of seventy ten-by-ten kilometre quadrats of the road system in south-western England. The general frequency distributions (Fig. 2.33-A) show a random distribution for first-order pairs, but stronger evidence of Manhattan structure in third-order pairs. Examples of extreme cases are given in Fig. 2.33-B.

An alternative approach to network spacing has been through the study of dimensional similarities. Bifurcation, length, and area ratios for stream channel networks developed on bedrock lacking pronounced structural control, have prompted speculation that erosional drainage

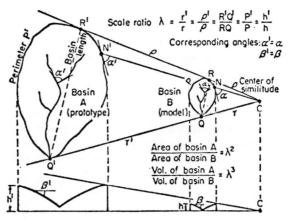

Fig. 2.34. Geometrical similarity of two drainage basins. Source: Strahler, 1957, p. 914.

basins in different environments show a close approximation to geometrical similarity when mean values are considered (Strahler, 1957, p. 913; Morisawa, 1962, p. 1045; Chorley and Morgan, 1962). This concept of the geometrical similarity of certain geometrical network properties suggests a more universal significance for the laws of morphometry (Chorley, 1957, p. 149). For basins to be similar in this way, their parts having a similar ratio, all length measurements between corresponding points must bear a fixed linear scale ratio (λ) and all corresponding angles must be equal (Strahler, 1957; 1958) (Fig. 2.34). Comparative data is given in Table 2.10 for length, drainage density and area measurements for a number of homogeneous basins in the Smoky Mountains of Tennessee and the Verdugo Hills of California. It is evident that some doubt regarding network geometrical similarity exists, and Hack (1957, pp. 63–4; see also Langbein, 1947, p. 145; Gray, 1961) has shown for a number of fourth-order basins (0·01–100 square miles in area) that the relationship between mainstream length (L') and basin area (A) is:

$$L' = CA^{n'}$$

where
$$C = 1·4$$
$$n' = 0·6$$

Now, if all watersheds are geometrically similar, regardless of area, n' would be 0·5; Hack (1957) assumed that a value of 0·6 means that as basins get bigger geometrical similarity is lost, because they tend to get

---

Table 2.10. Geometrical similarity of stream networks

Parameter:	Smoky Mts.	Verdugo Hills	λ
L_1 (miles)	0·115	0·062	1·85
L_2 (miles)	0·185	0·116	1·59
D (miles/square mile)	14·16	26·17	1·85
A_1 (square miles)	0·01321	0·00383	1·86
A_2 (square miles)	0·0591	0·0203	1·71

Source: Strahler, 1958, p. 293.

---

longer and narrower (Strahler, in Chow 1964). Smart and Surkan (1967) have pointed out, however, that geometrical similarity implies that *both* basin shape and mainstream sinuosity are constant with changes in basin size. They described sinuosity (s) as:

$$s = \frac{L'}{L_s}$$

and basin shape (S) as:

$$S = \frac{L_o}{A^{\frac{1}{2}}}$$

where L' is the length of the mainstream, L_s is the straight-line distance between the ends of the stream, L_o is the distance from the stream mouth to the highest point on the perimeter of the first-order basin containing the mainstream source, and A is the basin area. Thus basin shape approximates unity for square or circular shapes, and increases for elongate basins. Adjusting mainstream length and sinuosity to mesh length (L'_μ) and mesh sinuosity (s'_μ) by continuing the measurement to the divide, Smart and Surkan (1967, p. 972) showed that length and sinuosity increase with area, but that shape does not change so radically; they concluded that increases of sinuosity with area seem to make a significant contribution to the deviation of n from 0·5.

Part Two : Evaluation of Structures

Handeln vom Netz, nicht von dem, was das Netz beschreibt. (L. WITTGENSTEIN, Tractatus Logico-Philosophicus, 1922, 6, p. 35.)

Part Two / Evaluation of Structures

In the first part of this book, networks are treated as inert spatial structures. In this anaesthetized form it proves possible to isolate some of the major spatial parameters and to explore some of their characteristics and interactions. If, however, we are to make sense of the anatomy revealed in Chaps. 1 and 2, then we must bring the network back to life by reintroducing its fundamental function, namely, to conduct or impede flows. By flows we encompass not only a range of transported commodities (e.g. water, telephone messages, rail freight), but also the energy flows that lead to the creation of tensional networks. A general discussion of the relationship between flow efficiencies and network design (1), is followed by a case study of flow adjustments in stream-channel systems (2). Since network management hinges on the ability to predict future flows on the network and likely link loadings, a final section (3) raises the question of flow forecasting.

I. FLOW AND NETWORK EFFICIENCY

The primary purpose of a network has been described by Werner (1967, p. 3) as ' . . . to service transportation demand which can be represented and mapped as a set of locational vectors indicating the origin, size and destination of each individual transport demand.' If we identify three basic components—a set of *origin nodes*, a set of *destination nodes*, and a set of *vectors* joining the nodes—then we can set up three separate demand cases: (i) single origin to single destination (Fig. 3.1-A); (ii) multiple origins to single destination (Fig. 3.1-C); and (iii) multiple origins to multiple destinations (Fig. 3.1-E). For each of these cases we can set up

Table 3.1. Schematic components of highway costs

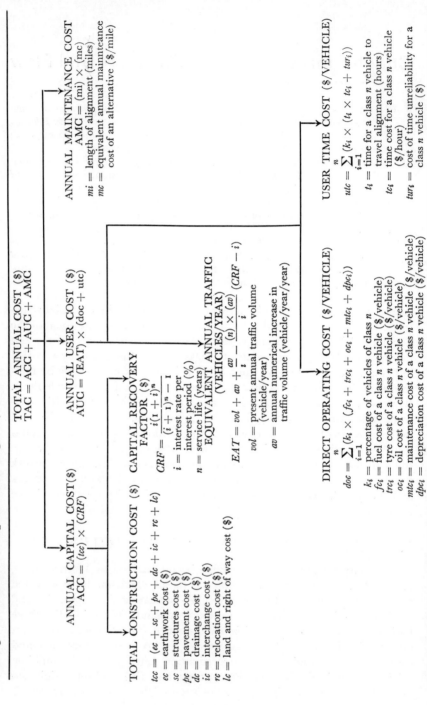

TOTAL ANNUAL COST ($)
TAC = ACC + AUG + AMC

ANNUAL CAPITAL COST($)
ACC = (tcc) × (CRF)

ANNUAL USER COST ($)
AUC = (EAT) × (doc + utc)

ANNUAL MAINTENANCE COST
AMC = (mi) × (mc)
mi = length of alignment (miles)
mc = equivalent annual mainteance cost of an alternative ($/mile)

TOTAL CONSTRUCTION COST ($)

$tcc = (ec + sc + pc + dc + ic + rc + lc)$
ec = earthwork cost ($)
sc = structures cost ($)
pc = pavement cost ($)
dc = drainage cost ($)
ic = interchange cost ($)
rc = relocation cost ($)
lc = land and right of way cost ($)

CAPITAL RECOVERY FACTOR ($)

$$CRF = \frac{i(1+i)^n}{(i+1)^n - 1}$$

i = interest rate per interest period (%)
n = service life (years)

EQUIVALENT ANNUAL TRAFFIC (VEHICLES/YEAR)

$$EAT = vol + av + \frac{av}{i} - \frac{(n) \times (av)}{i} (CRF - i)$$

vol = present annual traffic volume (vehicle/year)
av = annual numerical increase in traffic volume (vehicle/year/year)

USER TIME COST ($/VEHICLE)

$$utc = \sum_{i=1}^{n} (k_i \times (t_i \times tc_i + tur_i))$$

t_i = time for a class n vehicle to travel alignment (hours)
tc_i = time cost for a class n vehicle ($/hour)
tur_i = cost of time unreliability for a class n vehicle ($)

DIRECT OPERATING COST ($/VEHICLE)

$$doc = \sum_{i=1}^{n} (k_i \times (fc_i + trc_i + oc_i + mtc_i + dpc_i))$$

k_i = percentage of vehicles of class n
fc_i = fuel cost of a class n vehicle ($/vehicle)
trc_i = tyre cost of a class n vehicle ($/vehicle)
oc_i = oil cost of a class n vehicle ($/vehicle)
mtc_i = maintenance cost of a class n vehicle ($/vehicle)
dpc_i = depreciation cost of a class n vehicle ($/vehicle)

Source: Roberts and Suhrbier, 1962

a series of network configurations to service the flows. Although the designs chosen in Fig. 3.1 are arbitrary, they fall into the general classes of paths (Fig. 3.1-B), trees (Fig. 3.1-D) and circuits (Fig. 3.1-F), and underline

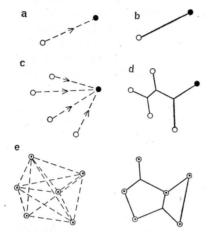

Fig. 3.1. Hypothetical demand and supply networks.

a general correspondence between the functional and topological character of network design (see Chap. 1.I(2)).

1. Scale bunching of flows

Inspection of the demand diagrams and possible networks in Fig. 3.1 suggests that one principal characteristic of a transport network is its merging of different flows in order to move them over a common transportation link. This merging reduces the total length of the network to be constructed but forces flows between nodes to take circuitous routes, thereby increasing the total flow costs. Clearly, different flows may use the same network link, and the orientation of the links may not necessarily correspond to the direction of the flow vectors.

(a) Basic cost elements

Transportation costs, whether measured in specific monetary terms in economic systems or energy terms within physical systems, are composed of two main elements: (1) Fixed costs attributable to the *length* of the route may be termed 'construction costs' or 'highway costs'; (2) Variable costs attributable to both the length of the route and the magnitude of the flow may be termed 'flow costs' or 'movement costs'. Since both types are based on different time spans it is necessary to reduce them to a common basis. Construction costs may be incurred once only, while travel costs may be incurred on each trip.

Table 3.1 shows a highly schematic picture of highway costs in which capital costs, user costs and maintenance costs are reduced to a common yardstick in terms of total annual costs (Roberts and Suhrbier, 1963). The detailed appraisal of route construction costs lies outside our immediate purpose (see Tinbergen, 1967; Friedlander, 1965, and Bos and Koyck, 1961, for some alternative econometric formulations), and we assume that network costs are clearly identifiable. Construction costs for unit lengths of the network (*a*) and user costs 'on' and 'off' the network (*b*, *c*) are therefore taken as given in the following discussion. It is, however, axiomatic that (1) transport costs on the network are less than off the network (i.e. $b < c$), and that (2) construction costs 'off' the highway (i.e. cross-country) are assumed to be zero. Clearly if axiom (1) were not true then networks would be barriers to movement.

Although the relationships between the four cost elements are highly complex, an important general linear relationship holds:

$$k(f) = a + b(f)$$

Here $k(f)$ is the total cost per unit flow; *a*, construction costs per unit length; *b*, flow costs per unit flow per unit length; and *f*, flow volume (Beckmann, 1967, p. 97; Werner, 1967, p. 4). All elements are reduced to a common time base. The equation makes clear that total cost per unit of flow decreases as total flow increases, and that substantial scale economies may be gained from merging different transport flows on to a common transportation link.

For a representative road system, a rural interstate highway in Rhode Island, U.S.A., Beckmann (1967) has provided the following cost estimates. Highway costs (*a*) are reduced to a common annual base for a one-mile section, based on original *construction* costs ($1,000,000)—maintenance costs ($4,550), *amortization* costs over a thirty-year lifetime ($35,000), and *annual interest* at five per cent ($25,000). Total annual fixed costs are, therefore, $65,000. Transportation costs per vehicle-mile (passenger) are estimated at $0.05 to $0.10, and for trucks the cost per revenue ton-mile for hauls of one hundred miles and over is estimated at $0.10. Beckmann assumes an annual flow of one million vehicles along the highway (about 3,000 per day) with one-fifth of the flow made up of trucks with an average load of ten tons. Thus the annual variable costs (*bf*) are around $180,000, and $a = b(f)$ for volumes of around 1,000 vehicles per day.

Empirical evidence for similar scale economies in transport networks is available. For example, studies of route costs on British Railways (British Railways Board, 1965, p. 8) show that routes with forty-five million ton-miles per annum have costs of around 0·07 pence per ton-mile whereas routes with only fifteen million ton-miles per annum have costs over twice this amount. Cost per unit carried falls rapidly as traffic density over a route increases. Similarly, Cookenboo (1955) has evaluated pipeline costs composed of capacity costs (diameter), energy costs (pumping), and throughout costs (related to flow). Fig. 3.2 summarizes these relationships by plotting unit movement costs (in cents per barrel per thousand miles) against the throughput in thousands of barrels (i.e. flow) for four ranges of capacity. The curve for the twelve-inch pipeline shows a typical U-shaped cost curve. It is in the range of the curve for a given capacity where it lies below other average cost curves, that the cost is optimum for the range of flows. Again the general fall in optimum costs with line size is apparent.

The cost elements in economic transport systems have analogues in the energy relationships in physical systems. For stream systems transporting water and debris from the watershed through river channels to the outlet, the stream cross-section is the critical element in efficiency. When

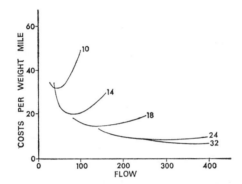

Fig. 3.2. Efficiency of channels in relation to flow volumes. Costs per barrel of crude oil for trunk pipe lines of varying diameter plotted against throughput. Source: Cookenboo, 1955.

one compares the cross-sections of different natural channels (e.g. Brandywine Creek, Fig. 3.3) great differences both in scale and form emerge, due largely to the wide range of bankfull discharges and channel roughness characteristics encountered. The most valuable measure of channel cross-sectional form is the *hydraulic radius* (R), defined as the cross-sectional area (A) divided by the wetted perimeter (P) and having the dimensions of L. This ratio gives a measure not only of channel shape but of the relationship between the cross-sectional area, through which flow occurs, and the perimeter of bed and banks, along which frictional drag occurs. The hydraulic radius is therefore a measure of channel efficiency, and approximates to the mean depth in wide shallow channels. More than ninety-seven per cent of the frictional energy loss in fluid flow is the result of interactions between the fluid particles themselves, and therefore any factors which increase turbulence also increase effective *roughness* (n, dimensionally equal to $L^{\frac{1}{6}}$), or resistance to flow. The Manning coefficient

(n) therefore appears prominently in the equation for mean velocity of flow ($\bar{V}$) (using f.p.s. units):

$$\bar{V} = \frac{1\cdot486}{n} R^{\frac{2}{3}}S^{\frac{1}{2}}$$

where R is the hydraulic radius and S the energy gradient (i.e. the loss of head of flow due to friction: in practice being approximated by the slope of the water surface). The coefficient n is a very complex parameter influenced by the roughness of the channel bed and banks, the type of fluid flow, and the shape and pattern of the channel. Values range from 0·012 for smooth concrete channels to up to 0·060 for mountain streams

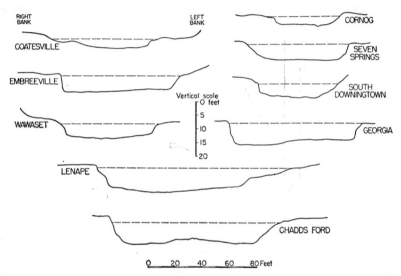

Fig. 3.3. Profiles of cross sections of Brandywine Creek at the principal measuring sections. Dashed line shows bankfull stage at each station. Source: Wolman, 1955.

with rocky beds and variable sections. From studies of Brandywine Creek, Wolman (1955) showed that as discharge increases downstream roughness decreases, owing to the floodplain bank materials being finer, the deeper downstream channel being usually straighter and better aligned, and to the submergence of protuberances in the deeper water.

(b) Network implications of cost ratios

The balance between the fixed route costs (*a*) and the variable transport costs (*b*) is of critical importance in determining the spatial pattern of a network. Fig. 3.4 shows extreme network solutions to a problem in which flow-demands exist between five nodes that are both sources and destinations. In the first case (Fig. 3.4-D) fixed costs are very low and only flow costs are minimized. The resulting network is one of straight-line connections between the nodes with no bundling of flows involved. This solution is clearly that with least cost to the *user* (i.e. it is the shortest and most convenient to and from any of the five nodes), but has the highest construction costs.

In the second case (Fig. 3.4-E) fixed costs are dominant and flow costs may be ignored. The resulting network is one which minimizes construction costs only and is least cost to the *builder*. This least-length network is neither trivial nor necessarily unique but must inevitably be a tree in graph-theory terms since ' . . . from any loop included in the system at least one link could be omitted without separating any pair of sources'

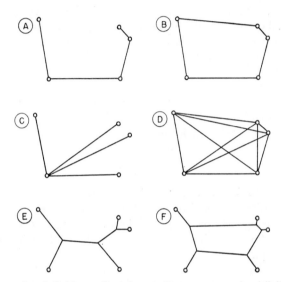

Fig. 3.4. Alternative definitions of minimum-distance networks. (*A*) 'Paul Revere' network, (*B*) 'Travelling Salesman' network, (*C*) Hierarchic network, (*D*) Least cost to user network, (*E*) Least cost to builder network, (*F*) Topological framework. Source: Bunge, 1962, pp. 183–9.

(Werner, 1967, p. 7). The general problems of shortest-path problems are discussed later in this volume (Chap. 4.I).

Beckmann (1967) has considered the effect of varying highway and transport costs for a highly simplified transport situation, viz. a single source for the flow of goods over a large, homogeneous and symmetrical plan. In this network equivalent of Thunen's *Der Isolierte Staat*, Beckmann considers the form that a supply network might take. Clearly the symmetrical flow field will be radial about the source, and any network must involve some flow moving longer distances than for unchannelled flow. In this situation, how many trunks will emerge from the centre, where will they terminate, how will they branch, and what design will the optimum network follow (Wasiutynski, 334)? Some general conditions for resolving these problems are given.

The *optimal number of trunks* may be given by the criterion:

$$F \eqsim \cfrac{a}{\cfrac{c}{\pi} \sin \cfrac{\pi}{n} - \cfrac{b}{n}}$$

where F denotes total flow from the centre, n is the number of trunks, and a, b, and c, are the cost elements as before (Beckmann 1967, p. 114). Wherever the costs of transport on the network are equal to or less than half the costs off the network (i.e. $b \eqsim c/2$), and cases $n = 1$ and $n = 2$ do not occur, for if it pays to have any trunks from the centre it pays to have three. For the Rhode Island cost data, Beckmann shows that the relation between number of trunks and flow is convex, with 27·6 million cars per annum needed to support four trunks, 36·3 for six trunks, and 46·5 for eight trunks.

Providing the flow field is symmetrical it is assumed that the trunks will be symmetrically arranged about the origin node. The termination of trunks is determined from an equilibrium of tension approach. The optimum terminus must be located where:

$$f = \frac{a}{c \cos \gamma - b}$$

where f is the trunk flow, and $\overline{\cos \gamma}$ is the weighted average of the cosines of the angles between the direction of the trunk and of the flows leaving, the weights being the flow levels (Beckmann, 1967, p. 113). Again, with a symmetric flow-field it is assumed that all trunks will have similar optimal termini.

The *branching of the trunk* poses more complex problems. Beckmann proposes that the opening angle between the two branches at a symmetric junction (angle γ) will be given by the expression

$$\cos \gamma = \frac{a + b(f)}{2a + b(f)}$$

When transport costs are very low (say, $b = 0$) all trunks are straight lines and all roads join at angles of 120°. In the continuous model with a single source and where the supply region is homogeneous but very large, Beckmann (1967, p. 116) conjectures that 'the solution is a symmetrical honeycomb with holes placed in such a way that the network remains simply connected (a tree)' (Fig. 3.5-A). An alternative case, when the density of both origins and destinations are uniform, gives a closed honeycomb as the optimum network (Fig. 3.5-B).

A number of practical extensions of Beckmann's model are apparent. Chisholm (1962, pp. 136–8) has discussed the optimum location of farm roads and boundaries in the reclaimed Dutch polders. Tanner (1967B) has conducted a theoretical study of how road systems in large plantations might be laid out in order to minimize transport costs. It was assumed that vehicles were confined to a permanent road system and that harvesting roads would consist of blocks of straight, parallel and equally

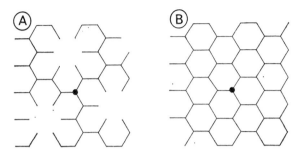

Fig. 3.5. Optimum network for homogeneous region. (*A*) Single origin with uniform terminal density. (*B*) Uniform origin and terminal density. Source: Beckmann, 1967, p. 116.

spaced harvesting roads to which crops must be carried by hand. Fig. 3.6 shows some alternative structures for the road layout and Fig. 3.6-D an optimum radial system when the ratio of operating costs on and off the

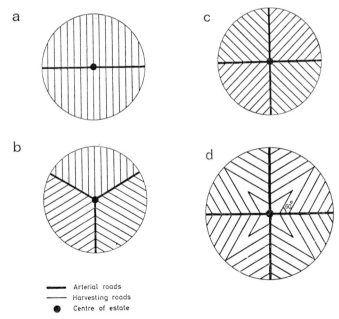

Fig. 3.6. Alternative arrangements for harvesting roads on a circular plantation. Source: Tanner, 1967B, pp. 13–14.

arterials is in the ratio 1 : 2. The spacing between harvesting roads (d) is given as:

$$d = \sqrt{\frac{2v}{wn}\left(\frac{C}{k} + x\right)}$$

where v is the walking speed of labourers (distance per hour), w is wages per hour of labour for carrying, n is the number of loads carried from each unit area at each cropping, k is the number of croppings per year, C is the cost per year of unit length of road and x is the cost per unit distance of operating collecting vehicles. It is suggested that systems based on three or four arterials will give appreciably shorter distances to the centre than two, but that systems using more than four radial arterials are not likely to be required. It was found to be usually preferable for harvesting roads to meet arterials roads obliquely; the junction angle varying with costs of operating on and off the arterials (Tanner, 1967B, p. 11).

2. Efficiency of alternative network designs

Alternative plans put forward for transport networks range from the radial, circumferential or grid roads for traditional towns with high-density centres, through to more complex hyperbolic and spiral networks for more experimental linear or circular cities (e.g. Miller, 1967). Different geometric patterns of road construction are associated with different patterns of relative accessibility: these patterns are discussed here in terms of *undirected* patterns in which flow in either direction is allowed and *directed* patterns in which one-way systems are incorporated.

(a) Undirected patterns

Much of the fundamental work on the efficiency of road networks has been conducted by Smeed and his associates (Smeed, 1968). Fig. 3.7 shows the average distance (d) travelled on some imaginary and real road networks in terms of the area within the boundary (A). Since Smeed was concerned with problems of congestion and therefore travel at peak periods, it was convenient to measure journeys between points on the outskirts of the central area of the town and scattered points inside it, and to assume first that origins equally distributed amongst the points at which the road leading into the central area meet its boundary (the broken line in Fig. 3.7); and second that destinations are distributed either uniformly along the sides of the roads of the central area or uniformly within the area.

Under these assumptions Smeed was able to calculate the average distance travelled on the roads of any town centre assuming journeys are made by the shortest route. The mean value for the realistic cases studied was:

$$d = 0.87 \, A^{\frac{1}{2}}$$

with variations between 0·70 $A^{\frac{1}{2}}$ and 1·70 $A^{\frac{1}{2}}$. Similar results were gained for a sample of journeys in Central London: external-internal journeys 0·70 $A^{\frac{1}{2}}$, internal journeys 0·87 $A^{\frac{1}{2}}$, through journeys 1·22 $A^{\frac{1}{2}}$, and overall average 0·87 $A^{\frac{1}{2}}$ (Smeed, 1968, p. 4).

A simple model for the distributions of trip lengths over alternative road designs in an urban area has been developed by the Road Research Laboratory (1965). Two types of urban area are assumed: the first has a uniform distribution of work places within a circular city; the second

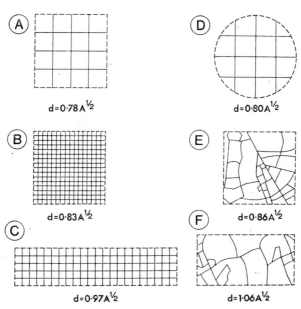

Fig. 3.7. Average distance travelled (d) on some imaginary and real road networks. A = area within boundary (broken line). Source: Smeed, 1968, pp. 5, 23.

has a distribution in which the density of work places is inversely proportional to the distance from the centre of the circular city. For both the 'uniform' and the 'clustered' distributions three alternative routes are examined by which workers move from the perimeter of the city to their work places within it: (1) *direct* routes follow a straight line from the entry point on the perimeter (γ) to the destination (β); (2) *radial* routes follow a straight line to the centre of the town (δ) and then another straight line on from the centre to the destination; and (3) *ring* routes follow a clockwise or anti-clockwise route around the circumference of the city and then follow a radial route inwards to the destination. In all cases the location of residences (e.g. α in Fig. 3.8) is assumed to be outside the perimeter of the city or the city-central area.

Average distances travelled from points on the circumference of the central area to points inside it are shown in Table 3.2. Radial routes are longer than direct routes, and ring routes longer than both. The same order applies to both types of urban areas, although the ring road is at a greater disadvantage with the more realistic assumptions of clustered workplaces. The ring road does, however, have the compensation that distance travelled *within* the central area (as distinct to distance around the perimeter) is substantially less. Clearly use of the ring road can

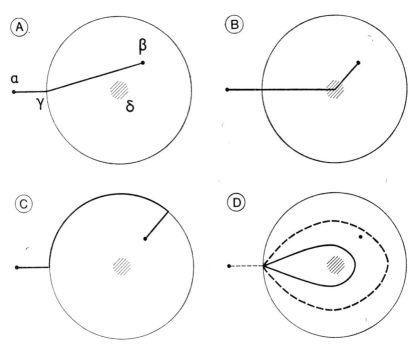

Fig. 3.8. Alternative routes between peripheral origin (α) and urban destination (β) (*A*) Direct. (*B*) Via city centre. (*C*) Ring road. (*D*) Areas that can be reached more rapidly by ring route than direct route. Source: Road Research Laboratory. 1965, pp. 254, 258.

substantially reduce the amount of travel in the congested central area, but since the route is longer it must be compensated by higher speeds. Fig. 3.8-D shows destinations which it is quicker to reach by ring route than by direct route from the entry point (γ) for two speed ratios: when the ring road speed is two times and five times that of speeds within the central area.

The concept of route factors, i.e. the ratio of the road distance to the direct origin-destination distance, has frequently been used in the evaluation of the efficiency of hypothetical network systems (see Chap. 2.I(1a)). Holroyd (1966) showed that the average route factor for long trips on a

Table 3.2. Average distances travelled from point on the circumference of a hypothetical urban area to destinations inside it

Type of route:	Direct	Radial	Ring	Ring (Urban Section)
Average distance travelled:				
Town with uniform distribution*	1·13R	1·67R	1·90R	0·33R
Town with clustered distribution*	1·09R	1·50R	2·07R	0·50R

Source: Road Research Laboratory, 1965, p. 255.
    * R = Radius of the urban area.

square network is 1·27, on a triangular network 1·10, and on a hexagonal network of 1·27. Tanner (1966) has investigated the efficiency of hexagonal systems at greater length. For short distances Timbers (1967, p. 394) has examined the distance between (i) equally-spaced points on the sides of a cell in the network, and (ii) points at intervals of up to four times the length of the cell sides. The results (plotted in Fig. 3.9) show that in the case of the square and hexagonal networks the resultant ratios oscillate about the theoretical value, while in the case of the triangular network the results are generally above the theoretical value. In all three cases convergence increases with length of journey. The relationship between theoretical systems and actual route networks has shown that route factors within British towns approximate the value for square grid systems (1·27), while the inter-urban network (1·17) may be derived from the combination of a basically triangular system (1·20) with a road deviation ratio (1·04 to 1·08) (Timbers, 1967, p. 401).

Fig. 3.10 shows twelve different routing systems for circular cities studied by Holroyd (1966). Clearly some of the systems make use of internal or external ring roads while others rely on radial, rectangular or other regular lattices. The efficiency of the twelve networks is evaluated for three types of movement: (i) *internal* movement when both origin and destination are inside the city (as in Fig. 3.10), (ii) *cross-cordon* movement when one end of the journey is inside the city and one outside, and (iii) *through* movement when both origin and destination lie outside the city. Table 3.3 shows the results of Holroyd's work in terms of average distances between random pairs of points. This allows the determination of optimum positions for circular ring roads. Extensions of the work to include realistic distributions of origins and destinations and average speeds on links is being conducted at the Road Research Laboratory.

The search for efficient geometrical structures for regional transport networks is of long standing. Kohl (1850) developed an intricate tracery of branching networks for his system of cities (Fig. 3.11-A), and Lalanne

(1863) developed Buffon's 'ratio of reciprocal obstacles' to give a triangular lattice modified to the needs of French prefecturial and canton structure. A century later Christaller (1933) developed an optimal transport network to serve his own central-place hierarchies (Fig. 3.11-B). Modifications of the Christaller system by Lösch (1954) in terms of sector rotation and specialization, and by Isard (1956) through agglomeration economies each lead to characteristic transport networks (Fig. 3.11-C and -D). Other less formal schemes for regional optima were put forward by McLean using a rectangular grid system, and by Kehra and Comeya using modified

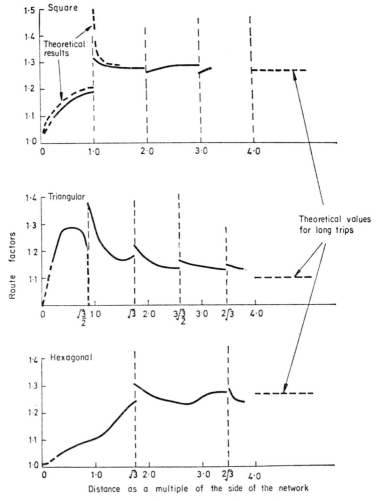

Fig. 3.9. Route factors for square, triangular and hexagonal networks. Source: Timbers, 1967, p. 394.

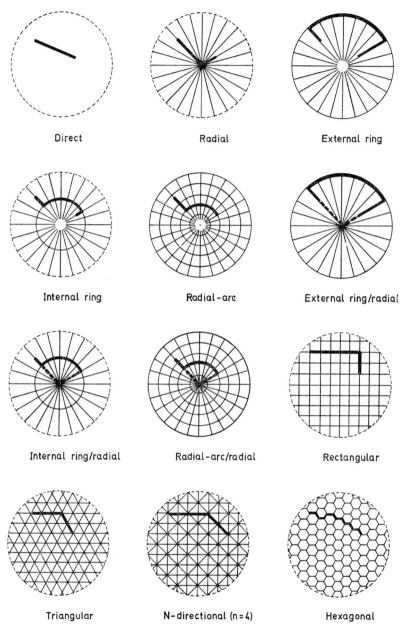

Fig. 3.10. Alternative routing systems within circular regions. Source: Holroyd, 1966, p. 22.

Table 3.3. Average length of journeys in a circular town*

Routeing system:	Internal journey	Cross-cordon journey	Through journey
Direct	$\dfrac{128a}{45\pi} = 0\cdot905a$	$\dfrac{32a}{9\pi} = 1\cdot132a$	$\dfrac{4a}{\pi} = 1\cdot273a$
Radial	$\dfrac{4a}{3} = 1\cdot333a$	$\dfrac{5a}{3} = 1\cdot667a$	$2a = 2\cdot000a$
External ring	$\dfrac{2a}{3} + \dfrac{\pi a}{2} = 2\cdot237a$	$\dfrac{a}{3} + \dfrac{\pi a}{2} = 1\cdot904a$	$\dfrac{\pi a}{2} = 1\cdot571a$
Internal ring	$\dfrac{4a}{3} - \dfrac{(4-\pi)b}{2} + \dfrac{4b^3}{3a^2}$	$\dfrac{5a}{5} - \dfrac{(4-\pi)b}{2} + \dfrac{2b^3}{3a^2}$	$2a - \dfrac{(4-\pi)b}{2}$
Radial-arc	$\dfrac{4a}{15} + \dfrac{4\pi a}{15} = 1\cdot104a$	$\dfrac{a}{3} + \dfrac{\pi a}{3} = 1\cdot381a$	$\dfrac{\pi a}{2} = 1\cdot571a$
External ring/radial	$\dfrac{4a}{3} - \dfrac{19a}{45\pi} = 1\cdot199a$	$\dfrac{5a}{3} - \dfrac{a}{\pi} = 1\cdot348a$	$2a - \dfrac{2a}{\pi} = 1\cdot363a$
Internal ring/radial	$\dfrac{4a}{3} - \dfrac{2b}{\pi} + \dfrac{2b^3}{\pi a^2} - \dfrac{19b^5}{45\pi a^4}$	$\dfrac{5a}{3} - \dfrac{2b}{\pi} + \dfrac{b^3}{\pi a^2}$	$2a - \dfrac{2b}{\pi}$
Radial-arc/radial	$\dfrac{4a}{3} - \dfrac{16a}{15\pi} = 0\cdot994a$	$\dfrac{5a}{3} - \dfrac{4a}{3\pi} = 1\cdot242a$	$2a - \dfrac{2a}{\pi} = 1\cdot363a$
Rectangular	$\dfrac{512a}{45\pi^2} = 1\cdot153a$	$\dfrac{128a}{9\pi^2} = 1\cdot441a$	$\dfrac{16a}{\pi^2} = 1\cdot621a$
Triangular	$\dfrac{256\sqrt{3}a}{45\pi^2} = 0\cdot998a$	$\dfrac{64\sqrt{3}a}{9\pi^2} = 1\cdot248a$	$\dfrac{8\sqrt{3}a}{\pi^2} = 1\cdot404a$
N-directional	$\dfrac{256n\tan(\pi/2n)a}{45\pi^2}$	$\dfrac{64n\tan(\pi/2n)a}{9\pi^2}$	$\dfrac{8n\tan(\pi/2n)a}{\pi^2}$
Hexagonal	$\dfrac{512a}{45\pi^2} = 1\cdot153a$	$\dfrac{128a}{9\pi^2} = 1\cdot441a$	$\dfrac{16a}{\pi^2} = 1\cdot621a$

a = radius of town
b = radius of internal ring-road

Source: Holroyd, 1966, p. 5.

* See Fig. 3.10

hexagonal systems (see review by Domanski, 1963, pp. 20–4). Chisholm (1962, pp. 136–8) has considered the lay-out of rural road systems on the reclaimed Dutch polders in relation to farm shape, farmstead location, and access costs.

(b) Directed patterns: one-way systems

One rather common solution to heavy traffic loads on part of a network is to convert links to one-way flow. The relative accessibility of points on networks before and after conversion has been studied from both empirical and theoretical standpoints. Four studies in urban areas reported by the Road Research Laboratory showed that increases in journey lengths

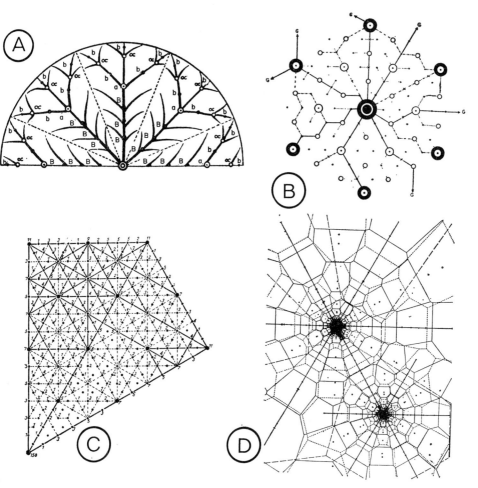

Fig. 3.11. Transport networks for theoretical settlement systems. (*A*) Kohl, 1850, (*B*) Christaller, 1933, (*C*) Lösch, 1954, (*D*) Isard, 1960. Source: Domanski, 1963, pp. 21–9.

of between seven and twenty-one per cent were either balanced by reduction in mean journey time (of from seven to ten per cent), or were offset by substantially unchanged or moderately changed journey times. A related parameter, accidents, showed reduction in some categories but increases in others. The equivocal nature of the evidence has spurred theoretical studies on both artificial networks and simulated urban networks. Bunton and Blunden (1962) showed that one-way working for random trips over a six-by-six grid would result in an overall increase in journey distance of about forty-five per cent. Orientation of links to meet major flows would reduce actual increases to well below this figure. Fairthorne (Road Research Laboratory, 1965, p. 338) showed an increase in number of turns with random trips over one-way grid system, but elimination of turns involving crossing opposing traffic flows. Calculations for a sample area of London streets (4·5 miles of network over a 0·125 square mile rectangle) showed a twenty-eight per cent increase in journey distances for all pair-combination trips for nineteen sampling points on the periphery and within the area. Since vehicle mileage and traffic flows also increased by the same amount, and since increased flow would result in levels high enough to slow average speeds over some links, the average capacity of one-way links would have to be increased by substantially more than twenty-eight per cent to attain the same average journey times.

3. Network efficiency under complex assumptions

Clearly the efficiency measures described in the preceding section (Chap. 3.I.2) refer to highly simplified conditions. Attempts have been made to specify network performance under more complex but realistic conditions (Peterson, 1961). Creighton, Hoch, and Schneider (1959) have studied the spacing of arterial roads within a grid-iron system in terms of six variables: (a) cost of construction, (b) value of travel, (c) trip density, (d) trip length, (e) speed relative to that on feeder roads and (f) absolute speed on the arterials. The relationship is positive with the first variable (i.e. as the cost of building arterials grows, their respective spacing decreases), and negatively with the other variables (i.e. as the value of travel increases, the optimum spacing of arterials decreases). Using assumptions of a regular cartesian grid of local streets spaced at intervals of 0·125 miles and empirical travel speed data from the Chicago Area Transportation Survey, the following formula for the critical spacing of expressways in an urban area was developed:

$$z = 2 \cdot 24 \left[C_z \Big/ \left(PK \left\{ \frac{1}{V_y} - \frac{1}{V_z} \right\} \left\{ \sum_p^{\infty} F_i \right\} \right) \right]^{\frac{1}{2}}$$

Here z represents the respective spacing of expressways in miles; C_z, the construction costs per mile for expressways; P, trip density or number of trip origins per gross square mile; K, a constant representing the value of travel time over a thirty-year period; V_y and V_z, the average speeds on arterial streets and expressways respectively; and F_i, the percentage of trips having any airline length i.

Although the relationships are not precise the formula does provide a logical statement of the ways in which a set of important factors affect spacing. Using data for Chicago, the formula suggests that arterials should be spaced about 0·65 miles apart in the higher density central regions, but 1·50 miles apart in the lower density suburban areas. In the former region the great number of trips forces the arterials closer together despite the considerable increase in construction costs.

Levinson and Roberts (1965) have investigated the transport performance of eleven American cities ranging in size from Chicago to Lexington. The cities were classified into three main network types, *grid*, *radial-grid* and *radial circumferential*. For each type the relation between maximum and average loadings on selected urban freeway systems was analysed. The ratio between the anticipated load point volumes and the average volumes approximated 2·4 for grid system in large cities, 4·0 for radial circumferential system in large cities, and 2·2 in medium sized cities. This analysis suggests that grid freeway systems appear to develop more equitable traffic loadings in large urban areas. However, because of of the intrinsic variability of the cities studied, it proved difficult to draw general conclusions from these results.

Simulation of traffic was studied for a hypothetical city using a symmetrical urban area containing 2·9 million people spread over 784 sq. miles in a series of five rings of density, decreasing from 15,000 to 1,000 persons per square mile with distance from the downtown area. Assuming conventional rates of trip-attraction and production and using a conventional gravity model, they were able to build up a picture of potential flows within the area. With assumptions of average speeds of fifteen miles per hour on local streets, twenty-five miles per hour on arterials and forty-five miles per hour on freeways, these trips were loaded on to a series of freeway configurations. The basic freeway systems considered are shown in Fig. 3.12-A. They are (1) attenuated grid, (2) symmetric grid, (3) radial grid, (4) semi-radial circumferential, and (5) radial circumferential. Using computer simulation, some 2·6 million trips were assigned to these hypothetical networks. From a review of the traffic flow patterns that emerged, Levinson and Roberts (1965, pp. 80–2) suggest seven basic similarities in the flow pattern. These are (i) an increase in loading as routes approach the centre; (ii) ratios of maximum load points to average volumes range from two to 2·7; (iii) loadings are generally comparable to those anticipated for most urban areas of similar size; (iv) loadings appear more sensitive to changes in system links than to changes in the original destination pattern of trips; (v) intermediate circumferential freeways play an important part in terms of number of trips carried; (vi) heaviest volumes occur in continuous routes; and (vii) volumes slightly exceed the flow that can be effectively carried on existing eight-lane urban freeways. The last conclusion suggests that even with optimum system configuration, freeways will need to be backed up with other transportation services.

Differences in specific loading patterns are shown in Fig. 3.12-B. These show that the heaviest volumes are found under the assymetric grid system (2), and that elimination of assymetry appears to develop the

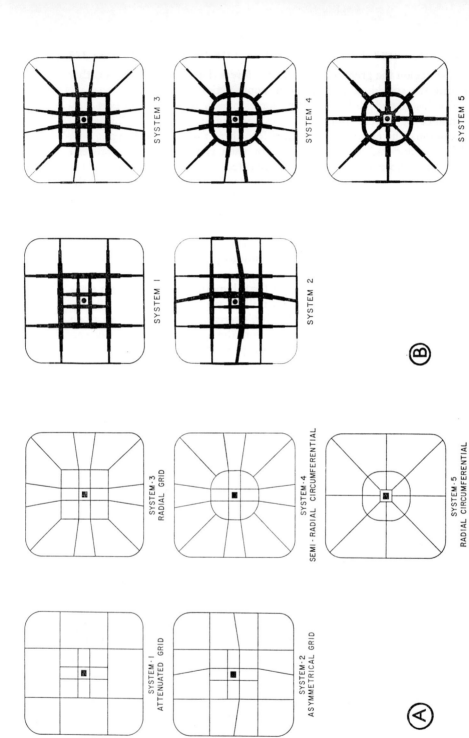

Fig. 3.12. Simulation of traffic flows on five alternative freeway networks for a hypothetical urban area. (*A*) systems tested, (*B*) Assigned daily flows. Source: Levinson and Roberts, 1965, pp. 78, 81.

most equitable loading pattern. The heaviest volumes on the inter-mediate loop occur with the intenuated or truncated grid system (1). The completely radial circumferential system (5) tends to develop the greatest extremes in radial loadings, and all systems with radial circum-ferential designs (3), (4), (5) tend to increase inner loop volumes by about ten per cent over that of the symmetrical grids. While a simulation study can only provide suggestive rather than definitive results, it does emphasize that loadings are more sensitive to adding, deleting or warping links than to downtown concentration as such. System continuity tends to maximize use, and demonstrates the importance of the intermediate freeway loop in large urban areas. It underlines the importance of avoiding route convergence, although the differences between radial and grid systems are less pronounced than suggested by Fisher and Boukidis (1963).

Evaluation of alternative road systems in newer expanded towns in the United Kingdom suggest that transport investment forms a very significant part of the total investment: for example, L.C.C. plans for Hook New Town show that £17 million out of the total anticipated cost of £157 million were allocated to the development of road transport. Farbey and Murchland (1967) investigated three trial schemes for a new town in the process of expansion from a population of 125,000 to nearly 300,000. The three alternative schemes are based on flows arising from (1) expanding the existing town centre, (2) providing a limited amount of development outside the centre and (3) high suburban and rural dispersal. The results of the detailed study were somewhat surprising. Although flow patterns show the expected variation with lower peak levels in the dispersed scheme, the user and builder costs were substantially the same in all the schemes. Extra mileage required in the undeveloped land in the dispersed scheme roughly balanced their lower demands on central area land. Experiments were carried out varying a number of the assumptions on the gravity model used, but without making significant changes in the costs.

The classic area for research on the efficiency of network structures under uncertain demand is telecommunications. Here channel capacity is rigidly constrained, but demands are unsteady or stochastic: too many messages arriving at the channel are reflected in message delays. Early work by Johannsen and Erlang on the delays of messages in the Copen-hagen telephone exchange was followed by a half-century of intermittent mathematical research leading to the present elegant theory of congestion and queuing (Benes, 1965). The nature of the design problem, albeit at a trivial level, is shown in Fig. 3.13. Here five major American cities are connected by a telegraph-line network arranged in three different con-figurations: the original 'diaper' net (Fig. 3.13-A), a reduced 'star' net (Fig. 3.13-B), and an augmented 'fully-connected' net (Fig. 3.13-C). The performance of the system is measured in terms of average message delay (in seconds) under varying conditions of load. Message delay is a function of both path length and traffic-flow concentration, and the trade-off between the two allows the sequence of optimal network topologies to be determined (Kleinrock, 1964, p. 31). For the example given the diaper network is always above the other two designs; at small network loads a

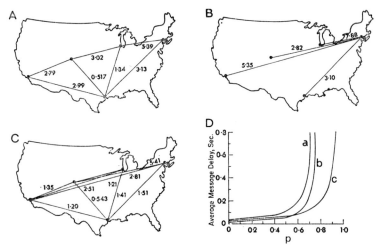

Fig. 3.13. Comparison of message delays in relation to network load (p) for three topologic arrangements: United States telegraph network example. Source: Kleinrock, 1964, pp. 23–31.

star-net centred on the main traffic centre (New York) gives the smallest delay, but as the load increases the curves cross and the fully-connected net is preferred (Fig. 3.13-D). Histograms of message delay from digital simulation allow evaluation of networks under a very wide range of assumed topologies and loads, for, despite the analytical progress, many complex telecommunications networks pose computational problems too massive for current computing capacity.

II. FLOW AND NETWORK ADJUSTMENT: THE CASE OF STREAM SYSTEMS

The relations between the form of a stream channel network and the character of throughput flows is a complex one and, in particular, the distinction between cause and effect is very much a function of time (Schumm and Lichty, 1965). For example, in the long term it is apparent that drainage density must be a function of some attribute of hydrology or paleohydrology, and Melton (1957) has shown that the Precipitation-Effectiveness Index of Thornthwaite 'explains' about eighty-nine per cent of the variation in drainage density observed in twenty-two drainage basins in the western United States. On the other hand, it is also clear that the flood peaks of individual storms falling on different drainage basins must be strongly influenced by the drainage density, in that the latter is a measure of the efficiency with which surface runoff can be gathered into channels and carried away. Carlston (1963) obtained a close correlation between drainage density and mean annual flood ($Q_{2.33}$) for fifteen basins in the eastern United States of quite wide variation in both area and relief. The control exercised by drainage density over flood peaks is obvious; but there are two ways in which an increase in the time span of our concern with stream networks forces us more and more to consider network form to be the product of process: firstly, through the cumulative effect of frequent processes of small or moderate intensity, and secondly, because the longer the time span the greater the probability that infrequent events of high intensity will be encountered. Wolman and Miller (1960) have shown that for many processes the combination of the approximately logarithmic-normal frequency distribution of magnitudes and the power function of the rate of their operation, means that the frequent events of moderate magnitude are responsible for the majority of erosional work. Thus they estimate that more than half of suspended sediment moved in stream channels is transported by flows which occur on average as little as once or twice a year. However, as they point out (p. 71), some erosional events, like gully initiation, must await the occurrence of infrequent intense storms, which, however, once formed, can be maintained and propagated by more moderate storms. In this way, an extreme event can have a lasting effect upon a channel system, particularly because the value of drainage density is very strongly controlled by the occurrence of fingertip tributaries (Kirkby and Chorley, 1967). This control may be counteracted to some extent, however, by the tendency for streamhead filling by slope debris to lead to some balance in the long term between channel extension and infilling (Kirkby and Chorley, 1967, p. 16).

Inputs of precipitation into a stream network produce two types of output: (1) The discharge of water and debris, with that of the latter closely related to the former. This discharge is a relatively immediate output, clearly related to a given input, and the prime concern of those who exploit the short-term operation of river systems. (2) Changes in the channel network itself. It has already been noted that changes in many aspects of channel geometry are the natural result of certain magnitudes and frequencies of runoff and discharge input. The larger the time interval under consideration the more important are these changes, either due to cumulative effects or to the increasing probability of encountering extreme events of high energy. Thus some attributes of the hydraulic geometry of self-formed channels (e.g. bankfull hydraulic radius) reflect events of relatively short recurrence interval (in this instance the bankfull discharge, $Q_{1.58}$, the most probable annual flood discharge), whereas other network attributes (e.g. the areal drainage network density) are more the product of events of larger recurrence intervals.

Because the competence of the velocities consequent upon different discharges to entrain material and alter channel form depends on the characteristics of the available material (e.g. calibre, shape, cohesion, etc.), it is obvious that channel form is the product of both discharge and debris characteristics. The temporal variability of the former makes its influence especially difficult to evaluate and no single definition of 'channel-forming' discharge is possible, since it has to be redefined for each phase of channel formation (Rzhanitsyn, 1964, p. 139). It is thus apparent that discharge events of different recurrence intervals are responsible for different aspects of hydraulic geometry, which is affected by a range of flows (Wolman and Miller, 1960, pp. 65–7). For example, channel cross-sections in non-cohesive material are almost instantaneously adjusted to changes of discharge, whereas the lateral cutting of cohesive banks of a small stream in Maryland is effected primarily by discharges occurring on average some eight to ten times a year (i.e. $\simeq Q_{0.1}$) which attack previously wetted banks. The most generally useful discharges for explaining channel form are the *mean annual flood* ($\bar{Q}_{2.33}$), the *mean annual discharge* ($\bar{Q}$ — equalled or exceeded about twenty-five per cent of the time and usually filling the channel to about one-third its bankfull depth: Leopold, Wolman, and Miller, 1964, p. 243), and the *bankfull discharge* (Q_b). Bankfull discharge, with a recurrence interval averaging some 1.58 years, completely fills the channel, and for this reason is considered the most important discharge in explaining channel form (Leopold, Wolman, and Miller, 1964, p. 241). This view is supported by the relationships which exist between Q_b and such channel features as meander wavelength (Wolman and Miller, 1960), and by the abrupt change of discharge habit which occurs at bankfull.

1. Flow and channel cross-section

To maintain the generally trapezoidal cross-section of a stable natural channel with movable bed and banks, two conditions must be satisfied: the flow must be transmitted, and the stability of the banks must be maintained (Leopold, Wolman, and Miller, 1964, p. 200). With generally non-cohesive material the change in the channel cross-section, both of depth and width, is closely related to changes in discharge (Fig. 3.14), such that at-a-station curves yield the following relationships:

$$w = aQ^b$$
$$d = cQ^f$$
$$v = kQ^m$$

where $b + f + m = 1$, because $Q = vwd$. These exponents also give an indication of channel geometry and resistance to erosion, in that a shallow dish-shaped channel would exhibit a big increase of w with Q (i.e. b would be large), whereas a rectangular channel with resistant vertical banks would have a small increase of w but a big increase of d with Q (i.e. b small and f large) (Leopold, Wolman, and Miller, 1964, p. 217). When discharge increases the water level rises, the velocity increases, and more bed and suspended sediment is entrained in the flow. The change of channel geometry and the general smoothing out of the small scale bed

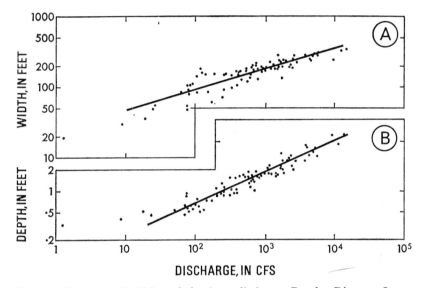

Fig. 3.14. Relation of width and depth to discharge, Powder River at Locate Montana, U.S.A. Source: Leopold and Maddock, 1953.

features by the movement of bed material have the effect of decreasing the flow resistance and allowing velocity to increase further.

The irregular initial arrangement and subsequent movement of the bed material may mean that at first the bed at a particular location may aggrade due to movement of material from immediately upstream. Eventually, however, an increase in discharge will lead to the scouring of the bed (Leopold, Wolman, and Miller, 1964, p. 227–32), though the maximum scour may not coincide with the maximum discharge (again because of irregularities in bed load movement). A decrease in discharge leads to a rapid deposition of stream load, although this deposition does not take place uniformly over the bed but as shoals ('riffles'), the location of which determine whether the bed at any particular point will aggrade appreciably as discharge decreases.

In many instances, however, the bank material differs from that of the bed in containing higher proportions of cohesive silt and clay material, and this places a constraint on the changes of hydraulic geometry associated with changes of discharge. Using data from forty-one locations on twenty-nine rivers of the northern Great Plains, Schumm (1961-B) showed the following relationship between channel width (w), the mean annual flood $(Q_{2.33})$, and the percentage of silt and clay ($<$0·074 mm) in the channel perimeter (M):

$$w = \frac{Q_{2.33}^{0.45}}{M^{0.36}}$$

A multiple correlation analysis showed that, independently, $Q_{2.33}$ explains seventy-one per cent of the variation of w, and M fifty-five per cent; whereas together they explain eighty-three per cent. The two independent

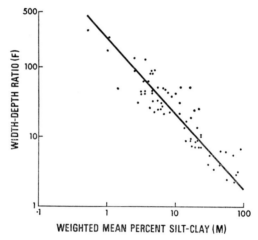

Fig. 3.15. Relation between width–depth ratio and weighted mean per cent slit-clay at different cross sections, northern Great Plains. Source: Schumm, 1960.

variables together only explained forty-one per cent of the observed variation in maximum channel depth. Calculating the width/depth ratio (F) at ninety locations in the northern Great Plains, Schumm (1960) demonstrated the relationship (Fig. 3.15):

$$F = M^{-1.08}$$

showing that poorly-cohesive sands and gravels are associated with channels which are relatively wide and shallow, the opposite being true for channels in more cohesive material. Neither $\bar{Q}$ nor $Q_{2.33}$ variations seemed to affect this relationship, although there was a clear control by discharge over the absolute values of depth and width. It was found that, as M increases downstream, depth increases more rapidly than width, with a consequent decrease of F. Where aggradation is occurring changes in F are also dependent on M, for, with cohesive material, deposition takes place on the banks as well as on the bed and tends to reduce the value of F. Aggradation where the banks are poorly-cohesive, however, is predominantly a matter of bed deposition, and consequently of an increase in F (Schumm, 1961-A).

2. Flow and channel plan

(a) Meandering channels

Where the bed material is composed of a range of grain sizes, the scour and fill associated with changes of discharge cause a sorting of debris which may profoundly affect hydraulic geometry. There is a tendency to bring the coarser debris to the surface of the bed, and to cause it to move in a discontinuous manner so that it forms bars or riffles along the

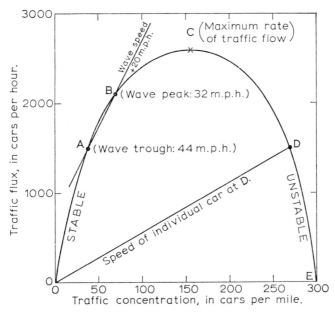

Fig. 3.16. Relation between traffic flux and concentration. Source: Langbein and Leopold, 1968.

stream course. These tend to be equally spaced at some five to seven stream widths apart along both straight and meandering reaches; they are most pronounced where the bed material is poorly sorted and most equally spaced where all the coarse load can be moved by the high discharges (Leopold, Wolman, and Miller, 1964, p. 203–6). Some authors have associated this with a wave of phenomenon of widespread occurrence, and taken it to indicate that the meandering tendency is present even in more or less straight streams (Leopold, Wolman, and Miller, 1964, p. 203; Langbein and Leopold, 1966), and will be re-established in a similar locational pattern even after the occurrence of large discharges (Leopold, Wolman, and Miller, 1964, p. 209).

Riffles, or bars, may exist independently of the individual particles composing them, and, as particles are progressively lost and replaced by others, the bar form may move upstream, downstream or remain stationary. Such a linear concentration of bed particles may result from the direct relationship observed to exist between the speed of moving objects and their spacing, as a result of interaction between them. This relationship has been explored most fully for moving traffic (Lighthill and Whitham, 1955), in which traffic flux (i.e. 'discharge' of vehicles) is at a maximum where traffic concentration and individual vehicle speeds are at medium values (Fig. 3.16). If a wave forms on the rising limb (e.g. A–B) it will move 'downstream' at a speed less than that of the individual vehicles within it, whereas a wave on the falling limb moves upstream as a shock. On the rising limb a local increase of traffic concentration causes an increase in flux characteristic of the stable condition of an 'uncrowded highway', but a similar increase on the falling limb will further increase concentration on the unstable 'crowded highway' (Langbein and Leopold, 1968, pp. 3–4). Experiments with glass spheres introduced into a narrow flume similarly show that speed of individual movement is inversely related to the linear bead concentration (Fig. 3.17-A) and that bead flux is related to concentration (Fig. 3.17-B). This shielding effect which a particle has on its neighbours downstream has been demonstrated for marked rocks in the bed of the Arroyo de los Frijoles, New Mexico, showing that, for a given discharge, the probability of a given particle moving is inversely related to the spatial concentration of particles of that size (Langbein and Leopold, 1968, p. 14). Another factor tending to concentrate locally stream bed particles of a given size is the combined effect of roughness and trapping produced by a bed surface of particles of the same size or larger. Thus reaches of high and low concentration of coarse channel debris tend to become accentuated, and regular debris bars are generated and maintained. Langbein and Leopold (1968, pp. 15–17), beginning with a random linear distribution (mean density 0·33 per unit length), randomly selected for each time interval a number of particles (probability of selection = 0·50) which either could not move if their paths were blocked by particles immediately ahead or which could overtake the next particle in front. Fig. 3.18 shows that after only a few time intervals this simulation begins to produce linear concentrations, the wavelength of which is proportional to the mean concentration and to the mean speed of all particles (even though all are not in motion at any given time). Riffles differ from traffic waves in having more locational stability and in exhibiting more constant wavelengths. The latter has

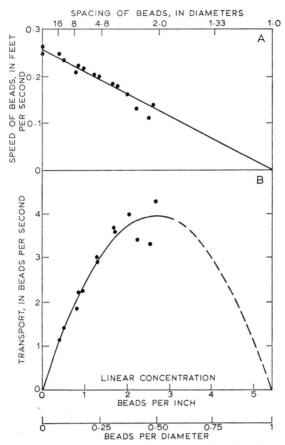

Fig. 3.17. Relation between speed (A) and transport (B) of beads and their linear concentration in a narrow channel. Source: Langbein and Leopold, 1968.

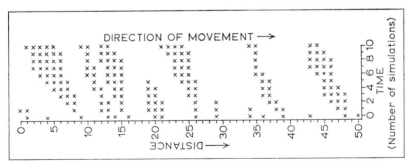

Fig. 3.18. Simulation study showing change from random linear distribution to regularly spaced concentrations in ten simulated steps. Source: Langbein and Leopold, 1968.

been ascribed to a supposed compromise between the tendency of increasing velocity to both increase wavelength directly and to decrease it
indirectly due to increased erosion (Langbein and Leopold, 1968, p. 19).

Riffles are associated with the crossings between swings, the intervening pools with the locations of the outer meander swings, and it is possible
that one of the mechanisms for initiating meanders in more-or-less straight
streams is the initial formation of the regularly-spaced riffles. Along such

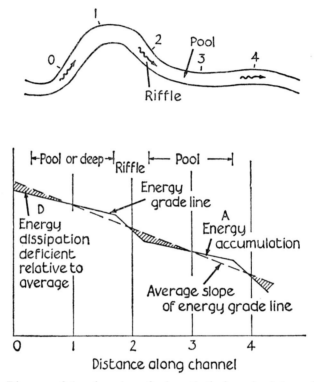

Fig. 3.19. Diagram of the plan view of a hypothetical reach of channel showing
location of pools and riffles, and longitudinal profile of total energy (energy
grade line) for same reach. Source: Leopold, Wolman. and Miller, 1964.

streams the energy dissipation becomes very unequal, being greater where
the channel is rougher and shallower over the riffle (Leopold, Wolman,
and Miller, 1964, pp. 295–308; Langbein and Leopold, 1966), and less
rapid in the deeper and smoother pools (Fig. 3.19). In an attempt to
equalize this energy loss along a reach, the energy excess in the pools is
compensated for by being partly consumed by the forming of bends at the
pools, so causing an increase in energy loss owing to the changes in flow
direction and hence the greater internal interference and friction between
the water particles. According to this view the meandering pattern is

potentially more stable than a straight one, which is regarded as probably a temporary state. Thus the inherent regularity in the fluvial processes is believed to be mirrored in the regularity of meandering patterns, and it has recently been shown that the deviations of a meandering river in progressing elemental distances along its course are normally distributed (Thaker and Scheidegger, 1968).

Considerable investigation has been carried out to determine the discharge frequencies which correlate best with meander geometry, and which therefore appear to be dominant in their formation and maintenance. Some of the most recent has been by Carlston (1965), which was based on measurements at thirty-one sites in the United States, showing that meander wavelength (L_m) correlates well with $\bar{Q}$, Q_b, and $\bar{Q}_{mm}$ (the mean discharge of the month of maximum discharge, which in this instance produces 16·6 per cent of the total annual discharge) (Fig. 3.20). Because the better correlation coefficients are given by $\bar{Q}$ and $\bar{Q}_{mm}$, although their marginality is questionable, Carlston (1965) suggests that discharges of less than bankfull are most significant, particularly $\bar{Q}_{mm}$ for it operates in cutting previously-wetted flood-plain banks (see also Wolman and Miller, 1960, p. 65). He concludes, however, that there may be a range of effective discharges between $\bar{Q}$ and Q_b responsible for controlling meander geometry, especially those with a falling stage attacking wetted and weakened banks and promoting meander development. This conclusion is supported to some extent by observations of the effect of changing discharge on channel conditions, for it has been noted that at $\frac{3}{4}Q_b$ the surface slope of a meandering stream is virtually uniform and the pools and riffles 'drowned out', whereas that of nearby straight reaches is still irregularly stepped (Langbein and Leopold, 1966) (Fig. 3.21). However, even if these ideas have validity, they do not provide a complete explanation of the phenomenon of meandering flows, which also occurs occasionally where the bed material is very uniform, where no bed material is involved (as in melt water channels on ice surfaces), and when one fluid flows through another (as with the Gulf Stream). Other mechanisms which have been invoked are harmonics in the flow itself, the irregular movement of uniform bed material, systematic bank caving, the helicoidal tendency, the Coriolis effect, etc., and it is clear that there is much still to explain with respect to the formation of meandering channel networks.

Sinuosity can be distinguished from meandering either on the basis of a lack of regularity in the windings of the channel, or as having a sinuosity (P) of less than 1·5 (Leopold, Wolman, and Miller, 1964, p. 281) (see Chap. 2.I.1(a)). Data from fifty locations on rivers in the Great Plains gives a good inverse correlation between sinuosity and the width–depth ratio (F), and a good direct correlation between sinuosity and the silt–clay percentage (M) (Schumm, 1963, pp. 1092–3). Sinuous streams tend to be narrow and to develop where the banks are resistant,

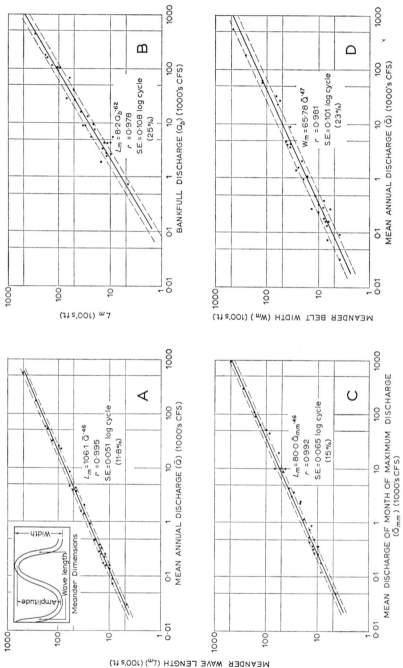

Fig. 3.20. Relation of meander wavelength to (*A*) mean annual discharge, $\bar{Q}$, (*B*) bankfull discharge, Q_b, and (*C*) mean discharge of the month of maximum discharge, $\bar{Q}_{mm}$. (*D*) relates meander-belt width to $\bar{Q}$. Source: Carlston, 1965.

owing either to a high silt–clay content or to bank vegetation (Schumm, 1963; Brice, 1964). They are characteristic where bed material lacks a significant coarse fraction which might collect in riffles, and seem

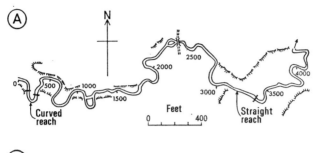

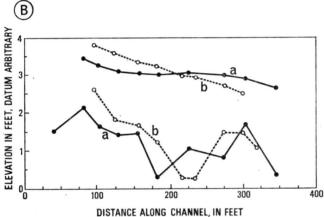

Fig. 3.21. Plan and profiles of straight and curved reaches, Baldwin Creek near Lander, Wyoming, U.S.A. (*A*) Planimetric map showing location of reaches; (*B*) Profiles of centreline of bed (*below*) and average of water-surface elevations (*above*) on the two sides of the stream; curved reach is dashed line and straight reach is full line. Source: Langbein and Leopold, 1966.

to have a lower gradient than straight streams with the same discharge. Clearly the factor of discharge alone does not seem to determine whether a stream will be sinuous, and the influence of bank resistance and uniform fine bed material seems to be paramount.

(b) Braiding channels

The more erodible the banks, and consequently the larger the proportion of available bed material, the more likely the channel is to widen during extreme discharges and to braid (Schumm, 1963; Brice, 1964). Braided or anastomosing channels are therefore associated with rather well-sorted non-cohesive material such as sand and fine gravel, where the shifting of large amounts of available bed material over a wide channel

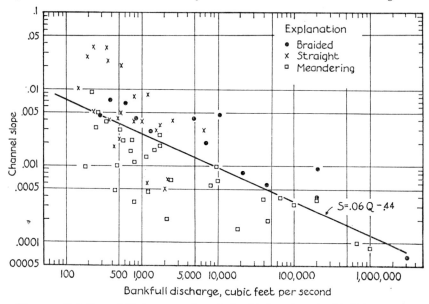

Fig. 3.22. Relation of discharge to slope in braided and nonbraided rivers. Line separates data of meandering from braided channels. Source: Leopold and Wolman, 1957.

during high discharges forms moving bars. These are exposed at low flows and may be turned into islands if vegetation gains hold enough to fix them and to trap abnormal sediment amounts. A comparison of braided, meandering and straight stream reaches on the basis of Q_b and channel slope shows that, whereas straight streams occur under a wide variety of conditions, braided channels tend to be steeper than meandering ones for a given Q_b, and to have a larger Q_b for a given slope (Fig. 3.22) (Leopold, Wolman, and Miller, 1964, p. 284–95). The effect of braiding is to increase the overall channel width and to decrease its depth, although individual braids may be deeper and have increased velocities (though increased turbulence in these divided reaches probably decreases the effective downstream velocity: Morisawa, 1968, p. 148). Braided channels appear to have a geometry which offers more flow resistance than straight ones, which is compensated for by an increase of bed slopes. Braiding might be viewed merely as an extreme case of high sinuosity, and Brice (1960) has suggested a braiding index (BI), where:

$$BI = \frac{2 \text{ (sum of length of islands and bars in a reach)}}{\text{length of the reach, measured mid-way between the banks}}$$

Because most islands and bars are long and narrow their perimeter is approximately equal to twice their length, and thus the braiding index is a measure of the increase in bank length resulting from braiding. Braiding has long been associated with stream aggradation and, although by no means all braided streams are depositing and raising their beds, the most impressive examples of the splitting of channels into braids are found where local deposition is forming alluvial fans and deltas. In both situations a progressive decrease in velocity causes deposition: on alluvial fans because infiltration decreases the discharge, and in deltas because the standing body of water checks the flow into it. Alluvial fans are characteristically covered with a network of abandoned channels and debris flows of very very variable age (Beaty, 1963) (Fig. 3.23). A detailed analysis of a fan

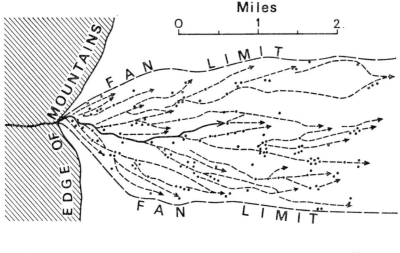

Fig. 3.23. Map of the Rock Creek fan, California, showing location of large boulders and former flood channels. Traced from an aerial photograph. Source: Beaty, 1963.

surface in terms of the assumed tractive force necessary to entrain the particles observed on it (tractive force = specific weight of the transporting medium × depth of flow × slope of the energy gradient; this has been simplified by omitting the specific weight, putting depth of flow equal to the maximum particle diameter, and substituting actual local gradient for the energy gradient), shows that the most tractable particles are arranged in a bifurcating pattern, presumably reflecting the pattern of the last material to be deposited at the termination of the most recent flood (Lustig, 1965). Deltaic deposition has been referred by Bates (1953) to the flow from a free jet emanating from a slot into standing water, producing deposition along its flanks and in the form of a transverse bar at a distance of more than four times the slot width. The channel is split as breaks are

made through the lateral levees and the transverse bar during high discharges, such that each break can become a slot for a new jet.

(c) *Channel plan and bed gradient*

The generally-observed decrease of channel gradient with distance
from the source has been traditionally explained as the combined effects
of the inverse relationship between bed slope and discharge, and of the
progressively finer material assumed to occur downstream as the result
of attrition and sorting. As has been noted in the case of pools and riffles,
however, this generalization breaks down on anything more refined than
a regional basis, largely because of the detailed disposition of the coarse
bed material remaining from the last flood capable of moving it. Combining all measurements from a region involving sixteen streams in central
Pennsylvania, Brush (1961) showed that this general decrease of channel

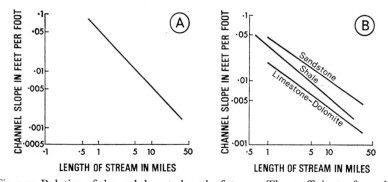

Fig. 3.24. Relation of channel slope to length of stream. The coefficients of correlation are sandstone −0·76, shale −0·87, and limestone −0·78. Source: Brush, 1961.

slope varied with distance from the drainage divide (Fig. 3.24-A). The
same author showed a similar tendency for channels within each geological
formation (Fig. 3.24-B), as did Hack (1957) for some channels in Virginia
and Maryland. It is interesting that although the absolute slopes for
channels on each type of bedrock are different (in the case of Brush's
data, significantly at the one per cent level), presumably owing to differences in the calibre and shape of the bed materials shed by different
types of bedrock, some downstream rates of change of slope were similar
for different types of bedrock.

Attempts to disentangle the effects of discharge (operating through its
control over velocities near the bed) and calibre have indicated that bed
gradients can vary significantly even where discharge does not, and for
a number of rivers Hack (1957) was able to show that, holding discharge
constant within a given lithology, bed slope is a direct function of median
bed grain size raised to the power 0·6. However, although bed particle
size shows a regional tendency to decrease with distance from the divide
on most (but not all) lithologies, and although there is a similar regional

tendency for bed slope to be directly related to particle size, data from individual channels show a rather poor correlation between calibre and bed slope. Brush (1961) showed that some of this lack of correlation might well be due to special processes occurring in both the headwaters (e.g. more active valley-side slope movements) and in the lower reaches (e.g. backwater effects), by demonstrating that in central Pennsylvania the best correlations between mean bed particle size and bed slope are found in the middle reaches of streams, irrespective of lithology.

A further refinement has been highlighted by Broscoe (1959) who found that most longitudinal profiles can be best approximated by logarithmic plots fitted to individual reaches, and that most profiles show marked segmentation, particularly by order (as predicted by Horton's law of stream slopes: Horton, 1945, p. 295; Schumm, 1956, p. 605), probably as the result of changing discharge–load relationships, channel characteristics and, perhaps, calibre of bedload (Mackin, 1948, p. 491; Woodford, 1951, p. 819; Yatsu, 1955, p. 657).

3. Flow and channel organization

(a) The basic Horton model

Of all the morphometric features connected with stream networks none has been held to be more indicative of flow conditions than drainage density. Horton (1945) not only defined drainage density, but developed an attractive flow model—the infiltration theory of runoff (Horton, 1945, pp. 306–31)—that explained how a given set of controlling conditions could bring about an equilibrium drainage density, which imparts to a given physiographic region so much of its morphometric character. The Horton model of infiltration and runoff assumes that, for a prolonged storm of constant intensity, a continuous exponential decrease in the infiltration capacity of the surface occurs (due to compaction, the breakdown of the soil structure, the filling of the pores with water, the washing down of the fine particles, etc.) until a constant low value is reached over the whole drainage basin. If the infiltration capacity (f) falls below the rainfall intensity (i), the rainfall excess causes overland flow to begin all over the slopes which, if it persists long enough to achieve a steady state will produce a flowing layer the depth of which will increase downslope according to a power law.

At a critical distance from the divides (X_c), depending on the runoff intensity (Q_r), the depth of flow becomes large enough to produce velocities which give shear stresses sufficient to overcome the surface resistivity (R_i), to entrain the surface material and to cause erosion. Downslope of this point first rills and then, perhaps, streams develop; thus it is implicit in Horton's reasoning that the width of the belt of no sheet erosion (X_c) sets a limit to the continued development of the drainage network, and therefore to drainage density. The width of the belt was held by Horton

(1945, pp. 320–4) to be principally controlled by R_i and Q_r, although, apart from noting that 'most of the work of valley and stream development by running water is performed during floods' (1945, p. 284), he did not suggest the magnitude of the runoff events which might be expected to control X_c.

Horton, however, did identify many of the significant controls over drainage density (D), and it has become recognized that this depends on factors concerned with climate, vegetation, rock and soil type, rainfall intensity, infiltration capacity and relief (i.e. possible stage of development, expressive of land slope) (Smith, 1958, pp. 1001–2), among others. Low values of D seem, therefore, to be favoured by highly-resistant or highly-permeable surface material, dense vegetation and low relief; whereas the opposite conditions will produce a high drainage density (Strahler, in Chow 1964, pp. 4–52).

Strahler (1958, pp. 288–90; 1964, pp. 4-69–70), in an attempt to rationalize these controls, has employed dimensional analysis to produce key combinations of variables which may focus attention on the dynamic relationships involved. Drainage density (L^{-1}) is proposed as a function of the following variables: (1) *Runoff intensity* (Q_r), the volume rate of flow per unit area of cross-section. This combines the factors of rainfall intensity and infiltration capacity, and is dimensionally equal to $\dfrac{L^3 T^{-1}}{L^2} = LT^{-1}$.

(2) An *erosion-proportionality factor* (K), which is the mass rate of removal per unit area divided by the force per unit area, is dimensionally equal to
$$\frac{ML^{-2}T^{-1}}{ML^{-1}T^{-2}} = L^{-1}T$$

(3) *Relief* (H), which represents the potential energy of the system, and is a function of slope and time, is dimensionally equal to L. (4) *Density of the fluid* (ρ), dimensionally equal to ML^{-3}. (5) *Viscosity of the fluid* (μ), dimensionally equal to $ML^{-1}T^{-1}$. (6) *Gravity* (g), dimensionally equal to LT^{-2}.

The above variables can be reduced to four dimensionless terms by applying the *Pi* theorem: (1) A *Ruggedness Number* (N), equal to $\dfrac{1}{HD}$, which is greatest when slopes are both long and steep, and least when they are short and gentle. (2) A *Horton Number* (N_H), equal to $Q_r K$, and expressing the relative intensity of slope erosion processes in the drainage basin. (3) A form of the *Reynolds' Number* (N_R), the ratio of the force of inertia to the force of viscosity, in which Q_r (dimensionally equal to velocity) takes the place of velocity, and H of length. (4) A form of the *Froude Number* (N_F), the ratio of the force of inertia to the force of gravity. Combining these four solutions, the following relationship can be expressed:
$$D = \frac{1}{H} f\left(Q_r K, \frac{Q_r \rho H}{\mu}, \frac{Q_r^2}{Hg} \right)$$

The most common work relating to drainage densities, however, has been that involving their comparison, either from maps or in the field, in areas which have been selected to eliminate the possible effects of some factors and to highlight the effect of one or more others.

(i) The effect of *rock type* within otherwise rather similar regions has been often noted, such that clay and shale are commonly associated with high values of D, whereas sandstones more usually give lower values (Morisawa, 1968, p. 158). In southern Indiana Coates (1958) recognized that some lithological differences, but not all, are paralleled by differences in D, with sandstones and limestones being associated with values of D of about eight and massive sandstones of about fourteen. The most complete lithological analysis of D was undertaken by Melton (1957) for a large number of third- and fourth-order basins in the western United States. Statistical tests, employing the Scheffé method, showed significant differences at the 0·05 level between the mean values of D for shale and schist, also between the values of shale and schist, on the one hand, and of the remaining four rock types, on the other. However, it has become clear that even if other significant factors are held more or less constant in such an experimental design, drainage density does not always reflect differences in bedrock. One of the most striking features of the morphometry of lowland Britain, for example, is that, despite great lithological variation, the values of drainage density are remarkably similar. Contrast with this the great differences which appear between lithologies in more arid regions, where the vegetation cover is less important. As Carlston (1966, p. 68) has suggested, increasing aridity tends to accentuate the differential effects of bedrock upon drainage density, and Smith (1958) mapped a difference in drainage density between two shales in the South Dakota badlands, showing the Brule Formation with a drainage density of 258 and the Chadron Formation with one of 77·6.

(ii) Of course, one of the most important ways in which the bedrock influences drainage density is through the *infiltration capacity* of the surface. This formed the corner-stone of Horton's analysis when he pointed to the differences between the densities of the drainage network in a sandy basin (Fig. 3.25-B) and of one developed on less permeable bedrock, even with the mean annual rainfall thirty per cent greater on the former (Fig. 3.25-A). Although in the list of drainage densities previously given from Melton (1957) there was not a clear correlation with measured infiltration capacity, it is apparent that shale and schist have both the highest values of drainage density and the lowest mean values of infiltration capacity (f). In a recent analysis of drainage density, Carlston (1963; 1966), relying on one-square-mile quadrat sampling from 1 : 24,000 maps of thirteen basins in the north-eastern United States with areas between five and fifty-five miles, has employed the notion of 'terrain transmissibility'. This embodies the combined effects of infiltration, vadose water movement and ground-water flow in streams. Carlston showed that as drainage density is primarily a function of the amount of storm runoff, it must be inversely related to terrain transmissibility (i.e. inversely related to base flow).

(iii) Other workers have examined *rainfall intensity*, the other part of Horton's rainfall excess equation, as a control over drainage density. A world-wide study by Peltier (1962) led to the conclusion that the lowest 'topographic texture' occurs in true deserts, the next lowest in 'moderate' climatic regions, the next in semi-desert regions (where Langbein and Schumm (1958) have shown high erosional rates to occur due to the

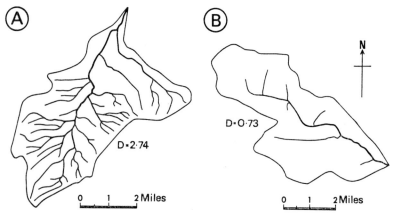

Fig. 3.25. Drainage patterns in well-drained (*A*) and flat, sandy and poorly drained basins (*B*). Source: Horton, 1945.

combination of quite high rainfall intensities with sparse vegetational cover), and the highest values in certain tropical areas. Cotton (1963) attributed the fine texture of many New Zealand landscapes to the relatively high intensities of winter precipitation (with little evaporation loss), comparing them with similar textures found in certain Mediterranean climates, and with the high drainage densities (approximately 53— determined from 1 : 25,000 maps) in south-west Honshu believed to result from the high rainfall intensities during the monsoon and typhoon seasons. Chorley and Morgan (1962) compared the drainage densities of two granitic areas of high relief (Dartmoor, England and the Unaka Mountains in the south-eastern United States), using fourth- and fifth-order basin data derived from 1 : 25,000 and 1 : 24,000 topographic maps. Dartmoor and the Unakas exhibited significantly different mean drainage densities (3·45 and 11·18, respectively), and this difference was attributed mainly to the contrast between the intensities of rainfall in the two regions (these differences operating over a long timespan), in so far as both regions have a complete vegetation cover.

There is some question remaining whether vegetation amount is an important control over a wide range of occurrence or whether, as seems more likely, once a complete cover exists, differences in the actual amount of vegetation per unit area have little further effect on drainage density, except in so far as its interception and transpiration influences surface runoff. However, these effects are so bound up with the effects of a thick surface soil cover that they had best be considered later, when a modification of the simple Horton runoff model is proposed. Another climatic control over drainage density that is difficult to evaluate is the magnitude

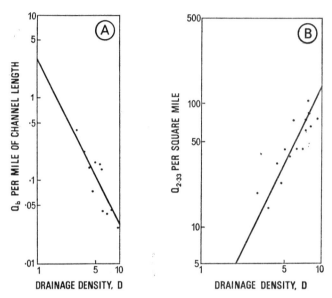

Fig. 3.26. The relation of drainage density to (*A*) base flow and (*B*) mean annual flood ($Q_{2.33}$). Source: Carlston, 1963, 1966.

of precipitation intensity (and therefore of runoff intensity), which is mainly responsible for initiating and maintaining drainage channels. Carlston's (1963) correlation between D and $Q_{2.33}$ (Fig. 3.26-B) led him to suppose that 'the drainage network is adjusted to the mean annual flood' (p. 7), although Kirkby and Chorley (1967) hold that for individual runoff events, the drainage density controls the flood peak. It would appear that drainage lines, once cut during infrequent storms of high intensity, may become quite persistent and even self-perpetuating features by controlling the locations of maximum erosion during the next storm (Chorley and Morgan, 1962, p. 29). This cutting of new gullies during exceptional storms, which once cut might continue to grow during more moderate storms, was recognized by Wolman and Miller (1960) as one of the relatively few extreme events which could have an enduring effect on the landscape. Even so, Kirkby and Chorley (1967) suggest that between-storm infilling of stream heads might cause reversions to lower drainage densities before the heads were excavated. Here one is encroaching the borderline between events which are extreme but 'normal' and those which represent a persistent change of input: the latter will be treated later.

(iv) The effect of *relief*, or slope, on drainage density has long been of interest, because it introduces a time ('stage') element from which evolution might be inferred. In this respect, however, observational results seem to be at variance with the theoretical assumption that steep slopes should give high runoff and a high drainage density. Coates (1958) could find no evidence in southern Indiana of any relationship between drainage density and relief or slope, although he did find that the most 'mature' basins (i.e. those with the largest upland area consumed) had the largest drainage density—but this increase seems inevitable throughout 'youth' if the basin area remains constant. Carlston (1966) also found that channel gradient seems to bear no relation to drainage density in the thirteen basins studied in the north-east United States. In contrast, Schumm (1956, pp. 612–17) found not only that drainage density increased with mass removed from the drainage basin throughout youth and maturity, but also that it increased with the relief ratio, or basin slope, for eight 'mature' basins of orders **4** and **5**. It would appear that when the surface is bare and of low resistivity, changes in basin slope may well affect drainage density, even though these effects are masked in vegetated basins. This matter of the effect of stage on the drainage network will be taken up again in the chapter on evolution.

(b) Elaborations of the Horton model

From what has been said so far it is obvious that of all the factors which have been suggested as controls over drainage density, some are much more important than others, and that even some of the important ones have effects of very varying significance depending on the magnitude of the other variables in conjunction with which they are operating. Attempts to evaluate the combined effects of the controls have commonly been of a qualitative nature. For example, Carlston (1966) pointed to the distinction between two nearby regions in the central United States: (A) The High Plains—having twenty inches mean annual precipitation, most of which evaporates, while the rest infiltrates into the sandy surface covered with short grass giving a mean annual recharge of less than one inch and virtually no surface runoff. Here the drainage density is extremely low. (B) The Badlands of South Dakota—having sixteen to eighteen inches mean annual precipitation mostly of fairly high intensity (the 10-year, 1-hour rainfall is 1·8 to 2 inches), of which almost all that does not evaporate, runs off the bare clay surface giving drainage densities of 200–400.

A quantitative attempt to attack the problem of controls over drainage density was made by Chorley (1957) who proposed a climate/vegetation index (I_c) which expressed the ratio of vegetation amount to precipitation amount and intensity, such that:

$$I_c = \frac{\text{Thornthwaite's P–E Index}}{\text{Mean annual precipitation} \times \text{mean monthly maximum precipitation in 24 hours.}}$$

The reciprocal of I_c showed some consistent relationship with the mean logarithm of the drainage densities calculated from topographic maps for

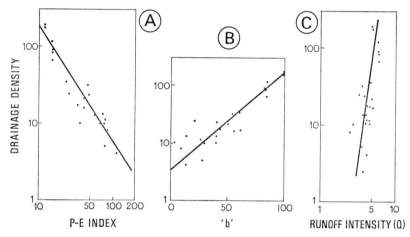

Fig. 3.27. Best-fit regressions of drainage density (D) as a function of P–E index, per cent bare area (b) and runoff intensity (q). Source: Melton, 1957.

three areas of massive sandstones in Exmoor, south-west England, central Pennsylvania and northern Alabama. The most satisfactory analysis of drainage density, however, was by Melton (1957), who carried out a multiple correlation of fifteen parameters relating to the morphometry, climate and surficial conditions from fifty-nine basins in the western United States (twenty-three of which were subjected to a full field study and the remainder were partly studied with the aid of 1 : 24,000 maps).

The correlation analysis showed that log D significantly correlates with log $(P–E)$ (Thornthwaite's precipitation effectiveness index), log b (per cent bare area), log f (infiltration capacity), and log q (runoff intensity), in that decreasing order, explaining, respectively, 88·9, 81·0, 62·9, and 43·3 per cent of the variance of log D (Fig. 3.27). These four variables together explained fully 93·2 per cent of the variance in log D, showing that climatic, hydrologic and surface properties of an area are of overwhelming importance in determining drainage density (Melton, 1957, p. 35). The analysis also showed that a combination of log $(P-E)$, log f and log q explains ninety-two per cent of log D, and that when these three are considered, per-cent bare area and surface roughness (M) are very minor influences. Perhaps the most interesting outcome of this phase of Melton's work was the dominance of the $P-E$ index, for when this had been considered the only other significant variable was log M, suggesting that 'the $P-E$ index influences drainage density through the agencies of infiltration capacity, amount of cover, and runoff intensity' (Melton,

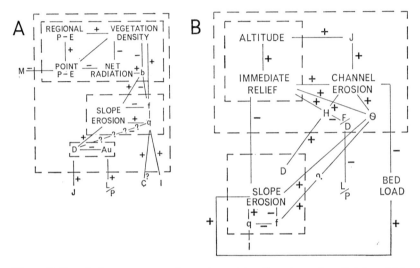

Fig. 3.28. Correlation structures of morphometric, climatic, and surficial variables as interpreted by variable-system theory. Source: Melton, 1957.

1957, p. 35). Melton (1958B) then mapped the variables which were significantly correlated into plane sets, and suggested directions of causality so that variable systems, commonly characterized by feedback loops, could be identified. Fig. 3.28 and Table 3.4 show such systems containing the climatic and surficial controls over the scale of the topography. It is interesting that an inverse relationship between D and q is necessary to provide a negative feedback to give D a limiting value, despite the positive relationship between the two noted from the field data (Fig. 3.28-B). For this reason Melton (1958B, p. 455) concludes that the commonly-observed regionally uniform value of drainage density results from 'a frequently realized mean value of runoff intensity capable of performing a certain amount of work on the surface in erosion and that, because the region has a certain mean surface resistance to erosion, a definite length of slope is needed before the accumulated runoff is sufficient

Table 3.4. Matrix of intercorrelations significantly different from zero

	θ	D	C	H	L/P	F/D²	f	P-E	Sw	SD	b	M	q	J	Au	
θ	1			+0·495		−0·294	+0·302						−0·236	+0·391		Maximum valley-side slope angle
D		1					−0·273	−0·376					+0·428	+0·285	−0·426	Drainage density
C			1	−0·160									+0·275		−0·216	Basin circularity
H				1	+0·253	−0·560	+0·362	+0·133					−0·308	+0·348	+0·178	Ruggedness number (relief times drainage density)
L/P					1	−0·322						+0·352			+0·506	Ratio of total channel length to basin perimeter
F/D²						1	−0·311	−0·225					+0·254	−0·197	−0·231	Ratio of stream frequency to drainage density‡
f							1	+0·341			−0·584		−0·737		+0·276	Infiltration capacity
P-E								1			−0·681	−0·307	−0·536		+0·366	Thornthwaite's precipitation effectiveness index
Sw									1	+0·496						Wet soil strength
SD										1						Dry soil strength
b											1	+0·278	+0·488		−0·499	Per-cent bare area
M												1				Roughness number (size of surface fragments)†
q													1		−0·518	Runoff intensity*
J														1		Relative January precipitation intensity**
Au															1	Basin area

Source: Melton, 1958B.

* Excess of precipitation over infiltration capacity for 5-year, 1-hour storms.
† Mean total length of all pebbles >½″ diameter in a randomly-selected 1-foot radius circle.
** Ratio of mean January precipitation to one-twelfth mean annual precipitation.
‡ A measure of the completeness with which the channel net fills the basin outline for a given number of channel segments.

to initiate a channel . . . Thus the regularity in the drainage density over a region is an expression of the regularity in climate, soil-forming processes, rates of plant growth, lithology, and structure rather than an expression of a hypothetical tendency for a particular drainage density to develop because it controls the regional elements that in turn control it.' This gives further support to the idea that under most conditions angle of land slope has little control over drainage density. The most ambitious variable system involving morphometric, climatic and surficial variables proposed by Melton (1958B) is given in Fig. 3.28-B. Negative feedback may operate in respect of drainage density probably where channel length is kept constant, providing that basin area, slope runoff rate and soil creep rate are themselves constant.

The Horton model of infiltration and runoff, on which most concepts of drainage density are based, has always been most applicable in areas of little soil cover and low infiltration capacity (e.g. badlands: Schumm, 1956), where a limited length of rainstorm is necessary to saturate the whole basin surface and to initiate unconcentrated runoff over all the slopes and concentrated flow in all the channels. For example, Gregory and Walling (1968) have shown that in two basins in south-east Devon, England, the total channel length contributing concentrated runoff varies with the basin discharge. Thus, over a wide range of discharges, from summer lows to winter highs, the contributing effective drainage density is not constant but varies with the discharge. This field work supports the view that every storm produces surface runoff from a limited 'contributing area' (Kirkby and Chorley, 1967). Gregory and Walling suggest that Carlston (1966) is incorrect in employing a constant drainage density, because the drainage density value applicable to peak flows do not apply when baseflow is making up most of the discharge. In soil-covered basins, throughflow (i.e. flow within the soil cover parallel to the slope) is much more important than in more arid regions, and under such conditions soil moisture is much more areally-variable over the basin than Horton assumed. Moisture content generally increases continuously down the length of a soil covered slope, and commonly approaches saturation only in a zone immediately marginal to the stream channels, although this zone will extend upslope as a given storm continues. Overland flow produced under these circumstances will therefore occur initially on restricted areas of the hillside at much lower rainfall intensities than would be required for universal Hortonian overland flow in the basin (Kirkby and Chorley, 1967). Thus, although Horton postulated that overland flow takes place universally in a basin when rainfall intensity exceeds infiltration capacity clearly it can occur very locally in soil-covered basins under rainfall intensities which are less than the overall infiltration capacity. These localities are those which are adjacent to flowing streams, those where lines of greatest slope converge, those where local concavities occur, and those where the soil cover is thin. With the exception of the last possibility, the favoured areas for overland flow and the initiation and maintenance of stream channels are valley bottoms and stream-head zones. Stream-head hollows are usually previously-eroded locations which are being infilled by slope processes, and periodically re-excavated when rare storms extend the drainage network (Kirkby and Chorley, 1967). The position of the stream heads, which is so important in determining the magnitude of drainage density, in humid soil-covered regions (as distinct from arid badlands) exhibits a wide variation in their distance from the divides: this indicates that the Hortonian concept of the limiting control over drainage exercised by the X_c zone is at best of only restricted application here.

4. Flow change and network change

From what has gone before it is clear that both the hydraulic geometry of stream channels and the patterns of drainage networks are capable of achieving an equilibrium state in relation to a number of controlling variables which have to do with climate and surface properties. A theoretical treatment of an aspect of this equilibrium was investigated by Strahler (1958, p. 295) for drainage density. Assuming that ρ, μ and g are virtually constant, it is possible to neglect N_R and N_F and to express D in terms of the Horton Number.:

$$D = \frac{1}{H}(Q_r\ K)$$

If we assume that the dimensionless valley-side slope $(\theta) = 2DH$, then

$$\frac{HD}{\theta} = \tfrac{1}{2}$$

This is a dimensionless 'geometry number'. If D and H are considered representative of length values in the horizontal and vertical planes, and θ is any slope representative of the mouthward inclination of the watershed, the geometry number shows by its small range in nature (Table 3.5) that it tends to be conserved near unity throughout a wide range of D. Thus, for example, a high value of D is compensated for by a low H and a high θ. The geometry number is a dimensionless group summarizing the essentials of landform geometry, just as the Horton number summarizes the intensity of erosional processes:

$$\phi\!\left(\frac{HD}{\theta},\ Q_rK\right) = 0$$

Strahler (1958, p. 295) concludes that 'conditions for a steady state within a drainage basin are these: for a given Horton number, that is, for a given intensity of erosion process, values of local relief, slope and drainage density reach a time-independent steady state, in which morphology is adjusted to transmit through the system the quantity of debris and excess water characteristically produced under the controlling régime of climate.'

Melton's (1957, p. 34) analysis led him to conclude that if, for example, the percentage of bare area is changed by some mechanism to a value which is not in agreement with the $P-E$ index, then the value of drainage density must change. Strahler (1958, p. 296) illustrated much the same point when he noted that devegetation sharply increases the Horton number (i.e. both Q_r and K increase), and that this leads to an increase in D, an increase in θ and a decrease in H (defined as the relief of the first-order basins). In other words, a new set of equilibrium forms are

Table 3.5. Rough estimates of ruggedness and geometry numbers, HD and HD/θ for a wide range of conditions

Locality*	Drainage density ($feet^{-1}$) D	Relief ($feet$) H	Mean valley slope side θ	Ruggedness number HD	Geometry number HD/θ
1. Gulf Coastal Plain (La.)	0·0009	25	0·06	0·022	0·38
2. Piedmont (Va.)	0·0013	75	0·17	0·097	0·57
3. Ozark Plateau (Ill.)	0·0028	125	0·54	0·35	0·65
4. Great Smokies (N. Car.)	0·0028	300	0·87	0·84	0·96
5. Verdugo Hills (Calif.)	0·005	200	1·00	1·00	1·00
6. Perth Amboy Badlands (N.J.)	0·12	6	1·10	0·72	0·66

Data for D and θ for localities 1–5 from Strahler, 1952, Table 1; for locality 6 from Schumm, 1956.
Data for H estimated from topographic maps.
* Localities arranged in order of increasing D.
Source: Strahler, 1958, p. 296.

produced having a smaller linear scale (Fig. 3.29), as when the destruction of grassland leads to the generation of badlands yielding sediment at a high rate. The manner in which the drainage density increases under such circumstances was explored by Melton (1958B) who noted that, because D is related to the stream frequency, as D increases not only do the existing channels lengthen but new channel segments are added as well.

The effect of a climatic change on the rate of erosion and the condition of the drainage net is very much dependent on the precipitation régime before the change (Langbein and Schumm, 1958, pp. 1083–4), such that an increase below twelve inches of effective precipitation may well produce increased erosion and channel extension, whereas the same change occurring above it may decrease them both. The fact that the highest world values both of drainage density (Peltier, 1962) and erosion rates (Langbein and Schumm, 1958) are found in semi-arid regions is no coincidence, and it may well be that an increase in precipitation in such a region causes a contraction of the drainage network.

The problem of what constitutes a significant and lasting change of input into a channel system is not a simple one, particularly since

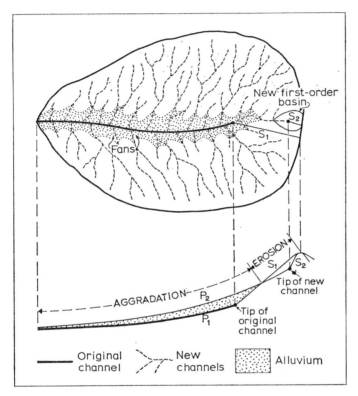

Fig. 3.29. Drainage density transformation. Source: Strahler, 1958.

Gumbel's (1958) theory of extreme values is based on the assumption that any hydrologic régime can experience very infrequent events of almost any magnitude. There seems to be an infinite gradation between the small continuous changes of channel discharge which continually modify hydraulic geometry and the wholesale network changes which accompany a significant climatic change. The effects of infrequent high-intensity processes on the stream network have been noted by Wolman and Miller (1960, pp. 70–3). Schumm and Lichty (1963) have shown how a major flood on the Cimarron River in south-western Kansas in 1914 increased the mean width of the river from fifty to over one thousand feet by evacuating a grassy floodplain, and that a series of further floods between 1941 and 1942 associated with deficient and variable precipitation further increased its width to 1,200 feet. During the following decade, floodplain construction, perhaps assisted by the increased erosion promoted by the extension of cultivation, caused the channel to narrow to some 550 feet in association with vegetation stabilization. A rather similar change of flood frequency on the upper Severn in Wales after about 1940 may have been responsible for the change in flood frequency noted at Shrewsbury (Howe, Slaymaker and Harding, 1967); but it is interesting that this may partly have been due to an increase in the rate of flood runoff owing to the effective increase of drainage density of large areas of the upper catchment as the result of the ploughing of drainage ditches some $3\frac{1}{2}$–$5\frac{1}{2}$ feet apart during afforestation. These ditches connect with natural streams and must affect the runoff rate for at least 30 years until complete crown cover has been achieved. It has been calculated that these ditches have effectively increased the drainage density of some areas of the upper Severn by 1·14 miles/square mile between 1961 and 1965 and, in view of the observed relation between drainage density and mean annual flood in the catchment, must have caused a change in the runoff régime.

The period of arroyo cutting in the south-west United States which began about 1880–5 and lasted until 1928 further illustrates the combined effects of climatic fluctuation and human intervention (Schumm and Hadley, 1957; Leopold, Wolman and Miller, 1964, pp. 442–5). The combined effects of low winter rainfall (supported by tree-ring evidence for the period 1870–1906) in decreasing the grass cover, together with the effects of overgrazing following the introduction of large numbers of cattle after 1870, allowed the probably more intense rainfall to cause considerable extension of the arroyos by headcutting. The Rio Puerco in New Mexico probably cut back an arroyo twenty-eight feet deep and 285 feet wide for well over one hundred miles during this period of relative aridity. Gully headcutting under such circumstances is a complex process involving seepage, piping and sapping, and producing gullies with very discontinuous longitudinal profiles (Schumm and Hadley, 1957), the headward break of slope being maintained during migration if the bed material has a greater resistance to shear stress than the flow stress, and if the flow is sufficient to transport the eroded material from the base of the nick (Leopold, Wolman and Miller, 1964, pp. 442–3). Laboratory experiments on headcut migration in cohesive material show that the rate of migration and the preservation of break of slope are dependent on the ratio of flow depth to initial height of the free face of the nick.

Arroyo trenching seems to be carried out in a cyclic manner of alternating cut and fill, producing the longitudinal breaks of slope so characteristic of semi-arid regions (Schumm and Hadley, 1957, pp. 170–2).

During erosion the downstream addition of sediment load increases faster than the discharge (owing to discharge loss through seepage and to the large amounts of material suitable for transportation which are usually available). This means that after the initial headward cutting at the lower parts of a drainage system (Fig. 3.30-A and -B), during which the flow concentration and runoff efficiency increase, the larger amount of sediment moved causes sedimentation downstream (Fig. 3.30-C and -D). This alluviation decreases owing to water loss by percolation, and discontinuous gullies begin to be cut in the especially oversteepened sections, each bounded by a headcut and yielding a fan downstream, the latter being itself susceptible to trenching. This complex operation results finally in the series of fans being flushed out to the main channel at the mouth of the gully system to form a fan. Trenching of this fan initiates a new cycle of semi-arid arroyo cutting.

Many assumptions regarding the possible effects of climatic change on

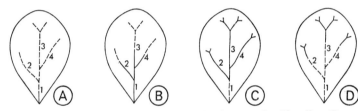

Fig. 3.30. The cycle of trenching and alluviation in a semi-arid valley. Dotted line indicates alluviation, within a channel. Solid line indicates trenching or a through channel. Source: Schumm and Hadley, 1957.

river networks derive from short-term observations concerned with the response of rivers to artificial regulation. River regulation and irrigation abstraction over the past sixty years, for example, have decreased the average width of the North Platte in western Nebraska from 2,800 feet to 500 feet, as $Q_{2.33}$ has fallen from 13,000 cfs to 3,000 cfs and $\bar{Q}$ from 2,300 cfs to 560 cfs. The South Platte river at Brule, Nebraska which had a wide braided course in 1897 was similarly narrowed by 1961 after reservoirs had reduced flood peaks and decreased the coarser sand load (Schumm, in Chorley 1969). Reservoirs on the Smoky Hill River in central Kansas have trapped much of the coarser sand bedload, and downstream, although discharge increases, the addition of silt and clay by the lower tributaries causes depth and sinuosity to increase, whereas width, width/depth ratio and meander wavelength decrease—all owing to the deficiency of bedload. Further downstream the Smoky Hill joins the Republican River to form the Kansas River, and the addition of bedload by the Republican causes width, width/depth ratio and gradient to increase abruptly, sinuosity to decrease, and depth to remain fairly constant (Schumm, in Chorley 1969).

Clearly imprinted on the landscape, however, is evidence of the important changes which stream networks have undergone as the result of past climatic changes, usually associated with the Pleistocene, although it is less easy to make estimates of a quantitative nature regarding runoff. It is apparent, for example, that fine-textured landscapes in New Zealand are currently being etched on to coarse-textured, subdued, periglacial forms (Cotton, 1963). Many systems of dry valleys in south-western United States (Dury, 1964A, p. 64) and the classic ones in the English chalk may be due to a reverse process involving a decrease of surface runoff. From what has been written above it is clear that, in general, an increase in discharge should increase the proportion of bedload transported, since this will be effected by width increasing more than depth

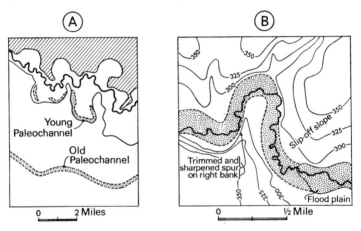

Fig. 3.31. Relationship of contemporary and former stream channels. (*A*) Murrumbidgee River, Australia. (*B*) Windrush River, England. Source: Schumm, in Chorley 1969; Dury, 1964A.

(so increasing the width/depth ratio); and that there should also be increases in the meander dimensions and a decrease of sinuosity. However, the effect on the drainage network depends very much on the conditions prior to an increase of discharge, and in less humid regions an increase in discharge can cause headwater erosion and channel extension to proceed simultaneously with deposition and loss of discharge by seepage downstream.

An excellent example of the presence of evidence in the landscape of changes which the channel system has undergone is provided by the Riverine Plain of the Murrumbidgee near Darlington Point, New South Wales (Schumm, in Chorley 1969). Here there is clear evidence of three stages of channel development: (1) The present river, having a sinuous, meandering course over a floodplain of silts and clays. (2) The young paleochannel cut in material similar to (1) and having both a channel slope and sinuosity similar to (1). However, width, depth and meander wavelength are longer, indicating a greater discharge than that responsible for the present channel. (3) The old paleochannel. This was cut in fine sand and gravel, and is straighter, wider and shallower than either (1) or (2). Its gradient is twice that of (1). These differences indicate its formation during a dryer climate than the succeeding ones, during which there was an abundance of coarse debris (Fig. 3.31-A).

The most classic example of stream changes that result from changes of discharge are provided by the so-called 'underfit' streams (Dury, 1953; 1964A; 1964B). These streams have obviously undergone a drastic decrease in discharge (sometimes owing to capture, but more usually to post-glacial changes of régime) and the most striking of them are the 'manifestly

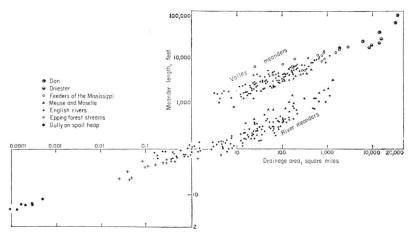

Fig. 3.32. Relationship between meander length and drainage area for rivers (lower set of data) and meandering valleys (upper set of data). Source: Dury, 1960; Leopold, Wolman and Miller, 1964.

underfit' streams which occur in more amply meandering valleys (Fig. 3.29-B). Their bedrock cross-sections show asymmetric bends, buried in alluvium on which the present rivers have meanders proportional to their present discharges. Although the ratio of wavelength to width of these meandering valleys shows a greater spread than similar measurements for meandering streams, it is clear that the wavelengths of valley meanders can be related to their assumed catchment areas, and that this relationship is similar in trend but distinct in magnitude from those of present river meanders—the former having about ten times the wavelength of present meanders for a given area (Fig. 3.32) (Dury, 1964A). The degree of underfitness of the present stream is expressed as the ratio of the wavelength of valley and present stream meanders (L/l), the Warwickshire Avon flowing in a valley cut by glacial overspill having $L/l = 10$; the streams of the north-east Ozarks having $L/l = 8$ upstream and 4 downstream; and the Salt and Cuiver Rivers, Missouri, having $L/l = 7\cdot5$ upstream and $3\cdot5$ downstream. It thus appears that meandering valleys in bedrock have a rather different hydraulic geometry from that of alluvial channels, and at present it is difficult to do more than speculate on the detailed processes responsible for their cutting. Meandering valleys are found widely throughout western Europe and north America and their magnitude suggested to Dury (1953) that they were cut by Pleistocene bankfull discharges of between eighty to one hundred times those in the existing streams (which, of course, does not imply that the precipitation was that much greater).

III PROBLEMS OF FLOW FORECASTING

Analysis of existing patterns of flow and their relation to the spatial structure of networks leads naturally to the critical problems of forecasting future variations in flow. In this section the general problems of sampling flows, the special problems of forecasting secular increases in flows, and the prediction of recurrent peaks in flow are examined. In each case extensive literatures within relevant applied fields of traffic engineering (e.g. Wohl and Martin, 1967) and hydraulics (e.g. Chow, 1964) have developed and only the more elementary models are outlined here.

1. Sampling and space-time distribution problems

Analyses of traffic flows moving over a wide range of channel networks, suggest that a small number of links within the total network carry a rather high share of the total traffic load; conversely, a high proportion of links are rather lightly loaded. For the British road system, Tanner,

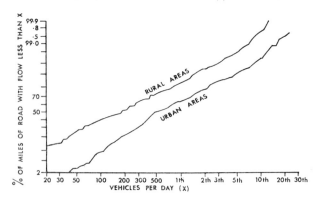

Fig. 3.33. Spatial concentration of motor-vehicle flow on urban and rural road networks. Great Britain, 1959–60. Source: Road Research Laboratory, 1965, p. 60; Tanner, Johnson and Scott, 1962, p. 42.

Johnson and Scott (1962) analysed a sample of flows at 1,000 sampling points distributed over all classes of road in Great Britain. Fig. 3.33 shows cumulative flow for movements in vehicles per day plotted against the length of road which carried less than this load. The horizontal scale is

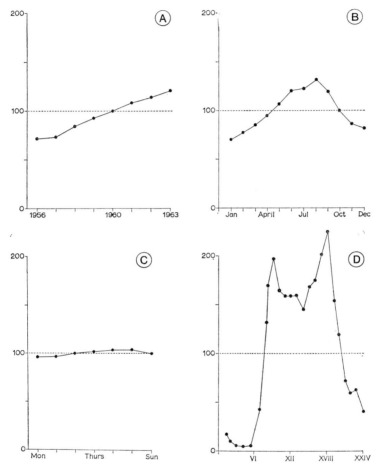

Fig. 3.34. Temporal variations in motor vehicle flow. Great Britain, 1956–63. (*A*) Secular trends. (*B*) Seasonal periodicities. (*C*) Daily periodicities. (*D*) Hourly periodicities. All flows expressed as percentage of average (100). Source: Road Research Laboratory, 1965.

logarithmic, but the vertical scale is calibrated to the normal probability distribution so that the general form of the distribution is clearly shown to be log-normal. Half the roads carried flows of less than 270 motor vehicles per day while the busiest ten per cent carried flows in excess of 2,300 and the busiest one per cent flows in excess of 11,600. For both urban and rural areas the curves bend upwards towards the right-hand end suggesting that there are fewer roads carrying very heavy flows than might have been expected. Broadly similar log-normal distributions are revealed by flow investigations on the United Kingdom railway system (British Railways Board, 1965).

The spatial concentration of flows on links is paralleled by concentrations over time. Fig. 3.34 shows a composite picture of flow variations in terms of four major time-cycles. In each graph the motor vehicle flow is shown as a percentage of the mean value (standardized as one hundred). The two main contrasts shown by the graphs are (a) the distinction between the secular trend shown by trends from year to year (Fig. 3.34-A) and the periodic or recurrent trends shown by the other three graphs, and (b) variations in the amplitude of the three recurrent trends. The greatest variation is shown in the pattern for hourly flows (Fig. 3.34-D): flows during the peak hour (the evening rush hour between five and six p.m.) may account for between 8 to 12 per cent of the total flow depending on the type of vehicle. By contrast the low hours in the early hours of the morning may carry less than one-half per cent of total flow. Month to month variations show a smaller range of amplitude between about seventy and 150 with the low point in January and the high point in August. Day-to-day variations show the weakest trend with limits of about ninety-seven and 103. Flows are generally at their lowest on Monday and Tuesday and build up in the later half of the week. Graphs here are based on the fifty-point traffic census of Great Britain (Road Research Laboratory, 1965, pp. 5–84).

The distribution of flows on stream channels in terms of magnitude and frequency is treated by Gumbel's (1958) theory of extreme values which states that, for such independent extreme events as the maximum instantaneous annual discharges of a river, there is a relationship between magnitude and occurrence (the latter commonly expressed as the average recurrence interval (RI) in years). This relationship is such that it appears as a more-or-less straight line plot on Gumbel's extremal probability paper, from which certain characteristic parameters can be obtained: e.g. the *most probable value* (RI: 1·58 years), the *median value* (RI: 2 years), and the *mean value* (RI: 2·33 years).

The space-time concentration of flows within networks poses severe sampling problems. For although spatial movements within networks are commonly mapped as flows (e.g. with the load on a link shown by its relative width) few flows are in fact continuous. Fig. 3.35 attempts to divide movement into four basic classes in terms of three criteria: (i) whether the flow is continuous or discontinuous, (ii) whether the discontinuous flow is channelled or unchannelled, and (iii) whether the channelled flow has a few specific terminals or unlimited terminals. Information on the first two of the four main classes is commonly gained

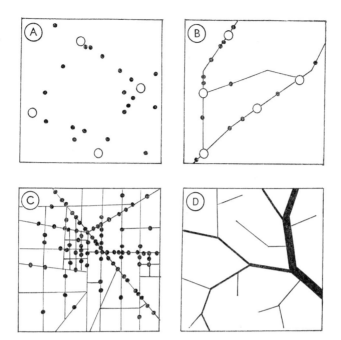

Fig. 3.35. Sampling problems posed by alternative types of flow. (A) Aircraft movements. (B) Rail movements. (C) Automobile movements. These examples of discrete movements are contrasted with the continuous movements in a stream-channel system (D).

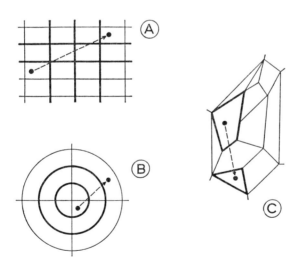

Fig. 3.36. Representative lay-out of spatial cordons. Source: Road Research Laboratory, 1965, p. 122.

by recording or sampling flow at specific points along the links of the network (see Fig. 3.35-c and -d); information on the second two classes is usually derived from the input and output of flows from origin or destination terminals (Fig. 3.35-a and -b). The distinction is not exclusive and often motor vehicle movements within a city will be gained both from road counts and from house-to-house OD surveys; in contrast data on

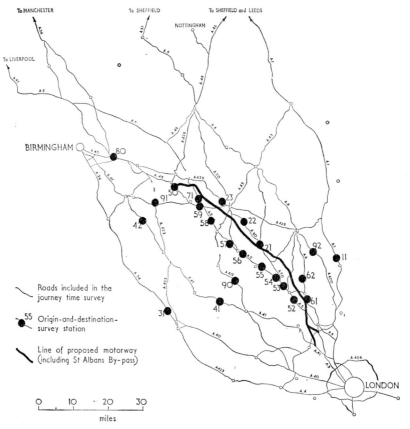

Fig. 3.37. Location of survey stations for assessment of economic effects of the construction of the London–Birmingham motorway. Source: Coburn, Beesley and Reynolds, 1960.

aircraft movements is recorded exclusively in terms of terminals (see Fig. 3.35-a). The wide range of data on transport movements is summarized by the Northwestern University Transportation Centre (Blaisdell, 1964).

Alternative arrangements of sampling points for determining the pattern of interregional flows within a network are shown in Fig. 3.36. Here the survey area is divided up into a number of cells by sampling lines or *cordons*. Cordons are chosen to 'catch' all major flows within the region and the sampling points are usually situated where a major link in the network intersects a cordon; minor links with small flows are not recorded. One arrangement uses a regular set of cordons set up orthogonally to form a rectangular sampling grid (Fig. 3.36-A), whereas an urban survey may use radial and annular cordons set up round the city centre (Fig. 3.36-B). Shorter cordons may also be used to divide the area into a series of irregular zones corresponding to important physical, commercial or administrative entities (Fig. 3.36-C). In all three diagrams certain cordons have been thickened. These are the critical sampling cordons that estimate flow from A to B. Total flow from A to B is estimated by the algebraic sum of the flows from A to B that cross the heavy cordons in the direction indicated. The best estimate is the average of the estimates obtained at each cordon (Road Research Laboratory, 1965, p. 122). Fig. 3.37 maps the location of twenty-three survey points used in an assessment of the economic effects of the London–Birmingham motorway in central England (Coburn, Beesley and Reynolds, 1960). For this purpose estimates were required of the proportion of traffic that would be diverted from existing roads and the consequent savings in travel times. Although sampling points were concentrated along the line of the existing major link (highway $A5$) some points were located on other highways to represent possible long-distance routes that might be switched to the motorway. Use of variable sampling and interception factors at each sampling station allowed a prediction to be made of *diverted* traffic using the motorway. Actual traffic volumes will of course include some traffic *generated* by the motorway (see Chap. 3.III(2)).

2. Forecasting secular changes : the traffic assignment problem

Planning a new transport network may involve the construction of a system *ab initio* but more frequently the modification and extension of an already existing system. Both types of problem commonly include the following basic design stages: (i) analysis of the origin and destination of existing traffic flows in the region (Chap. 3.III(1)); (ii) assessment of similar flows at some future date appropriate to the expected life of the new system; and (iii) assignment of the predicted flows to the existing network. This leads in turn to (iv) identification of stress in the form of over-loading on links; and (v) suggestions for improving the existing network. These improvements may be in the form of changes in the capacity of existing links, the addition of new links, (Chap. 4.I) or more fundamentally the creation of an entirely new network (Chap. 3.I(2)). The sequence of stages in forecasting future flows on links is shown in Fig. 3.38.

Assessment flows at some future date appropriate to the expected life of the new system may be estimated in two ways: (a) by simple expansion of the existing origin-destination matrix by appropriate constants, and

(b) by derivation of a mathematical model from existing flows with appropriate adjustment of parameters for future time periods. The expansion method multiplies each cell in the origin-destination matrix by a constant factor derived from observed trends in traffic flow. For example, in Great Britain the vehicle mileage has been expanding by

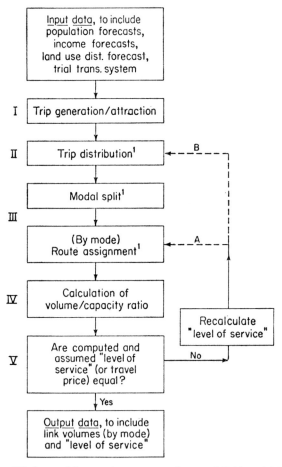

Fig. 3.38. Simplified travel-forecasting process. Source: Wohl and Martin, 1967, p. 130.

about eight per cent per annum compound, and this factor might be used to generate flows in future years. Regional variations between cells may be brought in (e.g. for central London the expansion rate is only about five per cent compound) and variations in land-use/traffic generation relationships allowed for. Information from planning schemes for a region with zoning proposals commonly gives information about future land-use. Since land-use is linked directly to traffic generation, it may prove possible to compute future traffic originating and terminating at zones. If we know the present flows from the ith to the jth cell (M_{ij}), the present traffic originating at zone i (O_i) and terminating at zone j (D_j), and the estimated future origins and destinations (O'_i, D'_j), then we may calculate future flow (M'_{ij}) using the formula:

$$M'_{ij} = \tfrac{1}{2}M_{ij}\left(\frac{O'_i}{O_i} + \frac{D'_j}{D_j}\right)$$

(Road Research Laboratory, 1965, p. 132). This average factor method may be improved by using an iterative method in which successive values of M'_{ij} are substituted in place of the original flows, revised estimates of O_i and D_j are obtained by row and column summation in the OD matrix.

A gravity model provides the most common mathematical form for estimating traffic movements between zones. The general form of the model may be given as:

$$M_{ij} = kA_iA_jf(Z_{ij})$$

where M_{ij} is the flow from zone i to zone j; k is a constant; A_i, A_j are measures of the attractiveness or size of zones i and j; and $f(Z_{ij})$ is a function of Z_{ij}, the distance or time or cost of moving from zone i to zone j. The term 'gravity' model has arisen from the similarity between the law of gravitation and inverse square function commonly fitted to Z_{ij}. A wide range of gravity models has been applied to interactions between zones; their major characteristics and performance are reviewed at length by Olsson (1965). More recently Wilson (1968) has provided a new evaluation of the model from the viewpoint of statistical mechanics. At the regional level, interregional flows have commonly been set within the framework of an interregional linear programming model and assessment of optimal flows predicted on this basis (Isard, 1960, pp. 413–92).

Assignment of the predicted flows to the existing network involves three separate stages of analysis: (a) selection of routes between zones in the network, (b) assignment of traffic movements to these routes and (c) adjustment of the loads on each link in the network in terms of modal splits and time-flow feedbacks. Selection of routes through the network implies some means of selecting the 'best' route or good routes where excellence is defined in terms of minimizing some distance or cost or time between the two cells. Methods for determining 'shortest paths' between vertices are discussed in Chap. 4.I(1).

Although rather simple measures of distance are frequently used in calculating shortest paths, some workers have gone to considerable lengths in developing realistic weights for links in a network. For example, Kissling (1966) used sixteen basic parameters in evaluating the regional highway network of Nova Scotia and New Brunswick provinces in eastern

Canada. These ranged from total mileage over the link through measures of the length of legal speed zones through to physical characteristics of gradient, curves, and passing opportunities as well as traffic itself. There is clearly an overriding need to replace simple measures of time or distance as a yardstick of the spatial separation of nodes. Quarmby (1967A and B) suggests that a useful estimate of generalized cost between nodes (C_{ij}) may be given as

$$C_{ij} = t_{ij} + e_{ij} + \alpha k$$

where t_{ij} is the travel time in minutes, e_{ij} is the excess time in minutes, α is the value of time calculated at twenty-four per cent of mean monetary income of the commuter, and k is cost per mile in pence (based on car mileage costs adjusted for parking and road pricing charges). 'Excess' travel

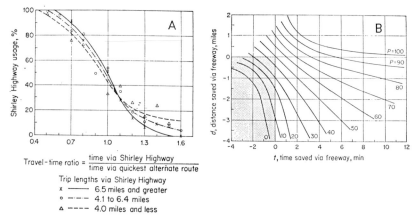

Fig. 3.39. Examples of diversion curves. (A) Diversion curve based on travel-time ratios. (B) Diversion curves based on time and distance saved by travelling via freeway route. Source: Wohl and Martin, 1967, pp. 134, 136.

time refers to waiting or walking time as opposed to 'in vehicle' time. The generalized cost function derived above is part of a group of studies of the choice of travel mode for the daily journey to work within urban areas; alternative formulations are needed for interregional movements.

Determination of the minimum path (however minimum be defined) allows the movement between two cells to be loaded on to the links making up that path. Empirical studies of road traffic suggest however that there may be a set of alternative routes available with only small variations in the journey time from season to season, and several methods have therefore been proposed which reduce the 'all or nothing' effect of minimum-path assignment. When two routes are equal in length it is assumed that the flow will be equally split, fifty per cent using both routes. As the difference in length changes the flow adjusts until one road is × 1·2 the length of the other when its share of the traffic falls to around twenty per cent; at $\times 2{\cdot}00$ the share has dwindled to almost nothing (Fig. 3.39). An alternative approach used in Manchester assigns traffic equally to all routes with a journey time not greater than $\times 1{\cdot}15$ the minimum value (Road Research Laboratory, 1965, p. 137).

Assignment of flows to a minimum path or series of minimum paths gives some measure of projected flow along each link in the network. The picture shown is complicated however by three additional difficulties: (1) traffic may be diverted from the mode under consideration to an alternative mode of travel; (2) flow along links itself changes the value of a link and therefore the pattern of shortest paths; (3) flow along links will lead to re-evaluation of the location advantages of vertices within the network and, in the long term, to consequent changes in land-use and traffic generation.

(i) *Short-term feedbacks.* When traffic flow is well within the design

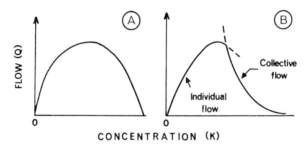

Fig. 3.40. Short-term feedbacks: self-induced network congestion. Diagrams showing the general character of flow-spatial concentration assumed by the (*A*) Lighthill–Whitham theory and (*B*) Prigogine theory. Source: Gazis, 1967, pp. 276–7.

capacity of the link, travel time is virtually independent of the volume of traffic. As volumes increase so does the spatial concentration of vehicles while speeds fall and travel times go up. Theoretical models of the relationship between volume and concentration are shown in Fig. 3.40 where the flow of traffic (e.g. in vehicles per hour) is plotted against its spatial concentration (e.g. vehicles per mile). The difference between the two statements is in the form of the flow-concentration function between these two points: the Lighthill-Whitham theory sees this as a continuous function (Fig. 3.40-A), while the Prigogine theory (Fig. 3.40-B) conceives a discontinuous function made up of curves for 'individual' and 'collective' flow. Despite the importance of the difference for wave behaviour in traffic and associated control problems, the overriding similarity of the curves suggests that sharp differences in link times are likely as concentrations pass the Prigogine transition point. The effect of this clustering on flow and capacity has already been discussed (see Fig. 3.16).

Attempts to build this consideration into assignment programmes may be illustrated by Smock's work (1963) on the future road network of an American town (Flint, Michigan). Under the Smock procedure travel times are determined on the basis of typical speeds, the shortest time path between every pair of intersections is identified, and all volumes are assigned to their quickest path. At the end of the first pass the ratio between assigned volumes and capacity for every link is calculated, and this quantity is used to compute a new travel time on each link according to the conditions of congestion built up. Where the ratio is high, speed is reduced below typical speeds and travel time is thereby increased. Where the ratio is low, speed is increased and travel time is reduced. The formula determining these increases and decreases is as follows:

$$T_i = e^{(R_i-1)} T_o$$

where T_i is travel time on a link for a given pass; e is 2·71828, R_i is the ratio of averaged assigned volumes (from previous passes) to capacity, and T_o is the original travel time on the link (Smock, 1963, p. 14). At the second and third passes the same procedure is followed and such passes are repeated until capacity-adjusted speeds come to approximate typical speeds. The two measures of speed will converge rapidly where the network provides sufficient capacity in the places where it is needed, but it will of course never happen if the network does not contain sufficient capacity for the volumes assigned to it. The network proposed for Flint in 1980 is shown in Fig. 3.41. Here the convergence or balancing process is demonstrated for the first five passes; links with either heavy congestion or low use are progressively altered to achieve balanced loads throughout the network.

(ii) *Long-term feedbacks.* Whatever the difficulties in forecasting the effect of short-term feedbacks in flow prediction, these are greatly overshadowed by the effect of long-term feedbacks. As Lachene (1965, p. 184) points out, once a transport link is established the low marginal cost of long-term transport will persist '. . . with the result that the economic effects accumulate with the passage of time; conversely the possibilities of making long-term economic forecasts are fairly limited particularly with respect to territories of small dimensions.' The persistence of transport routes, once established, and their crucial effect on the locational evaluation of regional resources are classic themes of geographic writing. For example Garrison in his *Highway development and geographic change* (Garrison, et al., 1959) has reported the effect of a number of route changes on the functional structures of a number of American cities. A typical example is the effect of a new by-pass on the trade and function of two small American towns Everett and Marysville, which lie some thirty miles north of Seattle, Washington state, on the highway (U.S. 99) which runs north to the Canadian border. Prior to October 1954, the main road ran through Marysville; after this date traffic was diverted around the town on to a four-lane limited-access highway. The effects of this new route on traffic were clear and expected. In the year following, the traffic through down-town Marysville fell to about a third of its previous flow (about 5,400 as against 14,000 vehicles a day). Less obvious were the effects on the func-tions of the town. Here Garrison found that the fall in through-traffic had

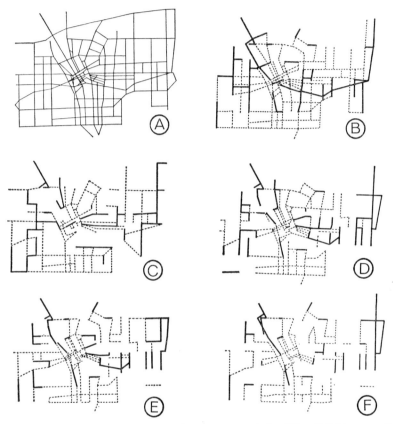

Fig. 3.41. Volumes assigned to the highway network for Flint, Michigan, 1980. (*A*) Network. (*B–F*) Assigned volumes as per cent of capacity for the first five runs of the assignment programme. Overloaded links (125 per cent capacity) shown by solid lines; underused links (75 per cent capacity) shown by broken lines. Source: Smock, 1963, pp. 19–24.

made Marysville so much more attractive as a local trade centre that sales of 'first- and second-order functions' increased to 121 per cent of pre-by-pass volumes. On the other hand, since it was now easier to get to Everett (a larger town) from the rural areas around Marysville, the higher 'third-order' functions of Marysville had dropped to eighty-three per cent of pre-by-pass levels (while the corresponding level in Everett went up). Against this, Marysville, minus its heavy through-traffic, became a pleasanter place to live in, rents of undeveloped sites went up and a residential boom appeared likely. Garrison's study illustrates clearly the effect of changes on route structure in reorientation and realignment of demands and supplies. People travelled further along the new highway for their higher-order needs at Everett, but Marysville had become a better local centre. Throughout the changes the urban system appeared to reorganize itself to the strains and stresses placed upon it in a way typical of open-system behaviour (*cf.* Ashley and Berard, 1965).

The immediate effects of new links on locational evaluation at the local level is clear. At the regional and national levels the picture is more clouded; for example the classical view of railroads as a critical factor in nineteenth-century economic growth is now challenged. Rostow (1960, p. 55) suggests that take-off (one of his critical 'stages' in economic growth) in the United States was sparked off by the rapid growth of the railway system in the period 1850–90. Railroads are seen as a leading sector setting off secondary growth in other sectors like coal, iron and engineering. This view has been challenged in a very detailed econometric study by Fogel (1964) which shows that the interregional savings from railroads in 1890 (as opposed to possible extension of the waterways-wagon road system) were surprisingly small—only 0·60 per cent of the gross national product of the United States. Fogel argues that the gains from railroads were far smaller than traditionally believed, that many were premature and un-economic, and that the railroads were part of, and not a pre-condition for, the American industrial revolution.

Despite the overall correlation between measures of transport invest-ment and economic development (Fig. 3.42) there is some variation in the 'lead' or 'lag' position of the transport sector. The lead position of the transport indices in Thailand stands in contrast to that of Peru where national income over the period 1948–60 advanced more considerably (Fig. 4.32) (Wilson, *et al.*, 1966). Kissling (1967) shows a close and balanced relationship between urban growth and accessibility for the maritime provinces of Canada, but Gauthier's (1967) study of São Paulo lends support to 'leading sector' hypothesis, i.e. improvements in transport structure over the period 1940–60 were generally in advance of changes in urban industrial structure. It is clear that transport networks alone cannot 'cause' economic development and transport investment may well run ahead of aggregate demand; alternatively there are situations in which shortage of transport facilities is serving as a constraint on growth. The need to develop systems models for the transport sector itself and for the whole development process is apparent; the moves already made by the Transportation and Economic Development group at Harvard involving simulation of alternative network strategies suggests one useful way to-wards this long-term goal (Roberts, 1966; Soberman, 1966; *cf.* Perle, 1964).

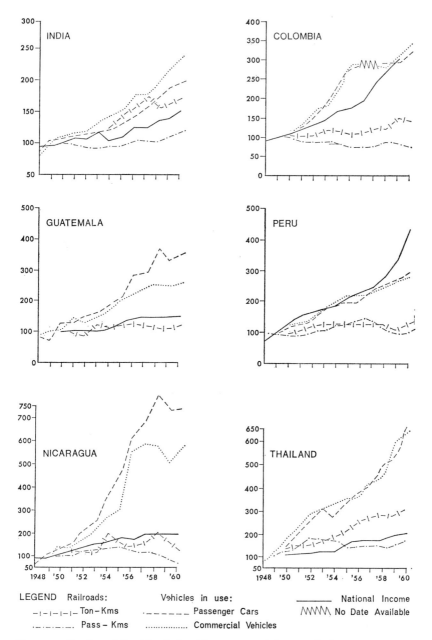

Fig. 3.42. 'Lag' and 'lead' relationships suggested by trends in transport and national income for selected countries, 1948–60. Source: Wilson, Bergmann, Hirsch and Klein, 1966, p. 15.

3. Forecasting recurrent changes: the flood problem

While the dominant problem in traffic forecasting concerns the *secular* increase in volume over time, the flood problem poses the difficulty of predicting *recurrent* patterns of change.

(a) Flows as hydrographs

Inputs of precipitation into stream networks produce an output which, when plotted as discharge versus time in the form of a hydrograph, appears as an attenuated curve (Fig. 3.43). This curve, although it varies between networks and between different storm inputs for the same basin, possesses certain characteristic components (Chow, 1964, pp. 14–18 to 14–19; De Wiest, 1965, p. 66): (a) The approach limb, prior to the first

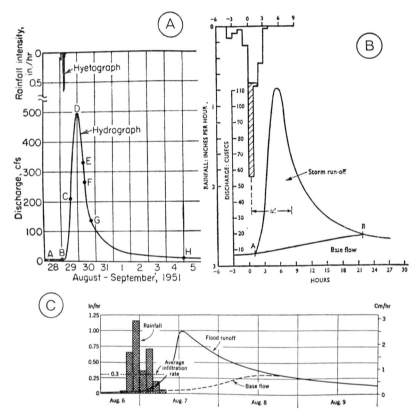

Fig. 3.43. Typical hydrographs with corresponding hyetograph. (*A*) Panther Creek, Illinois, U.S.A.; (*B*) Canon's Brook, Harlow, England; (*C*) Sugar Creek, Ohio, U.S.A. In the latter case only 3 inches of a total rainfall of 6·3 inches in twelve hours contributed to the hydrograph. Source: Chow, 1964; Nash, 1960; Strahler, 1965.

effect of the storm input at the point of rise (point B, Fig. 3.43-A). (b) The rising limb, which is variously defined as extending from the point of rise to the inflection point (point C, Fig. 3.43-A), to the crest or peak (point D, Fig. 3.43-A), or to some intermediate point. The form of this curve is largely a function of the time-area basin histogram (see later section) and of the duration and uniformity of the storm rainfall input (Linsley, Kohler and Paulhus, 1949, p. 394). (c) The crest or peak segment, representing the highest concentration of network runoff and, variously defined between the hydrograph inflection points (points C and E, Fig. 3.43-A). (d) The recession limb, usually more attenuated and less steep than the rising limb. This is sometimes defined simply as the curve following the crest segment, sometimes as the curve following the inflection point E (Fig. 3.43-A), but more usually as that following the time when all channel inflow from overland flow has ceased (Linsley, Kohler and Paulhus, 1949, p. 395). The last definition implies that the recession curve represents predominantly the withdrawal of water from channel storage, and its form is therefore almost entirely a function of the stream network characteristics. This curve merges into: (e) The ground-water or baseflow recession curve, a much longer term runoff originating from ground water.

Even in the most predictable precipitation régimes there is a wide range of storm area/intensity/duration characteristics, and in order to characterize the hydrological behaviour of a given stream network the concept of the *unit hydrograph* has been proposed. The unit hydrograph is the hydrograph of direct surface runoff resulting from a unit (commonly one inch) of effective rainfall (i.e. that in excess of losses, chiefly by infiltration) generated uniformly over the basin at a uniform rate during a specified time (for example, twenty-four hours) (Sherman, 1932; Wisler and Brater, 1959, p. 247; Linsley and Franzini, 1964, p. 54; Chow, 1964, p. 14–13). It is best adapted to runoff resulting from relatively short intense storms yielding a relatively-uniform effective rainfall of a duration not longer than the period of rise of the hydrograph, and producing a hydrograph with a well-developed single peak and of short time base in a basin of restricted area (Wisler and Brater, 1959, p. 247; Chow, 1964, p. 14–14). The concept of the unit hydrograph is based on the assumption that certain immutable physical characteristics of a given stream network effectively control the pattern of runoff (Foster, 1949, p. 318), such that all unit storms over a given basin which have the same antecedent conditions, regardless of their magnitude, will produce hydrographs of virtually identical pattern (Sherman, 1932; Linsley, Kohler and Paulhus, 1949, p. 444; Wisler and Brater, 1959, p. 247). This will be so because the time of concentration is constant for a given basin, and therefore the time base for the direct runoff hydrograph for all storms of the same duration is a constant for the basin (Linsley, Kohler and Paulhus, 1949,

p. 444; De Wiest, 1965, p. 79). These assumptions are questionable in that rainfall is seldom of uniform intensity in space or time (although basin and channel storage tend to have a smoothing effect on rainfall variations), and because relatively small differences in antecedent conditions, such as infiltration capacity and basin and channel storage, can affect not only the magnitude of the rainfall excess and therefore of the hydrograph but also its form.

The unit hydrograph is constructed, usually for basins of less than 2,000 square miles, by selecting a hydrograph of a uniform isolated storm with the required effective length (or of an average of storms of the required length, which is partly conditioned by the size and lag time of the basin), by separating the baseflow, and by scaling the discharge ordinates of the remaining direct runoff hydrograph to adjust for the required unit rainfall excess (Linsley, Kohler and Paulhus, 1949, p. 445; Wisler and Brater, 1959, p. 78; Linsley and Franzini, 1964, pp. 54–6; Chow, 1964, p. 17–17; De Wiest, 1965, pp. 79–80). Further assumptions are introduced by this procedure, for example, that the baseflow separation leaves an accurate estimate of surface runoff, and that the principle of proportional scaling of the ordinates is valid (Chow, 1964, p. 14–14). However, characteristic unit hydrographs have been constructed for many basins for different effective storms durations (commonly, six hour, twelve hour and twenty-four hour); the instantaneous unit hydrograph is assumed to result from the runoff associated with an instantaneous rainfall excess over the basin. Two important time spans used to characterize the spread of the unit hydrograph are the *time base*, the time interval between the point of rise and the end of all runoff originating on the surface (De Wiest, 1965, p. 70), and the *basin lag*, the time interval between the centre of mass of storm rainfall and the hydrograph peak (Linsley, Kohler and Paulhus, 1949, p. 395). A synthetic method for deriving the unit hydrograph was devised by Snyder (1938) for several Appalachian basins; this involved empirical determinations of (i) the basin lag (t_P, in hours):

$$t_P = C_t(L.L_c)^{0.3}$$

where $L =$ length (in miles) of the main stream to the divide, $L_c =$ distance from the mouth of the main stream along its course to a point opposite the basin centroid, $C_t =$ an empirical coefficient ($1 \cdot 8 – 2 \cdot 2$); and (ii) the standard unit duration of rain (t_R, in hours):

$$t_R = \frac{t_P}{5 \cdot 5}.$$

From these parameters were calculated the synthetic unit hydrograph peak discharge q_P (in cfs):

$$q_P = \frac{640(C_P)\ (A_d)}{t_P}$$

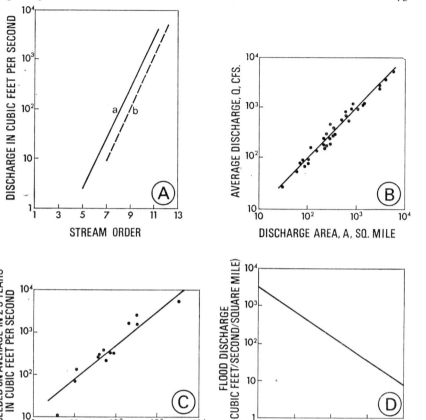

Fig. 3.44. Impacts of basin size on flow characteristics. (*A*) and (*C*) Two methods of estimation for ephemeral streams in New Mexico; (*C*) Potomac River; (*D*) Maximum flood discharge for basins in Colorado. Source: Leopold and Miller, 1956; Hack, 1957; More, in Chorley and Haggett, 1967.

where C_P = an empirical coefficient (0·56–0·69), A_d = the area of the basin (in square miles); and the time base T, (in days):

$$T = 3 + 3 \frac{(t_P)}{24}.$$

Knowing the time of the peak, the peak discharge and the time base Snyder (1938) constructed a valid synthetic unit hydrograph. The test of a unit hydrograph is the success with which it can be used to predict the hydrographs resulting from other than unit storms. This is most simply done for different intensity storms of similar duration by scaling the ordinates of the unit hydrograph in proportion to the required storm amount and superimposing them on the assumed baseflow curve. A more dubious operation is to adjust the unit hydrograph to storms of different duration by adding the ordinates of the required number of unit hydrographs, with the appropriate offsetting, and then rescaling the combined ordinates to the required storm amount.

(b) Factors controlling hydrographs

The 'immutable physical characteristics' of the stream network which were previously stated to control the form of the hydrograph are largely those which result directly, or indirectly, from its geometry. Basin area not only controls the magnitude of a given unit hydrograph (Fig. 3.44),

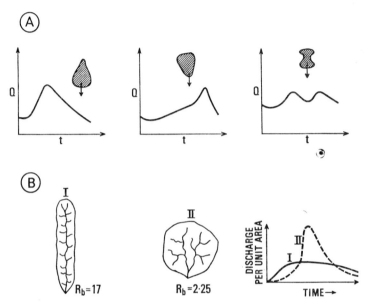

Fig. 3.45. (A) Influence of basin shape on hydrographs. (B) Schematic hydrographs for basins having high (I) and low (II) bifurcation ratios. Source: De Wiest, 1965; Strahler in Chow, 1964, pp. 4–44.

but has also been shown to control both the peak percentage discharge (i.e. the percentage of the runoff that occurs during the ten-minute interval containing the peak) and the period of rise of the hydrograph in the North Appalachian Experimental Watershed, Coshocton, Ohio (Wisler and Brater, 1959, p. 262). Basin shape is also an important control (Fig. 3.45-A), and Snyder (1938) found his parameter L_c to be a good indicator of basin shape. However, basin shape is usually only a function of the disposition of the channel segments and a more direct influence on the general form of the unit hydrograph is exerted by the bifurcation ratio (Fig. 3.45-B), where a high bifurcation ratio is associated with an elongate basin giving a low and attenuated hydrograph peak, whereas the more common lower bifurcation ratio of the compact basin produces a more pronounced peak (Strahler, in Chow 1964, p. 4–44). Drainage density is directly related to 'terrain transmissibility' (Carlston, 1963) and for fifteen basins in the eastern United States it was found to exercise an important control over runoff peaks, such that high drainage densities are associated with rapid drainage of overland flow into stream channels and therefore to more accentuated hydrograph peaks (Wisler and Brater, 1959, p. 47). The influence of basin slope is more difficult to evaluate in isolation because it is interlocked with other important variables such as surface detention, basin storage and channel storage. There is a tendency, however, for steep basins to be associated with more sharply-peaked hydrographs (Fig. 3.46).

There have been few attempts to evaluate the relative importance of the physical characteristics of the stream network in controlling the form of the hydrograph. Taylor and Schwarz (1952) found that for twenty basins of between twenty and 1,600 square miles in the north and middle Atlantic States, the rate of change of basin lag (t_{PR}) with the duration of the significant rainfall excess (t_R) (compositely expressed as m') was inversely related to $L.L_{CA}$ and that the lag of the instantaneous unit hydrographs (C') was inversely related to the square root of the equivalent mainstream slope (S_{st}: i.e. the slope of a uniform channel which has the same length as the mainstream and an equal travel time). Nash (1960) was also concerned with the lag of the instantaneous unit hydrograph— in this case the mean delay assumed to be encountered by a water particle between being precipitated in an instantaneous storm and appearing at the gauging station (m_1), being made equal to the time lag between the centres of gravity of the storm and the hydrograph. A multiple correlation analysis involving seven network geometrical variables showed that eighty-one per cent of the observed variation in m_1 for ninety British catchments could be explained either by a combination of basin area and the mean slope of the drainage basin, or by a combination of the length of the longest stream (carried to the basin perimeter) and the mean slope of the main channel. Following the work of Potter (1953), Morisawa (1959, p. 17; 1962, p. 1044) examined the factors controlling the peak ten-year runoff intensity for fifteen basins in the Appalachian Plateau Province of between 1·5 and 550 square miles. A multiple correlation analysis involving the logarithms of all values indicated that the frequency of first-order streams explained a significant 87·9 per cent of the peak intensity, and that when this was combined with the independently insignificant circularity and relief ratios the three factors together explained a significant 89·8 per cent. An analysis of the flood discharges of various recurrence intervals ($Q_{1.2}$ to Q_{50}) for 170 basins in New England of between 1·64 and 9,661 square miles and involving eleven basin parameters, as well as six climatic ones, showed that the flood discharges were

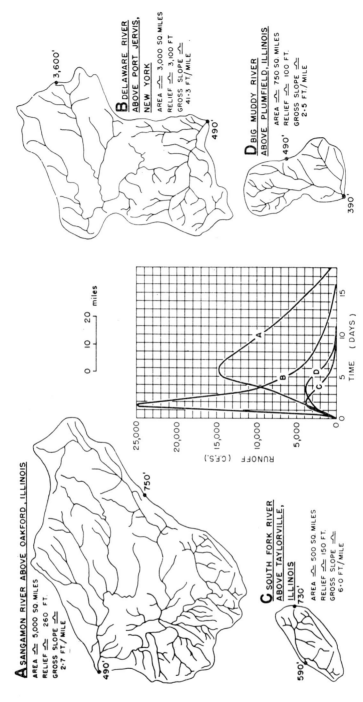

A SANGAMON RIVER ABOVE OAKFORD, ILLINOIS

AREA ≃ 5,000 SQ. MILES
RELIEF ≃ 260 FT.
GROSS SLOPE ≃ 2·7 FT/MILE

B DELAWARE RIVER ABOVE PORT JERVIS, NEW YORK

AREA ≃ 3,000 SQ. MILES
RELIEF ≃ 3,100 FT
GROSS SLOPE ≃ 41·3 FT/MILE

C SOUTH FORK RIVER ABOVE TAYLORVILLE, ILLINOIS

AREA ≃ 500 SQ. MILES
RELIEF ≃ 150 FT.
GROSS SLOPE ≃ 6·0 FT/MILE

D BIG MUDDY RIVER ABOVE PLUMFIELD, ILLINOIS

AREA ≃ 750 SQ. MILES
RELIEF ≃ 100 FT.
GROSS SLOPE ≃ 2·5 FT/MILE

Fig. 3.46. Unit hydrographs for (*A*) the Sangamon River, Illinois, (*B*) The Delaware River, New York, (*C*) the South Fork River, Illinois, and (*D*) the Big Muddy River, Illinois. These show clearly the effect of area on the magnitude of the unit hydrograph, and the effect of basin slope on its form. Source: Sherman, 1932; More, in Chorley and Haggett, 1967.

overwhelmingly predictable on the basis of the combined factors of drainage basin area (A) and the mean channel slope (S) measured between points ten per cent and eighty-five per cent along the mainstream length above the gauge (Benson, 1959). This was followed by a most ambitious analysis of mean annual flood $(Q_{2.33})$ discharges for ninety New England basins by Wong (1963) which showed by means of principal components analysis and multiple regression that, for ten morphometric variables plus one precipitation variable, fully eighty per cent of the variance associated with $Q_{2.33}$ can be explained by a combination of the logarithm of the length of the mainstream and the logarithm of the average land slope.

The attenuation and characteristic change of form to which each rainfall input is subjected in its translation into basin channel network output are primarily due to the differential patterns of travel time assumed by the various components of the basin hydrological cycle (More, 1969 in Chorley and Haggett, 1967). This differential is particularly marked between the direct or storm run-off from the surface and the groundwater runoff or baseflow.

A further group of factors responsible for the attenuation and characteristic peaking of the basin storm outflow hydrograph are those concerned with travel times of surface runoff. Obviously, inputs at different places in the basin will, suitably depleted by evaporation and infiltration, etc., take different times to reach the basin mouth and contribute to the hydrograph pattern. The simplest case is to imagine a regular, impermeable, unchannelled catchment from which unconcentrated runoff occurs to a single outlet. Fig. 3.47 shows runoff patterns from a semicircular catchment, the surface of which slopes such that lines joining points of equal travel time to the outlet (isochrones) enclose equal areas. Fig. 3.47-A shows the result when one inch of rainfall falls in one hour uniformly over the catchment, and introduces the concept of *time of concentration* (t_c), the time required for water falling on the most remote part of the catchment to reach the outlet (Linsley, Kohler and Paulhus, 1949, p. 391), at the end of which time (assuming long-continued, uniform rainfall on a small impermeable catchment) the discharge equals the rate of intensity of precipitation (Linsley and Franzini, 1964, p. 52). The concentration of runoff, as expressed by the hydrograph, is greatly affected by the distribution of catchment area with respect to the outlet, compact basins having more pronounced peaks than other basins. A measure of this compactness has been proposed (Langbein, 1947, p. 134) as Σal—the sum of each partial area in the catchment multiplied by the channel distance from the outlet to the midpoint of the main stream serving the partial area. Representative values for Σal for regular geometrical shapes are: (i) circular 'glory hole' with a central outlet, $0.375A^{1.5}$; (ii) an equalateral triangle with the outlet at one of the vertices, $0.94A^{1.5}$; (iii) a square with the outlet at one corner, $0.76A^{1.5}$; (where $A = $ the total catchment area). Langbein found that the average Σal value for a number of actual basins in the north-eastern United States is approximately $1\cdot2A^{1\cdot5}$, an indication that natural basins tend to be less compact than the above geometrical shapes. The addition of a channel network to the catchment introduces a further complication to the runoff pattern resulting from given inputs. Two examples of the results of basic channel inputs used in the construction of computer models to predict runoff are given in Fig. 3.48 (Crawford and Linsley, 1966, pp. 24–6). Fig. 3.48-B represents the hydrograph resulting from an instantaneous pulse of channel inflow into a simple channel reach, running off between times t_1

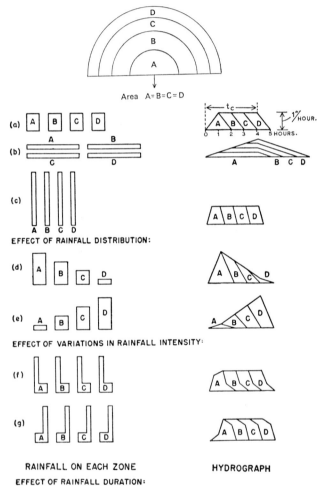

Fig. 3.47. Effect of rainfall duration, distribution, and intensity variations on hydrographs from a hypothetical semi-circular basin. Source: Linsley, Kohler and Paulhus, 1949.

and t_2 (this is over-simplified in that it ignores the attenuation to which this flood pulse is subjected as it moves towards the outlet—see below, Chap. 3.III.(3c). Fig. 3.48-c shows a similarly simplified hydrograph resulting from a converging inflow, duration t_2-t_1. If one takes a rectangular drainage basin over which input is a pulse of uniform, instantaneous rainfall, the hydrograph would record a trace similar to that in Fig. 3.46-D, the recession curve being an expression of channel storage. The effect of

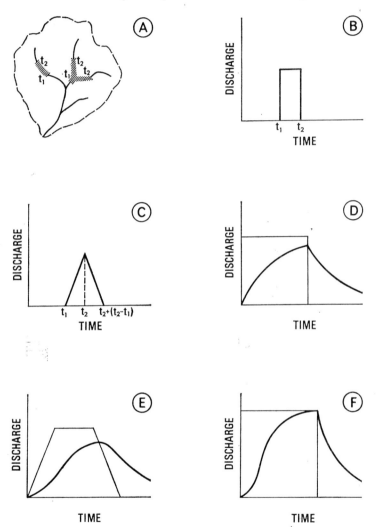

Fig. 3.48. Hydrographs for simplified drainage areas. The horizontal lines represent time of concentration (D), and time of tank filling (F). Source: Crawford and Linsley, 1966, Linsley, Kohler and Paulus, 1949; Clark, 1945.

channel storage is illustrated by the analogy with the free discharge from a tank in Fig. 3.48-F, which is filled at a constant rate for a unit time t_1–t_0. Here the area under the recession limb is an expression of the amount of water in the tank when the inflow ceases (Linsley, Kohler and Paulhus, 1949, p. 393). Returning to the rectangular drainage basin, a constant rainfall input over a unit period would produce an outflow hydrograph curve (Fig. 3.46-E) with features of both curves in Fig. 3.48-D and -F (Clark, 1945).

A further step is to consider the discharge resulting from a small instantaneous rainfall input over a drainage basin (area = a), divided into n subareas (each of area = Δa_j, where $j = 1, 2, 3, 4, \ldots n$) by isochrones (spaced at Δt) (Chow, 1964) (Fig. 3.49). The flow from each subarea would be considered as instantaneous into its channel segment and equal to $i_j.\Delta a_j$. Each subarea drains successively at the outlet, the hydrograph for which is given in Fig. 3.49-B, where the ordinates are

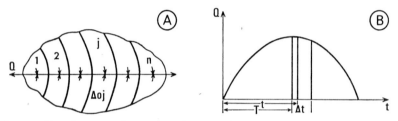

Fig. 3.49. Development of area-time diagram. (*A*) Drainage basin simulated by a linear channel, (*B*) area-time diagram. Source: Chow, 1964.

directly proportional to the magnitude of the drainage subareas such that the curve is termed a *time-area curve*. For an actual stream system, particularly one which has a high drainage density and relatively rapid inflow of overland flow into the stream channels, the first step in preparing a time-area curve is the construction of a distance-area curve for the channel system. Similar distances are measured along all streams from the outlet and these are connected to give distance isopleths, so that it is possible to plot the basin area beyond or within given distances to give a distance-area curve.

In much of the above discussion regarding the attenuation of runoff it has been assumed that the rainfall input is either a simple pulse or a period of uniform intensity. It is clear that the form of the hydrograph can also be controlled by the detailed characteristics of rainstorms affecting basin stream networks. Clearly, the amount of the storm rainfall influences the form of the runoff hydrograph, as does the areal and temporal pattern of rainfall intensity; thus the higher the initial rainfall intensity the steeper the rising limb of the hydrograph. Similarly, the position of the storm centre influences the runoff pattern (Fig. 3.50) such that in general the closer the storm centre to the outlet, the more rapid the rise of the hydrograph there. An exception to this might be where a storm occurred in a steep headwater area at a time when channel storage below it was very restricted: this might then give a steeper rise than that associated with an equal storm near the outlet falling on a floodplain area of low slope and large channel and floodplain storage capacity. An aspect of the affect of floodplain conditions on runoff is the concept of 'contributing area' (Kirkby and Chorley, 1967) where relatively high antecedent moisture conditions marginal to the stream channel promote the rapid achievement of overland flow there, such that the primary area contributing to surface runoff is only a part of the basin.

When the storm is of small areal size in comparison with the size of the drainage basin the direction of movement of the former can affect the

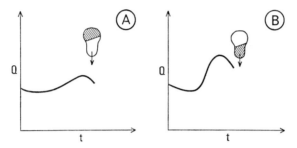

Fig. 3.50. Effect of storm location on hydrograph. Source: De Wiest, 1965.

form of the hydrograph, sometimes giving multiple peaks. Areal size of storms can also affect average runoff, in that although mean discharges naturally increase with basin order and with basin area (as do mean flood discharges), maximum flood discharges per unit area bear an inverse relationship to absolute basin size because the highest intensities are associated with storms of small area which give high runoffs per unit area from small catchments (Fig. 3.44-D).

(c) Flow routing

The time element can be introduced into hydrograph calculations in a number of ways, some very over-simplified and some involving complex routing equations. An example of the former is to assume virtually no time lag between the water falling as an instantaneous pulse on the surface and entering the channels, and to calculate the concentration of flow at the outlet by simply adjusting channel flow velocities according to the gradients of the different channel reaches. This involves the solution of the equation: time base (T) = duration of storm (t_r) + time of concentration (t_c). The time of concentration is that required for surface water falling on the most remote part of the basin to reach the outlet, and

Fig. 3.51 shows its simple calculation. Here the travel time along a channel reach is considered to be proportional to $\dfrac{l}{\sqrt{S}}$, where l is approximated by the horizontal distance and S by $\dfrac{\Delta z}{l}$ (Fig. 3.51-B). If the travel time from A to C (t_{AC}) is 3·48 hours, the equation in Fig. 3.51 gives the

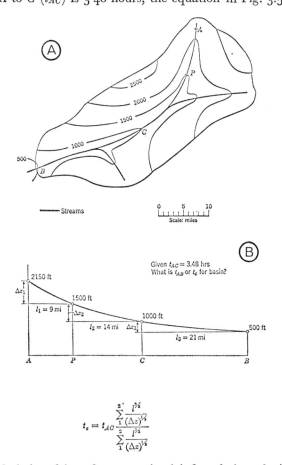

$$t_c = t_{AC} \dfrac{\sum\limits_{1}^{3} \dfrac{l^{3/2}}{(\Delta z)^{1/2}}}{\sum\limits_{1}^{2} \dfrac{l^{3/2}}{(\Delta z)^{1/2}}}$$

Fig. 3.51. Calculation of time of concentration (t_c), for a drainage basin employing average channel gradients. Source: De Wiest, 1965.

t_c for the drainage basin (i.e. t_{AB}) as eight hours (De Wiest, 1965). It is clear, however, that the many assumptions involved make this a very crude method.

A much more sophisticated method begins with some simple time-area curve and calculates the resulting hydrograph using routing methods, assuming, say, one inch of instantaneous runoff. While it is true that travel

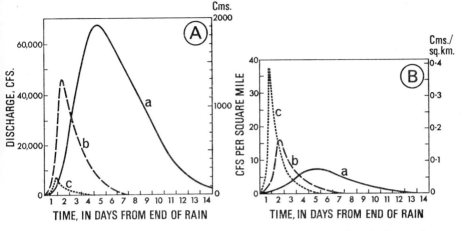

Fig. 3.52. The downstream progress of a flood wave is shown on these hydrographs of the Savannah River in South Carolina and Georgia. The area above the three points on the network are (a) 25,500 (b) 7,450 and (c) 526 square kilometres. Source: Strahler, 1965.

time of channel flow is a function of channel slope, bed friction, etc., it is more closely related to the variable storage capacity of the channel (Clark, 1945). The difference between unsteady flow in channels and pipes is that in the former there is a time lapse between the change of inflow rate and the change of outflow rate (Clark, 1945, p. 1432). This time lapse is virtually non-existent in rigid pipes, but becomes important in elastic pipes which have a variable capacity for storage (e.g. rubber hoses and the arteries of the human body) and in open channels. Thus the time-area curve is routed as if through a series of reservoirs that have wedge-shaped storage capacities equal to a constant times the outflow discharge. Each time-reach of the main stream is treated in this way (assuming, say, one inch of instantaneous inflow into the reach), employing a coefficient varying between zero and unity, depending on the form of the discharge hydrograph, to produce a routed time-area curve which can be translated into an instantaneous hydrograph. Some aspects of flood routing are more simply visualized by observing the behaviour of the downstream passage of a non-augmented flood crest. This results in the attenuation of the time base of the wave and a decrease in its peak discharge (Linsley and Franzini, 1964). This is partly due to the effect of channel storage on the inflow/outflow relations of the reach (Clark, 1945) and also to the fact that water particle velocities at all points on the wave are not identical (Linsley and Franzini, 1964). Where a flood peak builds up as the result of general rain over the basin its magnitude obviously increases downstream (Fig. 3.52-A); Calhoun Falls (b) is sixty-five miles below Clayton(c), and Clyo(a) a further ninety-five miles downstream, but when the effect of catchment area is removed by calculating discharge per square mile (Fig. 3.52-B) the contrast in the wave change is very marked.

The complexity of flood routing calculations, together with the desire to interfere with natural river networks by introducing artificial storage reservoirs through which flow must be routed, has led to the development of both analogue models and digital computer models of the operation

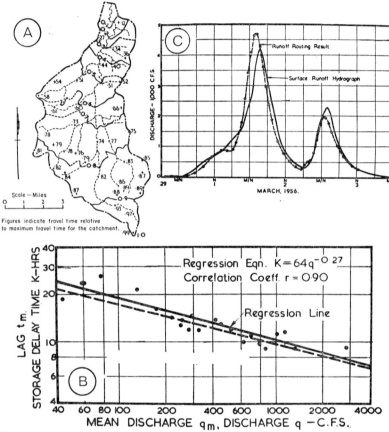

Fig. 3.53. South Creek, Australia. (*A*) Isochrones of storage-delay time. (*B*) Lag-mean discharge relationship. (*C*) Computed runoff (dashed), compared with measured runoff (solid). Source: Laurenson, 1964.

of flow in stream networks. An example of the former is the electronic simulation of the network draining the lower 7,000 square miles of the Kansas River (Harder, in Kneese and Smith 1966; O'Donnell, 1966). In this flood control model involving four reservoirs, channel characteristics and storage are simulated by resistances which operate as functions of discharge; rainfall inputs (together with reservoir release

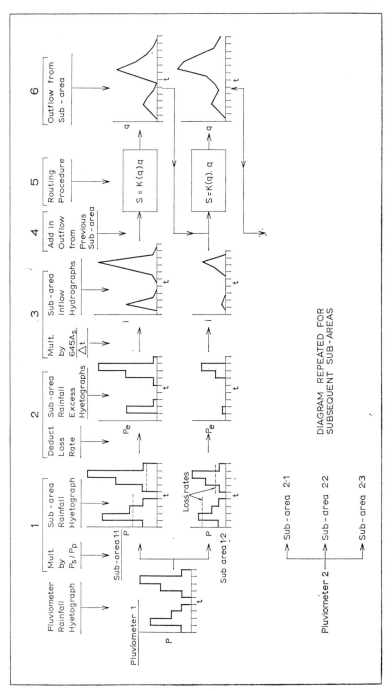

Fig. 3.54. Computations performed by runoff routing computer programmes. Source: Laurenson, 1964.

controls) lead to routing sequences through a number of reach elements simulating ten-day periods, the hydrographs for which are output on an oscilloscope. Rockwood (1961) describes a digital computer model designed to simulate flow in the Columbia River, in order to allow the optimum employment of reservoir storage for the production of hydro-electric power. An example of a computer runoff routing model involving surface runoff (i.e. not considering baseflow) is given by Laurenson (1964) for the thirty-five square mile South Creek, New South Wales. Recogniz-ing that storage delay times (i.e. inverse of travel times) are controlled by the discharge (Fig. 3.53-B), Laurenson calculated catchment travel times relative to the maximum travel time for the catchment (Fig. 3.53-A). Starting with the most remote of the ten subareas, the computer was programmed (Fig. 3.54) (1) to calculate the subarea rainfall hyeto-graph, (2) to deduct estimated infiltration losses to produce a rainfall excess hyetograph, (3) to convert to inflow by multiplying by an area/time constant, (4) to add to the outflow from the subarea upstream (in this case, zero), (5) to conduct a complicated routing procedure through the storage of the subarea, which was then used in the calculation for the next subarea downstream, and so on until the outflow hydrograph at the basin outlet was obtained. Fig. 3.53-c compares one of the better computer predictions with the observed surface runoff hydrograph. A more comprehensive programme, including baseflow and snow melt, has been proposed by Crawford and Linsley (1966).

In the approach to network structure in the preceding chapter the accent was on existing patterns and the balance of forces that constrain them. However, the creation of networks—both for the channelling and for the interception of flows—involves spatial design and raises questions of optimal location. It impinges on a major research area of optimization technique (notably through linear programming and associated programming methods) and the path followed here is to emphasize the nature of the spatial problem and the kinds of available solutions (where these exist) rather than develop full algorithms. The first part of the chapter considers the primal problems of the location of channel networks, and the second part the dual problem of barrier networks.

I. LOCATION OF ROUTES

In considering the balance of fixed and variable transport costs (Chap. 3.I) attention was drawn to the two extreme solutions when only flow costs and construction costs respectively were minimized. These are limiting cases of a general class of distance minimization or *shortest-path* problem. 'Shortest' distance is, however, an equivocal term. Bunge (1962) has illustrated six alternative shortest-path networks for a five-node problem. The first network (Fig. 3.4-A) shows the minimum-distance network for starting at a particular point and travelling to all the others in the shortest mileage: a solution described by Bunge as a 'Paul Revere' type of network. Fig. 3.4-B shows a similar distance problem, that of the shortest distance around the five points: the 'travelling salesman' problem. The next two definitions, shown in Fig. 3.4-C and 3.4-D, are for more complete networks —the first for a hierarchy connecting one point to all the others, and the second for a complete network connecting any point to all the others. If we examine this solution it appears to be the complete answer to our network problem in that it contains all the possible lines for the three solutions that precede it. But as Bunge points out the shortest set of lines connecting all five points does not in fact contain any of the elements shown in the previous diagrams. This shortest set solution is shown in Fig. 3.4-E. It can be found analytically or by analogues (Chap. 4.I(2)) and its intersections do not include any of the original points. Finally, Fig. 3.4-F shows the general topological case for a network of lines connecting five points as presented by Beckmann (Bunge, 1962, p. 189). Werner (1969) has provided a rigorous review of such 'minimum length' networks.

Examination of the final diagram shows that the two preceding cases—the completely linked network (D) and the shortest link network (E)—are but special limiting cases of Beckmann's general network. In this discussion solutions A to D are regarded as *fixed-node* solutions involving only the original nodes (Chap. 4.I(1)), and solutions E and F as *floating-node* solutions involving the creation of additional vertices. In Gould's terminology the original nodes are referred to as *pri-points* and the additional nodes as *sec points* (Gould, 1965). Both classes of shortest-distance paths assume that network solutions consist of straight line paths between arrays of pri-points and sec points. Where, however, the design problem is to route across a plane with variable resistance (where this may be measured in terms of construction costs, traffic generation, etc.) the optimum location for the path may not necessarily follow straight-line sections, but may be made up of arcs of varying radius. This more general problem of path location is termed a *geodesic* in this discussion (Chap. 4.I(3)).

1. Selection of fixed-node paths

(a) Link addition problems

With 'fixed node' problems, the shortest path is made up of links from the completely connected network. It would appear, therefore, that all shortest-path problems could be solved by evaluating alternative combinations of links and selecting the shortest. But although the shortest-route path through a network would seem for small graphs to be a simple matter of string and patience, it becomes surprisingly complicated and tedious where (a) very large numbers of links are involved, (b) the distance matrix is assymetric (i.e. some links are one-way only or longer in one direction than another), or (c) a number of vertices have to be joined in a single-shortest-route loop (the travelling salesman problem). Although the solution to this last problem is trivial for only five points, with larger numbers of points the computational problems become enormous. For example in Fig. 4.1 there are 47,900,200 solutions to the problem of the shortest cyclic line connecting all thirteen cities shown for the western United States: only one of these, that shown in Fig. 4.1-B is the optimum solution. The general relation is given by the formula $(N - 1)!$ where N

is the number of points, so that for one hundred cities the number of possible solutions rises to the astronomical figure of $9 \cdot 3 \times 10^{158}$. The practical significance of the derivation of such solutions (usually through high-speed computers) is for undertakings by, for instance, oil companies which have to ship products regularly by road to hundreds of local depots (Flood, 1956; Garrison, 1959–60).

The simplest method of evaluating alternative locations for adding links to an existing network is to treat the network as a planar or non-planar graph. Since each additional link in a graph (with a fixed number of vertices) will increase its connectivity, it follows that a comparison of the improved connectivity under experimental links will throw some light on their relative effectiveness. Examination of alternative measures of graph connectivity (see Chap. 1.III) suggests that the sum of elements in the shortest-path matrix (i.e. the *dispersion* of the graph) might provide a suitable overall measure of improved connectivity.

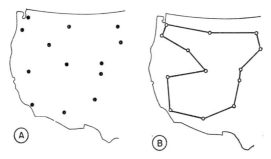

Fig. 4.1. Shortest cyclic path around thirteen major cities in the western United States. Source: Dantzig, Fulkerson, and Johnson, 1954, p. 219.

Burton (1962) examined the use of this measure in a study of the eastern part of Northern Ontario highway system in Canada. The network (Fig. 4.2) consists of thirty-seven links representing major highways (the King's Highways) and thirty-seven nodes representing either settlements with at least 3,000 inhabitants or intersections of the King's Highway system. The network, extending from Sudbury to Saulte St Marie, is of course connected to road systems in the remainder of the region, but these external links are sufficiently few to enable the network to be treated as a rational unit of analysis. The problem in route location facing the Ontario Highways Department was twofold: (1) Which of the seven links proposed (links A to G in Fig. 4.2) would have greatest impact on the total accessibility within the network? (2) What local effect would each new link have on the relative accessibility of individual nodes within the network? Answers to these problems were seen as providing some preliminary guide to both the regional case for an order of building priority and the local impact of individual links.

Changes in accessibility were measured by changes in the dispersion on the graph (Chap. 1.III). The dispersion of the original graph (8,151 links) was used as a control against which improvements resulting from new link construction could be measured. As Table 4.1 shows the greatest gains came from building new link *D* (from Folyet to Chapleau) which reduces the dispersion to 7,382 links, again of some 9·5 per cent in accessibility. Link *A* shows the second largest reduction in dispersion. Both links effectively short-circuit the whole network by joining one side of the 'star' to the other. In contrast links like *C* near the periphery have a very limited effect on overall accessibility. The local impact of each proposed new link is shown in Fig. 4.2 which plots the gains in accessibility for each node, differentiating link gain of ten or more links (small circle) from gains of fifty or more links (large circle). Inspection of the maps suggests that

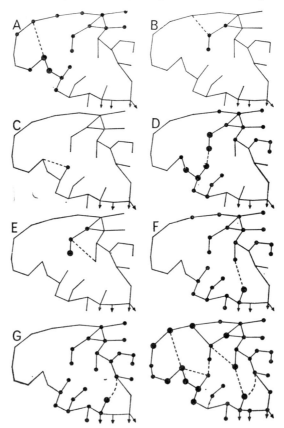

Fig. 4.2. Impact of projected road links (between lines) on the accessibility of nodes, eastern Ontario highway network, Canada. Improvements of five or more links (small dots) and of fifty or more links (large dots) in the shortest-path totals are indicated. Source: Burton, 1963.

the regional optimum may not coincide with a number of local optima: for example optimal link D has relatively little effect on the extreme western and south-eastern parts of the network.

The suggested sequence of construction is shown by the rank order in Table 4.1. There are, however, a large number of combinatorial possibilities, and if the highways authority could afford to build link pairs or link triads rather than individual links we would have no guarantee that the sequence shown for individual links would be the appropriate guide. The gains from building all links are a 26·1 per cent reduction in regional accessibility.

Since each link was given equal weight in the above analysis we can

Table 4.1. Gains in accessibility with link additions*

Network	Dispersion	Per cent gain	Rank	Rank†
Original network (N)	8,151			
N + Link A	7,466	8·4	2	4
N + Link B	8,003	1·8	6	6
N + Link C	8,095	0·7	7	7
N + Link D	7,382	9·5	1	5
N + Link E	7,906	3·0	5	3
N + Link F	7,559	7·3	3	2
N + Link G	7,721	5·3	4	1
N + Links $A, B \ldots G$	6,037	26·1		

Source: Burton, 1963.

* Eastern Ontario highway network
† Weighted index ($a = 0\cdot30$).

argue that the priority indications shown are rather severely constrained. In an attempt to outflank this assumption Burton has also computed a weighted measure of accessibility using the Shimbel-Katz index (Chap. 1.III). Following Garrison's (1960) study of the connectivity of the Interstate Highway System in the south-east of the United States, Burton uses a scalar of 0·30 (so that a one-link path is given the value 0·30) the two-link path 0·09 (i.e. $0\cdot30^2$), the three-link path 0·027 (i.e. $0\cdot30^3$), and so on. The effect of the index is to emphasize the impact of improvements in accessibility on near links and diminish their impact on more remote links in line with expected traffic-user behaviour. The changes in link order using the arbitrary 0·30 scalar is shown in Table 4.1; on this revised basis link G is now the preferred first-choice for highway construction.

Clearly graph theory can provide only a preliminary assessment of link priorities. Where the solution of a network optimization problem is defined in terms of building (1) or not building (0) a particular link, integer programming provides an appropriate analytical framework. Various formulations of the problem in these terms have been made by Burstall (1966), Quandt (1960), and by Scott (1967; 1969): most approaches involve some extensions, by algorithmic or heuristic means, of a systematic graph-searching procedure—the 'branch-and-bound' technique of Lawler and Wood (1966). The nature of the link-selection procedure is shown here by a very small sample network in which direct

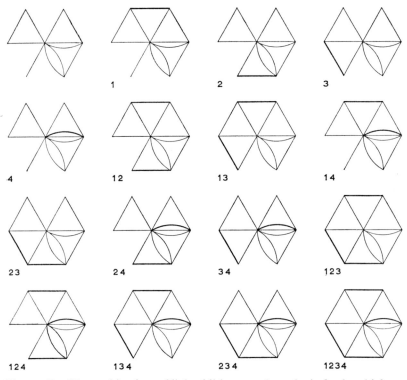

Fig. 4.3. Possible combinations of link additions to a hypothetical urban highway network. Source: Roberts and Funk, 1964, p. 25.

enumeration of alternatives is possible. Fig. 4.3 shows in graph form an oversimplified version of the freeway network for the city of Houston, U.S.A. (Roberts and Funk, 1964). With four links to be added there are sixteen different ways of implementing a highway programme including a choice of adding no links at all. In this case the authors built into their problem the unit cost of link travel, the capacity of each branch or link, and the cost of construction for the proposed link. Comparison of the alternative construction plans was based upon an economic analysis using the annual cost method and employing a twenty-five year life, a ten per cent interest rate and a value of time of three dollars per hour. Finally origin and destination flow requirements for each node were assumed from both the actual number of trips in the base year and the projected number at the end of a twenty-five year life. A uniform arithmetic increase in traffic volume was assumed over the twenty-five year period. For this sample problem, Fig. 4.4 plots capital expenditure against total cost.

In this case the construction of links 2 and 3 is indicated, since the construction of these two links would reduce travel costs plus construction

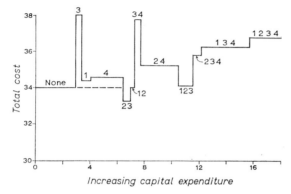

Fig. 4.4. Relationship of total costs to amount spent on links for hypothetical network shown in Fig. 4.3. Source: Roberts and Funk, 1964, p. 30.

costs to the lowest possible figure. If, however, the budget had been limited to say 500 units the optimum policy would have been to build no links at all. Clearly budget size is critical in determining which links are to be added to a network, though clearly different cost patterns in the example chosen could have produced a different situation. For problems of a realistic size and complexity total enumeration is not possible. If there are n projects to be considered and no limitations on the budget, then there are 2^n total possible combinations to be considered.

(b) Shortest-path problems

The problem of determining all shortest-paths in a network has attracted a large number of workers and a variety of approaches (see bibliography by Stairs (1965)). Pollack and Wiebenson (1960) in a review of the solutions put forward draw a general distinction between computational-mathematical solutions, such as the work by Moore (1959) at Harvard on the shortest paths through a maze, and analogue solutions like Rapaport and Abramson's (1959) model, in which electric or electronic 'timers' are substituted for distance and the shortest route is shown by a set of illuminated links. The basic problem in both is to reduce the long and expensive computations needed. For example, Little, Murty, Sweeney and Karel (1963) report that solutions for simple travelling salesman problems may become very costly in computer time when large numbers of cities are involved: using an IBM 7090 computer they found that while 'ten-city' paths could be derived in from one to three seconds, 'twenty-five-city' paths needed from four minutes to over one hour! (Fig. 4.5). Continuing research is producing more efficient algorithms and with the improving speed and capacity of both digital and analogue computers, more rapid solutions to complex shortest-route paths are emerging.

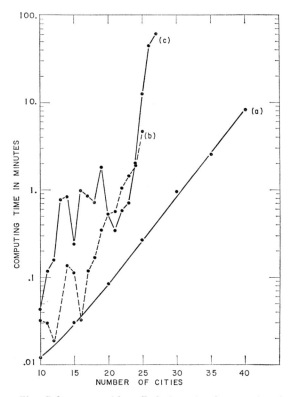

Fig. 4.5. Travelling Salesman problem. Relationship of computing time to number of cities for three sample cases. Source: Little, *et al.*, 1963, p. 987.

One of the simplest of the 'matrix' methods put forward is the *cascade* algorithm and its derivatives. Murchland (1965) has described a variant of the cascade method that needs only $n (n-1)^2$ simple operations for a graph with n vertices. As an illustration Fig. 4.6 shows a sample graph with four vertices *A, B, C* and *D* connected by directed arcs with specified lengths. The graph may be represented as a matrix (matrix *A* in Table 4.2) in which the distances are entered as off-diagonal elements and zeros along the main diagonal. Lengths which are unknown (e.g. from *A* to *C*) are replaced by infinity, or by any high value which is

greater than the sum of all the specified elements in matrix A. A value of
$*$ is substituted in matrix B. The rules for the solution of the shortest-path
matrix are given as follows: (i) successively choose one of the diagonal
elements as *pivot* until all have been used; (ii) choose in turn each element
not on either the main diagonal or on the same row or column as the

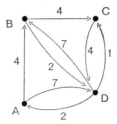

Fig. 4.6. Sample graph for cascade algorithm. Source: Murchland, 1965, p. 17.

pivot; and (iii) combine successive elements in the pivot row with successive elements in the pivot column and replace the element at their intersection by this sum if it is smaller (Murchland, 1965, p. 16). The successive pivots are shown in Table 4.2 by the starred diagonal elements in

Table 4.2. Stages in the matrix solution of shortest-paths*

(A)	To A	B	C	D		(B)			
From A	0	4		7			0	4	☆	7
B		0	4	2			☆	0	4	2
C			0	4			☆	☆	0	4
D	2	7	1	0			2	7	1	0
(C)	0	**4**	✱	**7**		(D)	0	**4**	(8)	(6)
	✱	0	**4**	**2**			✱	0	**4**	**2**
	✱	☆	0	**4**			☆	✱	0	**4**
	2	(6)	1	0			**2**	**6**	1	0
(E)	0	4	**8**	**6**		(F)	0	4	7	**6**
	☆	0	**4**	2			(4)	0	3	**2**
	✱	✱	0	**4**			(6)	(10)	0	**4**
	2	**6**	1	0			2	**6**	1	0

Source: Murchland, 1965, pp. 17–18.
* See network mapped in Fig. 4.6.

matrices C to F, and the revised off-diagonal elements in heavy type. Thus
in matrix C where the pivot selected is AA, the link DB has its value of
seven reduced to six, the combined value of row element AB ($=4$) and the
column element DA ($=2$). Inspection of Fig. 4.6 shows that it is indeed
shorter to proceed from D to B via A rather than following the direct
arc DB.

The cascade algorithm provides an extremely simple technique for finding all elementary paths in a complete directed graph. For symmetric graphs (i.e. where each link is matched by an oppositely directed link of the same length) the labour can be reduced by one quarter. None the less the matrix method makes heavy demands on the rapid-access stores of digital computers. Its restriction to rather small networks has been removed by the development of decomposition techniques (Foot, 1965; Mills, 1966A), in which the network is partitioned into a series of separate but interconnecting sub-networks. The cascade algorithm is run for each separate sub-network and the results aggregated by ingenious 'stitching' programmes to give shortest-paths for the complete network.

An alternative to the matrix method is to develop shortest-distances progressively an [origin node as a tree. Mills (1967) has suggested a heuristic approach to graph-searching in which a tree is built from the

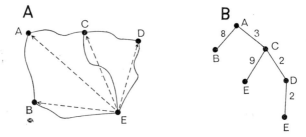

Fig. 4.7. Sample graph for heuristic graph-searching procedure. Source: Mills, 1968.

origin node, but a specific branch is developed only if it appears promising. 'Promise' is judged by approximate measures of the length of the shortest route from each node to the destination node which is assumed to be available. Mills' method, which builds on similar heuristics by Doran and Michie (1966) for a sliding block problem, may be illustrated from a simple example. In the five-node graph shown in Fig. 4.7-A the problem is to plan the shortest route from node *A* to node *E*. A tree is built out from node *A*. Intermediate nodes incorporated into the tree are given a value comprising the sum of the actual path length along the tree *from* the origin node *A*, plus the estimated distance from this node *to* the destination node *E*. The calculations are set out in Table 4.3 and the final form of the tree is shown in Fig. 4.7-A. The estimated distances to node *E* from nodes *B*, *C* and *D* are assumed to be 5, 6 and 6 respectively.

The utility of the method hinges on the availability of estimated path lengths. The crow-fly distances between nodes may be reasonable proxy road lengths. The benefit of the Mills heuristic is the considerable saving in computation, and the price, the failure to *guarantee* that the truly shortest-route is found. Heuristics are of special value where arc lengths

Table 4.3. Stages in graph-searching method for the solution of shortest paths*

Node being developed	Node added to the tree	Path length from A	Estimated distance to B	Sum	Preceding node on the path
A	B	8	5	13	A
	C	3	6	9	A
C	D	5	6	11	C
	E	12	0	12	C
D	E	7	0	7	D

Source: Mills, 1967.

         * See network mapped in Fig. 4.7.

can only be measured inaccurately or where the cost function can only be approximately specified. An extended trial of the method using a sample of six pairs of nodes from 'Network—90', a hypothetical urban road network with ninety nodes, gave encouraging results. In each case the overall shortest route was found and the savings (in terms of the number of nodes incorporated into the tree) ranged from seventy to eighty-eight per cent.

An alternative heuristic approach for getting good approximate solutions to problems of up to 145 points has been suggested by Lin (1965). The essential of the method is to make a large number of good approxi-

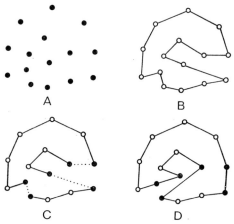

A   B

C   D

Fig. 4.8. Shen Lin method of solving a shortest-path problem. Source: Bell Telephone Laboratories, 1966, p. 295.

mations in a reasonably short time using a computer and selecting the best of these. Fig. 4.8-A shows the original distribution of fifteen points. To make one approximate path, the computer chooses a starting path at random (Fig. 4.8-B). Three links in this path are removed and the sections are reconnected in a different manner (Fig. 4.8-C). If this new path is not shorter than the original one, the process of breaking and recombining is continued. If, however, the reconnection does produce a shorter path, then this is taken as the starting point of a new starting path. An approximate path is completed when no further improvement results from this breaking and reconnection procedure. By using a large number of new starting paths, each beginning from a random distribution, many approximate shortest-paths are found. Lin reports that for a typical forty point problem, the experiments indicate that about one in sixteen approximations will be the actual minimum solution; for sixty points, about one in sixty-four. By employing a safety factor of around five, there is a high probability that one of the paths computed is in fact the shortest possible. If this is not so, it is at least short enough for most engineering purposes.

(c) Programming problems

The third group of methods for solving fixed-point problems involves mathematical programming and introduces a highly-developed and rapidly expanding field of applied mathematics. However, the programming of fixed-point problems, in which the selection of links for improvement is made from a finite set of links, may be simply illustrated from the network shown in Fig. 4.9. Here the problem may be set up in the form of a simple transportation problem in which supplies and demands, and the costs of flows between individual pairs of locations are known, and in which the objective function of the problem is to minimize the total cost or distance travelled. The problem may be set out algebraically as finding a set of flows X_{ij} such that:

$$\sum_i \sum_j X_{ij} d_{ij} = \text{Minimum.}$$

Here X_{ij} represent the unknown flows between the ith source and the jth destination, and d_{ij} the known cost or mileage between i and j. This objective function is subject to four basic constraints. The first states that the flows from the ith surplus region are equal to the surplus available there where a_i represents the amount of the commodity which is available or export from the ith surplus region:

$$\sum_j X_{ij} = a_i; \quad i = 1, 2, \ldots n.$$

Similarly, the second constraint states the flows into the jth deficit region is equal to the amount of the commodity demanded by the jth region (b_j):

$$\sum_i X_{ij} = b_j; \quad j = 1, 2, \ldots m.$$

The third constraint requires that total demand is equal to total supply:

$$\Sigma\, a_i = \Sigma_j\, b_j$$

The final and fourth constraint ensures that there will be no negative flows:

$$X_{ij} \geqslant 0 \text{ for all } i \text{ and all } j.$$

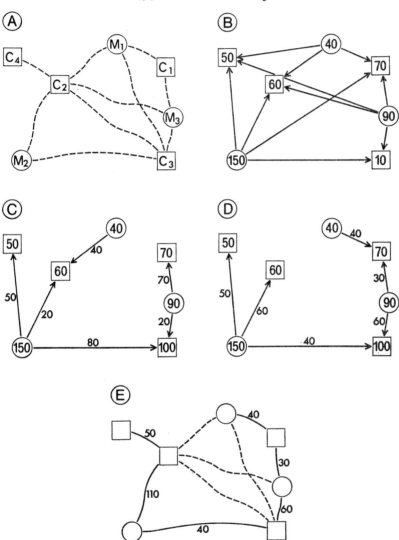

Fig. 4.9. Hypothetical problems in the allocation of coal flows between mines, M, and cities, C. The optimal allocation derived by a linear programming formulation is mapped in (*D*). Source: Cox, 1965.

More complex examples of the application of linear programming methods to locational problems are given in a number of standard texts (e.g. Dantzig, 1963).

Fig. 4.9-A shows a typical flow problem in a very elementary context. Here it is assumed that three mines (located at M_1, M_2 and M_3 respectively) and four cities (C_1, C_2, C_3 and C_4) are linked by a simple railway network. The production of each mine and the demand of each city is given in millions of tons in matrix (1) and the mileage cost (a rough surrogate of cost) is given in matrix (2) (Table 4.4). Here the basic locational problems are (i) to determine the optimum allocational flows so as to minimize the total ton miles of movement and (ii) to determine the flows of each link in the network.

The solution adopted here is based on the stepping-stone method of Charnes and Cooper (Cox, 1965, pp. 229–31). Six basic steps are involved: *Step 1:* Rank the mileage costs shown in matrix (2) in order from the cheapest to the most expensive (i.e. *1* to *mn*). *Step 2:* Take the cell with lowest mileage (i.e. at the intersection of M_3 and C_1). In this cell nine units are produced by mine 3, but only seven units are demanded by C_1. Enter the value of seven into this cell and adjust the row and common marginal totals accordingly. *Step 3:* With this reduced matrix for the cell with the lowest mileage (i.e. cell M_1–C_1), since the demands of this city column are exhausted, we search for the cell with the third lowest mileage. This iterative procedure is continued until we arrive at a basic feasible solution. This solution is shown in Fig. 4.9-C and in matrix (3) in Table 4.4. By multiplying the tonnage in each cell by the appropriate mileage costs, we can calculate the total transportation cost of this allocation scheme at 1,530 ton miles. We do not know as yet, however, whether this is the cheapest possible scheme of distribution. *Step 4:* To check whether this allocation provides a minimal transportation cost, shadow mileages for each row in common are computed, and the sum of these values used to compute the fictitious costs for each cell. To do this, shadow costs for M_1 are set at zero, implying, by summation, that shadow cost for C_2 must be four (see matrix (3)). Continuing this process, the marginal values are computed and the sums are entered in as heavy figures in matrix (3). *Step 5:* Shadow cost in the cells of matrix (3) are then compared with the true costs in matrix (2). Where shadow costs are less than the true cost, readjustment would lead to an increase in the total transportation costs, rather than a diminution; where shadow costs are equal to the true costs, readjustment is possible, but would not lead to any change in the total transportation costs; if, however, shadow costs are larger than the true costs, then the final solution has not been arrived at and a readjustment is required (Cox 1965, p. 56). Since two cells in matrix (3) (marked with an asterisk) fall into the third category, some readjustment is indicated. *Step 6:* Readjustments are made by a circuit

Table 4.4. Transportation problem

(1) Supply and demand totals

				4
				15
				9
7	6	10	5	28

(2) First assignment: A

4				4
2	8	5		15
7		2		9
7	6	10	5	28

(4) Adjustment Cycle

4			
+4	−4		
	2	8	5
	+4	−4	
7		2	
−4		+4	

(2′) Second Assignment: B

4				4
	6	4	5	15
3		6		9
7	6	10	5	28

(2) Transfer costs

2	4	7	5
10	6	9	8
1	11	3	20

(3) Shadow costs: A

(5)*	4	(7)	(6)*	0
(7)	6	9	8	2
1	(0)	3	(2)	−4
5	**4**	**7**	**6**	

(3′) Shadow costs: B

2	(1)	(4)	(3)	0
(7)	6	9	8	5
1	(0)	3	(2)	−1
2	**1**	**4**	**3**	

Source: Cox, 1965, pp. 229-31.

* $d_{ij} \geqslant d'_{ij}$.

which is alternatively horizontal and vertical (see matrix (4)) leading from the empty cell via cells which are occupied by flows in matrix (3). The value of the transfer is determined by the minimum value of any cell, which will be marked by a minus sign during the transfer cycle. Beginning with cells M_I–C_I, the circuit is traced by alternate horizontal and vertical movements with associated minus and plus transfers. Since the minimum value in any minus cell is four, the transfer takes the value of four units. The adjusted solution is shown in matrix (2'). The cost of transportation in the new solution is 1,410 ton miles as compared with 1,530 ton miles in the original assignment.

Since the solution method is iterative, we must now recompute shadow costs (see matrix (3')), and compare such costs with the true mileage costs as shown in matrix (2). Since none of the shadow mileages are larger than the true mileages, we may conclude that the adjusted solution is the final solution, i.e. no improvements are now possible. This final solution is shown in Fig. 4.9-D.

The optimum flows from the transportation problem may now be loaded on to the existing networks to give the pattern shown in Fig. 4.9-E. Clearly, the heaviest flows on any section of road occur on that connecting M_2 and C_2; conversely, five of the links in the original transportation system carry no flows. One useful by-product of the solution of the transportation problem is the information it gives about the comparative location advantages of the three mines and of the three cities. A full discussion of the dual is given by Dantzig (1965).

Applications of transportation methods to specific geographic networks may be illustrated by Gould and Leinbach's (1966) study of the projected assignment of hospital services in the western part of Guatemala. Here the population centres are treated as surplus nodes while the hospital capacities are defined as quantities to be filled. The transport costs for a person travelling to obtain hospital services were calculated from the most efficient and best service route available. Fig. 4.10-A shows the distribution of the three hospitals and the road transportation network of western Guatemala. To simplify the calculations, road distances were calculated from each hospital to every population point within the centre of eighteen hexagonal areas. Distance was measured over the most efficient route and took into account the quality of the road over which the journey had to be made. Distances over unpaved roads were weighted twice that of distances over paved roads. Starting with an optimal set of hospitals based upon equal capacities the solution was altered by changing hospital capacities so as to eliminate cross flows. Fig. 4.10-B shows the best solution obtained with the existing road services.

The investigators emphasize that they had no way of telling at the time whether this solution was the ultimate and final one. For although linear programming provided a more realistic estimate 'it must be remem-

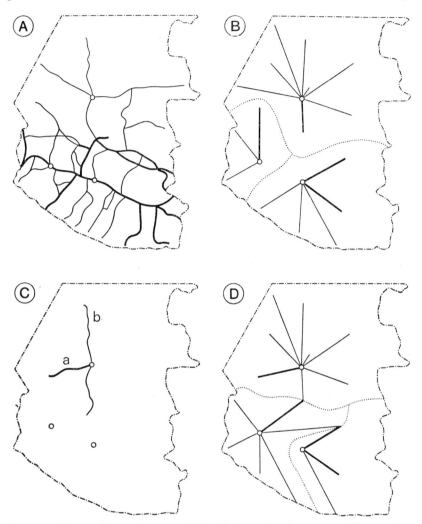

Fig. 4.10. Geographic assignment of hospital services in western Guatemala. (*A*) Original road network. (*B*) Optimal patient flows to three hospitals from eighteen population centres. (*C*) Projected improvements in road network. (*D*) Impact on patient flows. Source: Gould and Leinbach, 1966, pp. 203–6.

bered that in any truly general theory of location we would be in a very complex general system framework where probability constraints and queuing theory are involved' (Gould and Leinbach, 1966, p. 206). However, the model is useful in showing the impact of changes in the transportation network upon the hospital location problem. A road extension in western Guatemala (*a*) and an upgrading of the unpaved roads from the Northern border of Huehuetenango to Quezaltenango (*b*) (Fig. 4.10-c) are postulated. The marked effect of these highway changes on the pattern of patient flows and on the hospital collecting-boundaries is shown in Fig. 4.10-D. Although the example selected here was a very simple one and was used for largely expository purposes, the authors stress that time in medical practice may be a very crucial variable; 'minimum travel paths take on more than geometrical connotations when children's lives are at stake in emergency situations' (Gould and Leinbach, 1966, p. 203).

Solutions to the transportation problem were put forward by Hitchcock (1941) and Koopmans (1947) over twenty years ago. During the intervening decades a major body of programming techniques have been built up (see Dantzig, 1965). Work began with modifications of the basic Hitchcock–Koopmans problem to allow the introduction of (i) slack variables (where total production outstrips total consumption), (ii) preferential and inadmissible flows, and (iii) bounded flows where flow along a given link must not exceed or fall below some stated bounds. Further modifications were made by Orden (1956) in the form of *transhipment problem* where indirect flows are admitted and the *Beckmann-Marshak problem* where indirect transforming centres are introduced (Beckmann and Marshak, 1955; Maranzana, 1964). A review of programming in relation to its spatial applications is given by Garrison (1959–60). Methods have since been extended to meet variable costs, uncertain demands, integer forms, non-linear functions, etc., so that linear programming with its extensions probably represents one of the most important and analytical tools available for the analysis of flows in networks. Ford and Fulkerson (1962, Chap. III) in an outstanding survey, suggest that something like half the time spent on industrial and military applications of linear programming was concerned with this sub-set of problems. Because of their vast economic and military significance, a considerable body of sophisticated mathematics has been built up to structure these studies, and practical algorithms and computer routines based on both operations research and classical electrical-network studies (e.g. the Maxwell–Kirchoff problem) have been derived.

Fig. 4.11-A illustrates a simple problem in deriving minimal-cost paths through a network while obtaining maximum flow (Ford and Fulkerson, 1962, pp. 123–7). The network consists of a set of twenty-one links (edges) joining the two terminal points (*a, B*). Fig. 4.11-A shows the maximum flow capacity of each link, Fig. 4.11-B the unit shipping cost. If we assume our problem is to ship goods from *a* to *B* then it is possible to compute that a positive minimum-cost flow will get through the network to *B* by utilizing the paths shown by the arrows in Fig. 4.11-c. However, for the maximum flow through the network (given the cost and capacity constraints already referred to) almost all the links are utilized (Fig. 4.11-D). Two points worth noting are (i) that not all the links along the paths operate to their full capacity (links not fully 'saturated' with traffic are shown by broken lines) and (ii) that the direction of flow along some of the links may change as total flow through the network increases.

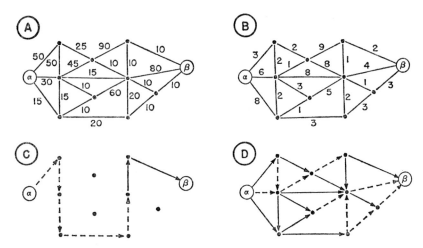

Fig. 4.11. Minimal-cost flow through a complex network: (*A*) Maximum flow capacity of each link. (*B*) Unit shipping costs along each link. (*C*) Initial flow pattern. (*D*) Maximum flow pattern. Links not fully 'saturated' with flow shown by broken lines. Source: Ford and Fulkerson, 1962, pp. 123–7.

2. Location of floating-node paths

Where builder costs are very important relative to flow costs, solutions are no longer entirely made up of links from the fully-connected network. The introduction of secondary junctions (*sec points*) makes the determination of the solution much more complex. Despite these complexities the importance of the answers, and the spatial economies to be derived from them, makes even partial solutions of interest. Some simple partial solutions for three-point, four-point, and N-point cases are discussed here; more complex empirical problems are explored in Chap. 3.I.2.

(*a*) *Three-point and four-point problems*

The simple problem of connecting three points on a plane by the shortest path has intrigued geometers since the ancient Greeks. Polya (1954) has examined the problem of the location of the shortest path between two villages which are situated near the bank of a stream which allows water to be collected en route. There are two solutions: (1) for a straight stream the solution is equivalent to that of a light-ray reflected in a mirror; (2) for an irregular stream the optimum water-collecting point is where the smallest ellipse that can be drawn (with the two villages as foci) touches the streambank.

Miehle (1958, p. 232) has drawn attention to Jacob Steiner's solution of a comparable problem, that of connecting three villages by a road network of minimum total length. Steiner found that study of the angles of the triangle formed by the three villages yielded the solution. If one of the angles of the triangle is equal to or greater than 120° then the villages at the other vertices should be directly connected to this vertex. If, on the other hand, all angles in the triangle are less than 120° the solution is to connect all three villages to a junction point so located that all three roads meet at 120° angles.

A more general approach by Beckmann (1967) allows the volume

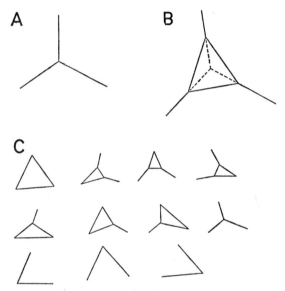

Fig. 4.12. Location of routes connecting three centres. Source: Beckmann, 1967, pp. 98, 101–2.

of flows to be taken into account. The condition of optimality for the triple intersection is given by:

$$\sum (a + bf_i) \frac{\eta - \omega_i}{|\eta - \omega_i|} = 0$$

where a and b are the highway and movement costs; f_i denotes the flows to point i; vector η denotes the location of the intersection; and vector ω_i the location of pri-points (Fig. 4.12-A) (Beckmann, 1967, p. 99). This represents an equilibrium of tension approach with the weights pulling from ω_i to η with strength $a + bf_i$; clearly the knot joining the three strings together will be pulled into location η.

The general case is shown in Fig. 4.12-B. If highway costs (a) are zero, the costs are minimized when movements costs are as low as possible and the triangle as large as possible; conversely if movement costs (*b*) are zero, then construction costs are minimized when the triangle is shrunk to a point. One implication of the linear cost functions assumed is that 'there should be no junctions of the kind shown in Fig. 4.12-B unless at least one corner of the triangle is occupied by a city (i.e. pri-point)' (Beckmann, 1967, p. 102). The eleven topological alternatives for efficient road networks between three cities is given in Fig. 4.12-C.

For four pri-points the number of topological opening possibilities rises to thirty-two. For larger numbers of points the number of opening possibilities rises rapidly to astronomical proportions. Adoption of local minima is no guarantee of achieving global minima, but Beckmann (1967, p. 104) suggests as a general heuristic that any closed loop of roads (e.g. triangle in the three-point case, pentagon in the five-point case, etc.) should contain at least one pri-point among its vertices. This provides some limit to the possible number of junctions.

(b) N-point problems

Miehle (1958) has suggested that the general case of the shortest route between a set of N points may be studied as an extension of Steiner's problem. For N points it may be shown that the maximum number of functions at which three lines meet at $120°$ is $N-2$. For a large set of points the mathematical solution involves the solution of sets of equations of the form:

$$\frac{\partial D}{\partial x_A} = m_{A1}\frac{(x_A - x_1)}{d_{A1}} + m_{A2}\frac{(x_A - x_2)}{d_{A2}} \ldots + m_{AB}\frac{(x_A - x_B)}{d_{AB}} + \ldots = 0$$

$$\frac{\partial D}{\partial y_A} = m_{A1}\frac{(y_A - y_1)}{d_{A1}} + m_{A2}\frac{(y_A - y_2)}{d_{A2}} \ldots + m_{AB}\frac{(y_A - y_B)}{d_{AB}} + \ldots = 0$$

where x_i, y_i are the co-ordinates of point i, where M_{ij} is the weighting factor in the link connecting the two points, where D is the total distance to be minimized, and where:

$$d_{ij} = \sqrt{(x_i - x_j)^2 + (y_i + y_j)^2}$$

(Miehle, 1958A, p. 237). Fixed points are designated by numbers and movable junctions by letters. Since the equations are non-linear an iteration method is used where the positions of the movable junctions ($A, B, \ldots$) converge on the minimum solution. Clearly the numerical analysis of the shortest-path problem is complex and becomes increasingly so when constraints (e.g. minimum separation distances) are placed on the junction locations.

Alternatives to the analytical solution are provided by a number of 'hardware' analogue models (see review by Morgan, in Chorley and Haggett, 1967, pp. 768–71). Three alternative approaches are considered here: *soap-film analogues* are based on the tendency of such films to minimize the surface area for the constraints imposed upon it. A scale model of the pri-points in the problem is constructed with small pegs connecting two parallel perspex surfaces. The model is immersed in a soap solution and ' . . . when the model is slowly and carefully withdrawn soap films are left perpendicular to the parallel plates and they gradually contract until minimum-length links are established with junctions of the film at 120 degrees' (Morgan, 1967, in Chorley and Haggett, p. 769). This analogue model is useful for demonstration purposes, but it has a number of serious disadvantages. Solutions are not unique since gravitational forces affect the orientation of the film walls, and with large numbers of points local solutions predominate. Miehle (1958) suggests that fifteen points probably provide the upper useful limit of the model. By breaking down a network into a series of sub-networks or by decomposing the network into 'nested' components, the size limitations can be overcome and partial solutions for large networks obtained.

Mechanical link-length minimizers represent the pri-points in the problem by pegs set in their correct location on a base-board, each peg carrying a small pulley. Sec-points are represented by movable pegs also with pulleys. Minimization is achieved by looping a thread around each peg in turn and tightening the thread until its total length is reduced to a minimum. Despite its bulky appearance and inconvenient operation large systems of points can be treated with integral weights assigned to individual links, and very rapid solutions of reasonable accuracy obtained. Silk (1965) made extensive use of a thirty-three point model in a study of optimal road designs for the county of Monmouthshire, Wales. Here fixed pegs located on a 1/63,360 base board represented the major urban centres within the county. Fig. 4.13 shows the evolution of the network as urban areas of successively smaller size are added to the problem. All solutions—mathematical, soap-bubble, and mechanical—show the characteristic 120° junctions inherent in Steiner's three-point case. Figure 4.14-C shows theoretical minimum link-length minimization for networks, including that for sixty-two weapon units in a military communications system (Miehle, 1958, p. 236). Here an original connecting scheme had a total path length of 121·4 units. This was reduced to only 68·8 units by using the Steiner solution shown (Fig. 4.14-A). More complex models using *electrical analogues* on the lines suggested by Enke (1951) are possible ways of improving the situation. Electrical circuits are a form of network and the step between the model and the real world is rather small; nevertheless, despite direct solution of fixed-point problems using electrical analogues (Road Research Laboratory, 1965, p. 352) the problem of variable point solutions remains.

(c) Special cases

Two special cases of floating-point problems are discussed by Beckmann (1967, pp. 106–8). The first case concerns the location of a single

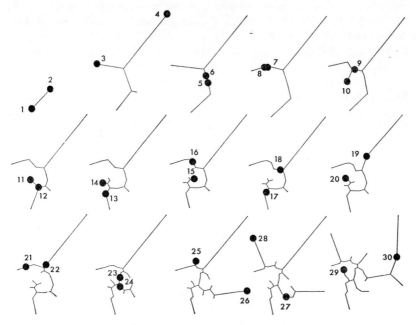

Fig. 4.13. Adjustment of shortest-distance network for the towns of Monmouth-shire, Wales. Successive addition of pairs of nodes is made in the order of decreasing population size. Source: Silk, 1965.

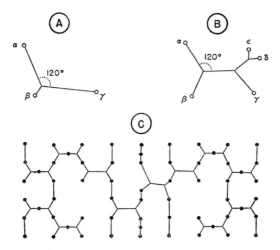

Fig. 4.14. Shortest-distance paths for (*A*) three nodes (Steiner's problem), (*B*) five nodes and (*C*) sixty-two nodes. Each node is assumed to have equal weight. Sources: Bunge, 1962; Miehle, 1958.

trunk between two termini but which would be accessible to feeder roads. No matter how slight the in-flow from the feeder roads, the optimal design will still have slight corners at these junctions (Fig. 4.15-A). Beckmann's design was anticipated in the practical context of railroad location by Wellington (1887). Wellington was a mining engineer who, employed on the planning of the Mexican railroad system, widened his findings to include general cases of route location. The practical problem as he saw it was to optimize the relationship between the total length of

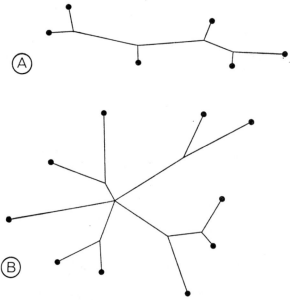

Fig. 4.15. Special cases of optimal location. (A) Trunkline connecting two terminals with feeder routes. (B) Network connecting satellite terminals to central origin. Source: Beckmann, 1967, pp. 106–7.

a path (i.e. the length of railroad to be constructed) and the number of points connected (i.e. the towns connected by the path). Clearly the objective in the first case was to shorten the line, and in the second to connect as many places as possible (and thereby to carry their traffic). From the data he was able to assemble on contemporary railroad practice, Wellington put forward three basic propositions: (1) that if all points intermediate between the two terminals of a railroad were of equal freight-generating capacity and if they were equally spaced, then the traffic varied as the square of the number of points served; (2) that if the intermediate points were 'small country towns' without competing alternative railroads, then the effect was to reduce the traffic by ten per cent for every mile that the station was removed from the town; and (3) that if the intermediate points were 'large industrial cities' with competing railroad facilities, then the loss would be still more abrupt—a reduction of twenty-five per cent for every mile that the station was removed from the town. Although these figures have now only antiquarian interest, and may have been quite inaccurate for contemporary conditions, they do represent an interesting early attempt to wrestle with the problem of path deviation between points. Indeed Wellington's attempt should be seen against the present state of research which still lacks a rule ' . . . relating to the deviation of paths between major places in order to pass through inter-mediate places' (Garrison and Marble, 1962, p. 85). Where the ratio of fixed/variable costs is greater than one then we might expect that the track mileage would be kept as low as possible (i.e. little deviation of the path to connect intermediate points) and vice versa. Bunge (1962, pp. 183–9) has extended this idea with reference to regional contrasts in the railroad pattern in the United States.

Beckmann's second special case (Fig. 4.15-B) concerns the central city with connections to outlying cities. This forms the discrete case of the continuous model examined in Chap 3.I(1b), and will not be discussed here. It is worth noting, however, that one implication of equilibrium conditions is that the angles between converging paths tend to become more acute as the central node is approached (*cf.* Chap. 3.IV). Beck-mann's general approach has analogies in the early attempts to devise optimal regional networks (see Fig. 3.11).

3. Geodesic paths

In the discussion of shortest-path problems (Chap. 1.II) all links were assumed to be straight-line vectors connecting nodes directly, viz. secon-dary nodes (sec points). Inspection of empirical examples of regional networks shows some examples where straight-line links are relatively common (e.g. short-haul airline routes), but a majority where deviations from the straight-line occur. If we are to treat such indirect paths as— in aggregate—rational, then we must assume that they represent some form of geodesic, or least effort path. Geographers, always profoundly conscious of place-to-place variations in environment, have found no difficulty in providing explanations for the devious routes followed by particular transport links (see review bibliographies of transport geo-graphy by Ullman, in James and Jones, 1954; Wolfe, 1961; Siddall, 1964). Classic studies like Vance (1961) on the contrasting lines followed by the Union Pacific Railroad across the central section of the Rocky Mountains, U.S.A. (Fig. 4.16); Momsen (1963) on the changing role of the Serra do Mar in deflecting least-cost paths between Rio de Janiero and São Paulo in south-eastern Brazil; or Appleton (1962) on the complex routing of competing railway lines in nineteenth-century Britain, show the intricacies of the problem. There remains, however, the task of build-ing these case studies into a more general model of route 'deflection'.

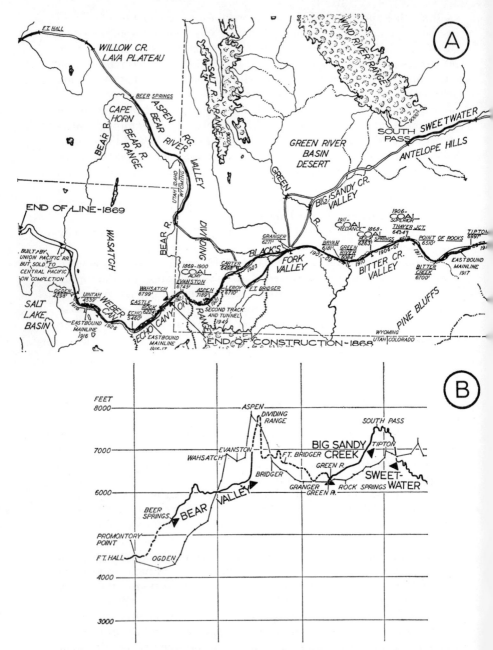

Fig. 4.16. Alternative line of routes followed by the Oregon Trail and Union Pacific Railroad in the Rocky Mountains section, western United States. Source: Vance, 1961, pp. 362, 370.

While no such general model exists, limited advances have been recorded in extending concepts of refraction to transport models. The attempt to extend ideas of refraction to transport routes is of long standing. Ratzel (1912, p. 347; cited by Werner, 1966; 1968, p. 28) observed that ' . . . traffic crossing areas that are difficult to traverse use routes consisting of straight segments that veer towards areas less difficult to traverse.' In the thirties, German economists von Stackelberg (1938) and Lösch (1954, translation) found that Snell's law for the refraction of light could be extended to transport routes.

(a) Discrete cases

A very simple application of a refractive model is shown in Fig. 4.17-A. Here the problem is to locate a route from P_1 to P_2 which has to cross

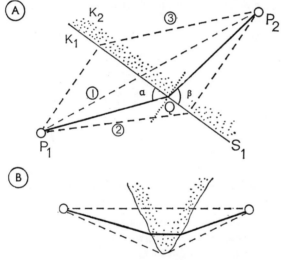

Fig. 4.17. Refractive indices in route location: elementary cases. Source: Werner, 1968, p. 33; Lösch, 1954, p. 186.

a straight boundary line (S_1) separating two areas of different unit transport costs, K_1 and K_2. The straight line solution, 1, fails to adjust to differences in transport costs while routes 2 and 3 represent extreme cases where $K_1 = 0$ and $K_2 = 0$ respectively. The optimal path, 4, via the corner point Q, is given by the ratios of the two unit costs, K_1 and K_2, and of the angles of entry (α) and refraction (β):

$$\frac{\cos \alpha}{\cos \beta} = \frac{K_2}{K_1}$$

(Werner, 1968, p. 33). In practical terms, Q might represent a coastal port separating an area of lower-cost ocean transport (K_2) from an area of higher-cost land transport.

Fig. 4.18-B shows a more complex case of the same refraction principle with the problem of a route between P_1 and P_2 which has to cross a mountain range (stippled). Again the cost per mile of the route across the plains is much less than the cost through the mountains, so that the direct route is not the cheapest. The higher the cost of traversing the mountain area (or the greater the 'refractive index' in Lösch's analogy) the more the least effort route will be deflected southwards. Again the final compromise location will depend on the construction and running costs over the two mediums of plain and mountain. Lösch reminds us of the 'deflection' of a great deal of nineteenth-century trade between the east coast of the United States and California via the Cape Horn route, a diversion which added some 9,200 miles to the direct distance overland across the United States. An equally direct parallel occurs in this century with the planning of a trans-isthmian canal across central America. Of the

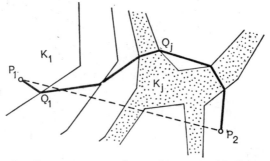

Fig. 4.18. Route location across a complex environment with discrete cost regions.
Source: Werner, 1968, p. 39.

two major routes considered (the Nicaraguan cut and the Panama cut), the sea distance between the eastern and western United States would have been most strikingly reduced by the northern route but this saving was insignificant compared to the saving in construction costs on the shorter Panama cut. Again it is the ratio of the costs that is important. Had the cost of ocean transport been much higher, the advantages of a more northerly route might have been decisive. Since other than United States shipping used the canal, the decision was far less simple in reality, but the basis of Lösch's idea remains valid. Researchers have drawn attention to ferry costs and resultant route 'bending' around Lake Michigan while small-scale examples that make the point as well are the orientations of bridges across estuaries, railway lines, or similar barriers. Unless a road is of very major importance the bridge spans the obstacle at or near a right angle, deviating from the general direction of the road on either side of the bridge. Lösch would describe this as a result of the very strong refractive or bending power of the bridge-construction costs on the alignment of the route (Haggett, 1965, p. 64).

Extension of the refraction law to multivariate cases has been made by Werner (1966, 1968). The plane across which the route is to run is subdivided into a number of sub-regions (cost regions) (F) which are assumed to delimit contiguous polygonal areas of constant unit cost per mile (K). (See Fig. 4.18.) The appropriate expression for total cost is:

$$C = \Sigma_j K_j l_j$$

where 'the set of j's is summed over the sequence of cost regions F_j with per unit costs K_j through which the path passes, and l_j is the length of the path in F_j' (Werner, 1968, p. 31). The minimum-cost path set consists of a set of straight-line segments connecting the nodes P_1 and P_2 through

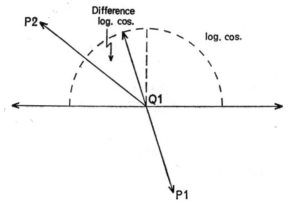

Fig. 4.19. Graphical procedure for locating optimum path. Source: Werner, 1968, p. 38.

an intervening set of corner points $(Q_j = (x_j, y_j))$ located on the boundaries of cost regions. The total cost K for a path with the corner points $Q_1, Q_2 \ldots, Q_n$ is given by Werner (1968, p. 31) as:

$$K = \sum_{j=0}^{n} K_j [(x_j - x_{j+1})^2 + (y_j - y_{j+1})^2]^{\frac{1}{2}}$$

where the first and last corner points, Q_0 and Q_{n+1} are identical with P_1 and P_2.

The vast variety of combinatorial and permutational possibilities for feasible paths and the complex arithmetic needed to derive the optimum may be sidestepped by using a geometrical instrument devised by Werner. The principle of the instrument is that the location of a corner point Q_1 on a regional boundary s_1 approaches optimality when the ratios of the cosines of the entry angle α_1 and the refraction angle β_1 approach the reciprocal ratio of the unit transportation cost K_{i-1}, K_i for the two contiguous cost regions F_{i-1}, F_i^-. The instrument (Fig. 4.19) is moved along the boundary s_i until either (1) the two vectors (A, B are pointing at the two neighbouring corner points Q_{i-1}, Q_{i+1}) indicate a difference of K_i/K_{i-1} on the log cosine scale, or (2) the centre of the instrument reaches one of the end points of s_i. Using successive iterations local and absolute minimum can be found, even where cost boundaries are not polygonal but curved. Fig. 4.18 shows a hypothetical case for a path between P_1 and P_2 over fourteen intermediate sub-regions. As Table 4.5 shows the optimal

Table 4.5. Cost assessment of alternative routes*

	Length in miles	Total transport costs ($\times$ $1000)
I. Straight-line connections	22·7	744·0
II. Alternative path	23·7	623·0
III. Optimal path	28·6	512·0

Source: Werner, 1968, p. 39.
* Mapped in Fig. 4.18.

path is about six miles longer than the overland path but over $200,000 less expensive.

Difference in unit costs over sub-regions have their parallels in movement along lines of different cost. Wardrop (in Herman, 1961, pp. 57–78) has examined the problem of the shortest distance between two points on a plane. If we assume that costs of movement from the first point

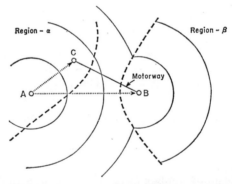

Fig. 4.20. Impact of a new motorway on shortest-distance paths. Source: Wardrop, in Herman, 1961.

(A) are directly proportional to distance and isotropic, then the shortest distance to B is the straight line AB (Fig. 4.20). If however a low-cost path (e.g. a motorway) is introduced (line CB), then the 'shortest' path across the plane in terms of cost will be line ACB. By making reasonable assumptions about relative costs of travel it can be shown that A lies within a region bounded by a hyperbola (Region α), where it is cheaper to use the motorway in travelling to B than to use a direct overland path. A similar region (Region β), also bounded by a hyperbola, can be demarcated to indicate those points that can be most cheaply reached from A by paths using the motorway CB. Wardrop goes on to develop more complex cases on rectangular grids, and Coburn, Beesley and Reynolds (1960) to apply more realistic analyses to the effect of the London–Birmingham motorway on optimum paths from points in the general vicinity of the new highway.

(*b*) *Continuous cases*

Refraction in the bivariate and multivariate cases with two to N homogeneous cost regions leads on to continuous cases where unit costs, K, vary continuously over the region. Here costs can be shown as an

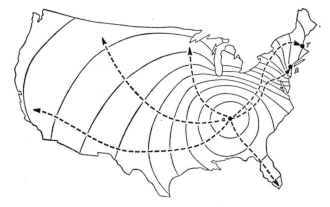

Fig. 4.21. Geodesic paths on an iso-cost surface. Source: Warntz, 1965, p. 17.

isarithmic surface. Warntz (1965) has shown that the general rule for least-cost paths (geodesics) is to follow the trajectory orthogonal to the isocost lines on a surface. Fig. 4.21 shows the isocost surface integrated about a Tennessee town, Murfreesboro (α), where the contours indicate the total cost of acquiring land on a route from this point to any other given point in the United States. On this basis, six least-cost routes from Murfreesboro to other parts of the country are shown. Note that the routes are orthogonal to the contours and that the paths to New York (β) and to Boston (γ) follow divergent paths. The fact that, because of variations in land acquisition costs, the least-cost paths are not straight lines may be demonstrated by converting the surface shown in Fig. 4.21 into a three-dimensional plaster model with point α as the pit. If we place a ball on that surface it will roll 'down hill' towards Murfreesboro, and (as long as the mass of the ball is sufficiently small and the gradients on the surface

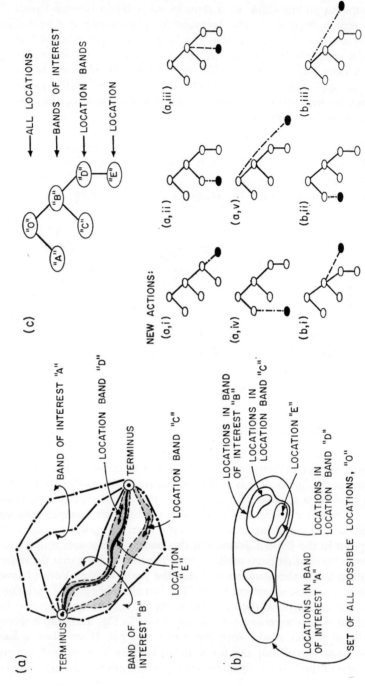

Fig. 4.22. Highway location as a Bayesian hierarchic decision process. Source: Manheim, 1964.

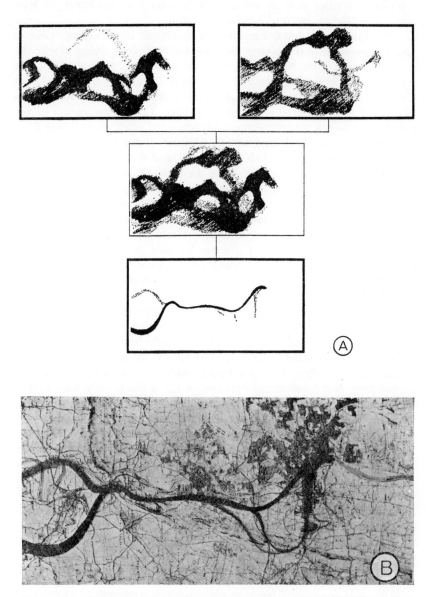

Fig. 4.23. Final stages in the hierarchic screening process for locating a twenty-mile stretch of highway in the Springfield–Northampton area, Massachusetts, U.S.A. Source: Manheim, 1964, p. 38.

sufficiently gentle) its path will accurately trace the least-effort (= least cost) paths shown by the lines on the map.

Geodesic paths have considerable practical implications. Warntz (1961) has studied the routes used by DC-8 aircraft flying across the north Atlantic from New York to cities in western Europe. Here the complex pressure systems encountered 'deflect' aircraft from the Great Circle (= least distance) paths to more complex and variable routes optimized in relation to time and fuel costs. The determination of the shortest path on a contour surface has attracted research attention from navigation to chemical engineering. Warntz (1965, p. 11) provides a graphical method for route determination, and Gould (1968) reviews a number of 'search' procedures when the form of the contour surface is unknown. Manheim (1964, 1966) has argued that route location may be viewed as a hierarchically structured sequential decision process. *Search*, the activity in which alternative locations are generated, is distinguished from *selection* in which a choice is made between locations. Fig. 4.22 shows a highly simplified history of a route location process structured as a decision tree. Each location is regarded as an 'experiment' and the results of alternative experiments evaluated through appropriate utility functions. The Bayesian formulation of the problem shows a number of analogies with Gould's search model of highway development in an underdeveloped area (see Fig. 5.26). Manheim illustrates his model with reference to the location of a twenty-mile stretch of highway in the Springfield–Northampton area, Massachusetts, U.S.A. Twenty-six locational requirements (ranging from earthwork costs, through pavement and sub-grade costs, to air pollution and duplication) were mapped on a black-white scale so that ideal locations were shaded in black and very poor locations in white. Thus on 'earthwork costs' flat areas on the map were recorded as white. The design problem was structured by grouping requirements in subsets and ordering the subsets into a hierarchic tree. By photographically merging the requirements into sub-sets an overall picture (on the same white-black) scale of each group of requirements was built up. Fig. 4.23 shows the final stages in this merging process with the line of optimal highway location. Problems of alternative weighing procedures for each requirement remain to be fully evaluated.

II. LOCATION OF BOUNDARIES

Design of systems of barrier networks raises, in a dual form, a number of the locational problems discussed in their primal form in the first half of this chapter. The partitioning of spatial arrays of points into sets of optimum networks is a direct analogue of the more familiar geographic problem of partitioning an area into sets of optimum regions.

1. Planar graphs and contiguous regions

In region-building studies a distinction is frequently drawn between methods which seek to classify or group spatial phenomena and those which implicitly include location as one of the grouping criteria. Bunge (1962, p. 16) stresses that it is the implicit inclusion of the category of location that makes for regional as against the purely classifactory approach to the earth's surface. Similarly Hagood and Price (1952, p. 542) argue for the importance of the locational category in their agricultural classification of the United States, viz. their statement that '. . . California cannot be put into the same category as New Jersey because of geographic separation.' This geographic separation is stressed in topological terms through linkage (i.e. contiguity), so that we may usefully approach the regional problem with its central concern with optimum location of boundaries through the medium of planar and non-planar graphs (Chap. 1.III).

(a) Districting problems

The most general form of the boundary problem occurs where a set of points is distributed on a Euclidean plane and the object is to establish some partitioning of the points into a predetermined number of groups to optimize an objective function (Scott, 1968). Fig. 4.24 illustrates the optimal two-fold partitioning of twenty randomly distributed points. All points in the problem are given equal weight and the objective function is to minimize the sum of distances from each point to the centre of gravity of the group to which it belongs. Since the number of points in this problem is finite the optimal solution can in principle be found by direct enumeration of all possibilities. However, in practice the number of solutions for the problem shown in Fig. 4.24 is a little over one million, so

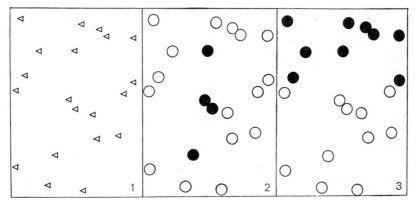

Fig. 4.24. Optimum partition of a point set (twenty nodes) into two groups evaluated by backtrack programming. (1) Original point set. (2) Sample from over one million possible non-optimal solution. (3) Single optimal solution.
Source: Scott, 1968, p. 10.

that for all but the simplest problems enumeration is not feasible. As a result, research has centred on the development of a number of algorithmic or heuristic solution methods; Scott (1968, p. 9) obtained the solution mapped in the figure by a backtrack programming algorithm, and a wide range of alternative approximation procedures are now available (Cooper, 1967).

One of the most interesting applications of the general point partitioning problem is in the drawing sub-regional boundaries for electoral districts. Here the principle is to assemble the sets of points (= counties or enumeration districts) into planar graphs (= contiguous or compact electoral districts) of equal weight (= equal voting population). The generally agreed criteria on which electoral boundaries should be drawn are thus: (a) *equality* of the voting population; (b) *contiguity* in the sense that the district should be in one piece and should be compact in terms of ease of communication between various parts of the district; and (c) *homogeneity* in that it should have common social, political and economic interests. It should be noted that no scheme can ever have complete equality in that the balance of population changes as time passes.

The drawing of such equitable political boundaries has become of greater interest since the historic Baker versus Carr decision in the United States supreme court in March 1962. In this, the court ruled that 'The apportionment of seats and districting of state must be so arranged that the number of inhabitants per legislator in one district is substantially equal to the number of inhabitants per legislator in any other district in the same state' (Silva, 1965, p. 20). The case was precipitated by a group of voters in Tennessee claiming that their votes were 'debased' by inequities in the State's existing apportionment districting. Tennessee had continued to elect its State legislators on the basis of an apportionment adopted in 1901, even though in the intervening sixty years there had been radical shifts in the distribution of the population from rural areas to cities and suburban areas. Thus, one vote in rural Moore county was equal to nineteen votes in Hamilton county, which contains the city of Chatanooga. The situation in Tennessee was paralleled by situations in other American states. In Vermont, the most populous district had 987 times more voters than the least populous district. Similarly, in Texas, one district had a representative in Congress elected on less than a quarter of the votes for other representatives in the same state; altogether, the courts found that thirty-seven of the American states needed some redrawing of electoral boundaries. Re-apportionment on the basis of 'one man, one vote' caused relatively few problems; but the equal population rule is not in itself sufficient to guarantee that economic, political and cultural sub-groups are adequately represented in the legislature. For example, Fig. 4.25 shows three examples of gerrymandering. The two main methods are for the governing party (i) to split up a stronghold of the opposition party and form districts, each of which has a majority of pro-government electors, and (ii) to concentrate the opposition's votes in a few districts where it is wasted on 'overkill'. New York city shows one of the most famous gerrymandering districts in Manhattan's 'Silk Stocking' district where boundaries wind around street corners to include the affluent residential sections and to exclude Negro and Puerto Rican neighbourhoods.

The need to reach decisions on electoral districts has led to a flurry

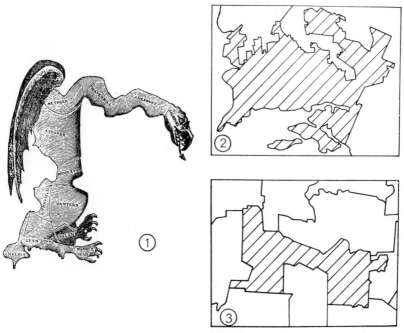

Fig. 4.25. Examples of electoral areas. (1) Original gerrymander of senatorial districts in Massachusetts depicted in the *Boston Gazette,* March 26, 1812. (2) Republican districting of the 24th congressional district in the Bronx section of New York city. (3) Democratic districting of the 30th congressional district in Los Angeles. Source: Silva, 1965, pp. 21, 23.

of proposals and counter proposals for reform. Some idea of the high costs of re-districting procedures is given by Nagel's (1965) estimate that one state alone, New York, had spent some $100,000 since the Supreme Court decision. In this context of urgency and high cost a number of schemes have been put forward for re-districting algorithms on pre-selected bases of equality, contiguity and homogeneity. The general form of such algorithms is shown in Fig. 4.26. The major steps are: (1) choose arbitrary or reasonable district centres for the *n* electoral districts; (2) compute the matrix of distances between tracts and the trial centres, solve the transportation problem (see Chap. 4.I) and, combine split districts; (3) compute the new gravity centre for each district (i.e. the gravity centre for all the tracts assigned to that district in step 2); (4) if the new centres differ from the trial centres, cycle back to step 2, if not, terminate. Since the solution is obtained by a 'local' rather than a 'global' optimum,

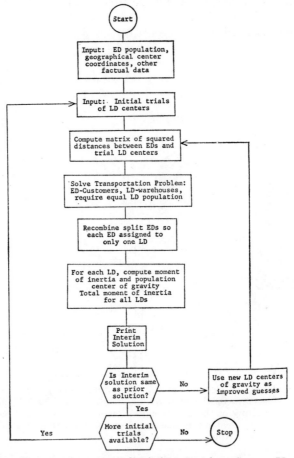

Fig. 4.26. Flow diagram of computer districting procedure. Source: Hess, Weaver, *et al.*, 1965, p. 1000.

some procedures allow re-cycling back to step 1 in which the researcher arbitrarily selects a new set of trial areas.

Operational definitions of 'equality' and 'contiguity' have been examined at length by Kaiser (1966). Where n_j is the population of the ith district in a set of k districts, then Kaiser proposes a value b where:

$$b = 1 - \sqrt{1 - \sqrt[k]{\pi r_j}}$$

Here r_j is the population ratio of district j, a value which ideally is one for all districts. The index b is taken as a stable measure of population equality between the districts. One additional advantage is that $b/2$ is a lower bound to the commonly used Dauer-Kelsay index (the minimum proportion of the population of the state capable of electing a majority of the representatives).

For a compactness index Kaiser uses a value v, where:

$$v = \sum (2\pi t_j / a_j^2)$$

where a_j is the area of the jth district and t_j is the moment of inertia of the districts area, viz.:

$$t_j = \iint (x^2 + y^2) dx \, dy$$

where the origin is at the centre of the mass of the area and the integration is over the area (*cf.* Reock, 1961).

To link together the measures of population equality (b) and of compactness (v), Kaiser uses a criterion function $f = b^w/v$ where w is a weight, $0 \leqslant w \leqslant \infty$. The choice of w depends on the relative weight we wish to place on population equality as opposed to area compaction. When w is zero the f criterion is concerned only with the latter, and where w is one both criteria are equally weighted. Choosing w very large will force population equality at the expense of geographically unwieldy areas.

Computer procedures for districting which use the twin bases of population equality and compactness have been proposed by Weaver and Hess (1963), Hess, Weaver, Siegfelt, Whelan, and Zitlau (1965), Nagel (1965), Kaiser (1966), Harris (1964) and Mills (1966B). The Weaver and Hess procedure is typical in that it proceeds iteratively from an initial input of population and gives the population, area, geographical coordinates and other factual data about each basic enumeration tract. The enumeration tracts are the smallest indivisible unit into which the population of a state is broken for electoral purposes. Fig. 4.27 shows the results of the first three stages on the re-districting of Sussex County, New Jersey, U.S.A. The maps show in sequence the distribution of tracts (enumeration districts), the initial centres for the six districts, and the first and second assignments of population to these centres. The The second assignment showed an improvement of from five to one per cent in the maximum deviation of population from the mean and a significant improvement in compactness. A similar re-districting of the twelve downstate congressional districts of Illinois, U.S.A., is reported by Kaiser. His weighting criterion function (w) was set at 1·00 to give equal greater weight for compactness and population equality, and 200 computer runs were made, each one allocating the 101 counties to twelve randomly selected initial districts. The present boundaries of the

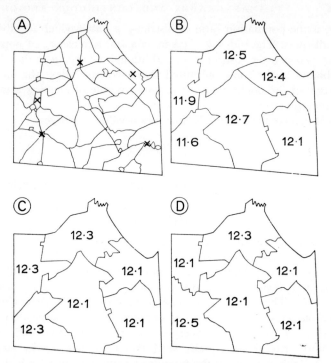

Fig. 4.27. Stages in the boundary optimization process for Sussex County, U.S.A.
Source: Weaver and Hess, 1963, p. 306.

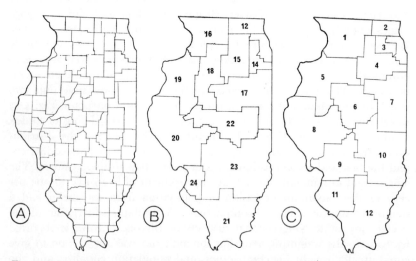

Fig. 4.28. Proposed reapportionment of electoral districts in Illinois, central United
States. (*A*) County system. (*B*) Existing congressional districts. (*C*) Proposed districts.
Source: Kaiser, 1966, pp. 210–12.

congressional district and the best of the 200 solutions ($f = 0.762$) are shown in Fig. 4.28.

Mills (1966B) has developed a linear programming-plus-heuristics approach, and has applied this to local government data within the urban area of Bristol, England. The basis of his approach is an iterative procedure in which (a) arbitrary ward centres are chosen; (b) the transportation problem is solved and split districts are combined; (c) a new centre of gravity is computed for the districts assigned to the ward in the second step; and (d) if the new centres differ from the old the programme goes back to step (b), but if not the programme is terminated at the local optimum. Different starting centres are used to generate different local optima, and human scanning of these optima is used to reject or to adapt non-contiguous solutions. As expected this heuristic procedure does not

Table 4.6. Convergence properties for a sample of computer runs*

| | | | End of Iteration | | | |
	0	1	2	3	4	5
Compactness	199	177	162	152	151	151
Population Variation	0.176	0.169	0.158	0.148	0.148	0.148
No. of Changes	—	34	12	8	6	0

Source: Mills, 1966, p. 13.
* Data for Bristol C.B., England, 1954.

guarantee unique optimum solutions, but it does provide a set of good alternatives. Mills was able to introduce considerable flexibility into his procedure by permanently assigning certain populations to certain wards so as to secure a natural boundary as a ward boundary (e.g. the River Avon in the Bristol case) or to avoid bisecting a uniform housing estate. Table 4.6 shows the convergence properties of a series of computer runs. Iteration O refers to the hand calculation used as the starting point of the runs. Although compactness and population variation show convergence, these do not always improve monotonically and there may be different iterations on the same computer run with much the same population variations and compactness. Occasionally runs started 'cycling' in that assignments 3, 4, 5, 6, 7 and 8 matched those in 9, 10, 11, 12, 13 and 14, and so on. Mills attributes this to instability in the rounding procedure. Solution to electoral districting problems through the use of mathematical assignment methods have direct parallels in other fields. Yeates (1963) has illustrated the use of transportation methods (see Chap. 4.I) to the delimitation of optimal school districts. Here the solution consists of minimizing the cost of shipping a product (children) from a set of sources (their homes) to a number of destinations (schools). Specifically

Yeates's problem was the cost of moving the 2,900 high school children of Grant county, Wisconsin, to the thirteen high schools in that county. The location of the schools and school districts for the central part of Yeates's study area is shown in Fig. 4.29-A. To reduce the size of the computing problem both schools and pupils' homes were assumed to lie in the centre of the square-mile section in which they were located; school boundaries were then redrawn on this simplified basis (Fig. 4.29-B) by including the whole square-mile section within a school area, if the greater

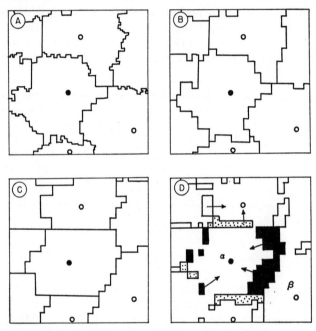

Fig. 4.29. Distance-minimization approach to the construction of optimum boundaries in a sample quadrat of Grant county, Wisconsin, United States. Source: Yeates, 1963, pp. 8, 9.

part was in that section. Since 754 sections were occupied by children from the thirteen schools the problem was reduced to a 754 × 13 matrix. With this data optimum boundaries were determined by computer such that: (i) total distance to schools was minimized; (ii) each school was filled to its capacity (in 1961). Subject to a number of restrictions the problem can be expressed algebraically as:

$$\sum_{i=1}^{n} \sum_{j=1}^{m} d_{ij} x_{ij} = \text{minimum}$$

where d_{ij} is the distance from the ith section to any jth high school, and

x_{ij} is the number of school-children in the ith section assigned to the jth high school.

The boundaries resulting from this minimum solution are shown in Fig. 4.29-C. Comparison with the boundaries of the formalized school districts shows that there is considerable change and overlap. The boundaries of the Lancaster high school, α, in Fig. 4.29-D show losses to the north and south (stippled) but large gains (black) from the neighbouring Platteville high school, β on the east. How important are the changes shown in Yeates's reallocation? Analysis is made more difficult by two factors (1) the fact that the theoretical boundaries were based on children's distribution in a single year (1961), while the actual school districts must remain static over a longer time; and (2) the difficulty of comparing the actual ground distance covered along roads by the children with the cross-country (direct) distance used in the theoretical analysis. Table 4.7 shows

Table 4.7. Distances travelled with alternative district boundaries*

High school:	Boscobel	Platteville
Actual school districts:		
Mean road distance to school, miles	6·7	6·4
Mean overland distance to school, miles	5·5	5·6
Theoretical school districts:		
Mean overland distance to school, miles	5·1	5·3
Estimated saving on overland distance, miles	0·4	0·3

Source: Yeates, 1963, p. 9.
* Grant county, Wisconsin, United States, 1961.

the results for a sample study of two high schools in the area, Boscobel and Platteville, where a comparison of the distances both by road and cross country suggests savings of the order of 0·4 to 0·3 miles.

Morrill (Garrison, *et al.*, 1959, pp. 244–76) used comparable approaches in an analysis of the effect of freeway construction on medical-care areas for selected areas within the United States. Fig. 4.30-A shows the medical service areas for a sample area in western Pennsylvania, 4.30-B additions to the highway network, and 4.30-C estimated shifts in boundaries after highway improvement (black). Stippled areas show unchanged boundaries despite the effects being felt in all interrelated counties and not just those through which the highway passes. Highway improvements brought a greater flow in services, but at a lower unit cost. Clapham (1964) was concerned with the effect of alternative distance functions on boundary demarcation when using transportation techniques.

In a study of the re-distribution of workloads for delivery routes (e.g. oil tanker distribution) he contrasted the effect of least distance (Fig. 4.30-D)

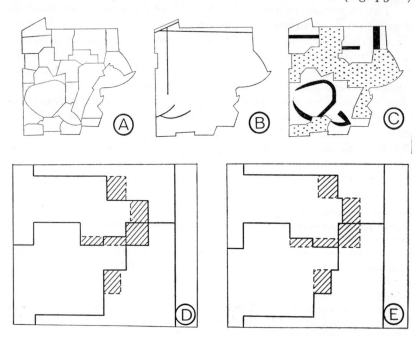

Fig. 4.30. (*A, B, C*) Impact of highway changes on optimum medical-care areas in western Pennsylvania, U.S.A. (*D, E*) Impact of alternative definition of distance on optimum workload areas for a hypothetical area. Source: Morrill, in Garrison, *et al.*, 1959, pp. 266–7; Clapham, 1964, pp. 72–81.

and least squared-distance (Fig. 4.30-E) on boundary solutions. The latter was found to give smoother boundaries with greater compactness of the resulting work areas.

(b) Boundary adjustment problems

Another set of problems concerns the drawing of boundary lines around sets of points under some definition of minimum distance, where the membership of groups is already defined. As with shortest paths (*cf.* Fig. 3.4) there is a considerable range in the definition of 'shortest boundary' and an attempt is made in Fig. 4.31 to illustrate some of the conflicting boundary patterns that may emerge.

(i) The bounding of a set of points by lines which divide the region into a set of areally homogeneous units has been variously studied under the terms *Dirichlet* regions (Dirichlet, 1850), *Voronoi polygons* (Coxeter, 1961) or *Thiessen polygons*. The procedure was used by the U.S. Weather Bureau in generalizing the rainfall of a given water catchment from a scatter of meteorological stations, and has been widely used in the study of market hinterlands in urban geography (Haggett, 1965, pp. 247–8). The method for drawing the boundaries is as follows: (1) lines are drawn joining a given centre to each adjacent centre; (2) each of these inter-centre lines is bisected to give the midpoint of the line; (3) from the midpoint of the line a boundary line is drawn at right angles to the original inter-centre line to give a series of polygons; (4) counties lying across the boundaries are included within the boundary of the centre within which the greater part of the county area lies. In practice the

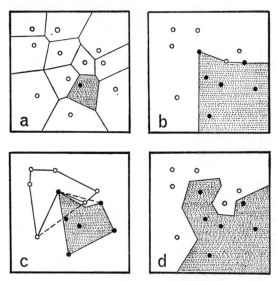

Fig. 4.31. Alternative definitions of 'shortest boundaries'.

drawing of Thiessen polygons is time-consuming, and there is usually some imprecision over the choice of diagonals to be used in drawing the boundaries of the polygon around a given centre. Kopec (1963) has reported an alternative method of construction in which arcs of circles of the same radius are drawn from adjacent points, and the side of a polygon is located by drawing a line through the points of intersection on the arcs. The argument for this method is that it eliminates the need to draw diagonals and reduces the chance of error from using inappropriate ones in construction.

Fig. 4.31-A shows the pattern of Dirichlet regions about a set of twelve points. All the area within the intersecting boundary lines of the polygon lies nearer to the enclosed centre than any other external centre. Whether the use of such polygons as proxy for the market area of an urban centre is valid is a matter for debate.

(ii) *Shortest fence* problems involve dividing a region into as many regions as there are sets of points by lines whose total length is minimized to enclose each kind of point within its own area (Bunge, 1962, p. 190). Fig. 4.33-B and -C illustrate two alternate concepts of the shortest fence problem when the twelve original points are partitioned into two sets. In the first or 'island' case the area is assumed to be bounded by the perimeter of the square. Bunge finds three relevant theorems from the mathematical literature are that the shortest bounding fence in these conditions will (1) consist of straight segments, (2) meet the perimeter of the region at right angles, and (3) minimize the second moment (i.e. define the most compact region) (see Fig. 4.31-B). There is some evidence that the problem is solvable only for four or less sets of points. In the

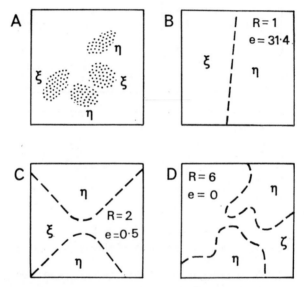

Fig. 4.32. Use of higher-order discriminant analysis to classify a mapped distribution (and) into discrete regions. Source: Sebestyen, 1962, pp. 69–71.

second or 'mid-continental' case the area is not bounded by the perimeter of the square, and the problem is to find the shortest total length of lines where each line encloses only one set of points. The problem has analogies in the literature of linear programming and the solution in two dimensions would be (1) to draw the convex hull around each set of points, and (2) to adjust for interpenetration (if it exists) by selecting the minimum deflecting path. Again the solution consists entirely of straight-line segments. In Fig. 4.31-C the original hulls are shown by solid lines, the area of interpenetration is stippled, and the minimal deflections by broken lines. (iii) One unusual boundary problem occurs when a line must be drawn across a region so as to maximize its distance from a set of points. Bunge

(1962) considers this as a class of 'obnoxious' boundary problems which might involve applications in the routing of motorways through existing urban residences or high-voltage pylon lines between rural settlements. Fig. 4.31-D illustrates one of the many cases in the group by a boundary line (made up of segments from the Dirichlet region solution) which separates two sets of points with a line placed as far as possible from points contiguous to it.

Sebestyen (1962, pp. 69–71) has illustrated the use of discriminant functions to specify boundaries between complex two-dimensional distributions of points. Using the same approach as Casetti (see Chap. 4.II.2c), he computed discriminant functions in terms of the two orthogonal axes (U, V). Fig. 4.32 illustrates the superimposition of successively more complex discriminant functions on the four clusters of points (sets η and ζ). Simple first-order polynomials appear as a straight line (Fig. 4.32-B); conversely a sixth-order polynomial (Fig. 4.32-D) appears as a complex line. The simple line misclassifies a third of the points, whereas the complex line accurately classifies both η and ζ distributions. Between the two lie intermediate boundaries of which the second-order polynomial (Fig. 4.32-C) appears particularly efficient in that it misclassifies only 0·5 per cent of the η and ζ distributions yet is a mathematically rather simple solution. This type of classification procedure might also be applied to the type of boundary problem in which we need to predict the probable course of a boundary across areas where information is either absent or very sparse.

(c) Statistical decision problems

The planar constraint of regionalization is seen in practical form in the many attempts which have been made to classify spatially-occurring phenomena into homogeneous but contiguous regions. This limitation is illustrated by an early study by Hagood and Price (1952), where the problem studied was to divide the conterminous United States into some six to a dozen contiguous groups of states, each group to be made as homogeneous as possible with respect to some 104 items taken from the 1940 censuses of population and agriculture. These items were equally divided into two major groups, agriculture and population, which in turn were divided into further sub-groups, fourteen in all. These groups varied in size from information on crops (twelve items) in the agriculture group, to information on race (five items) in the population group.

These items were used to draw up 'agriculture-population profiles' of each state. First, all the 104 items were standardized so that the mean value for the forty-eight states for each item was 50·0 and the standard deviation 10·0. Second, correlation coefficients (r) were calculated between the profiles of adjacent states. Correlation in a general sense denotes the association between quantitative or qualitative data. It is measured by correlation coefficients which vary between -1 and 1, with the intermediate value of zero indicating absence of correlation and the two extremes indicating complete negative or positive correlation. Kendall and Buckland (1957, p. 67) define a generalized correlation coefficient, Γ

$$\Gamma = (\Sigma\ a_{ij}\ b_{ij})/\sqrt{\Sigma\ a_{ij}{}^2\ \Sigma\ b_{ij}{}^2}$$

where, in two sets of observation $x_1 \ldots x_n$ and $y_1 \ldots y_n$, a score is allotted to each pair of individuals a_{ij} and b_{ij}, and Σ is a summation of all values of i and j from l to n. This general coefficient includes others such as Kendall's *tau*, Spearman's *rho*, and Pearson's *r*. Examples of the computation and application of correlation coefficients to geographical problems are given in Gregory (1963, pp. 167–84).

The resulting coefficients varied from very high values between like states (e.g. Alabama and Georgia had a coefficient of $+0.92$) to very low values between unlike states (e.g. Ohio and its southern neighbour,

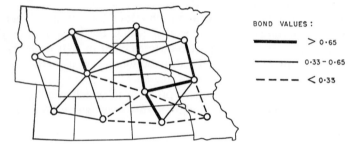

BOND VALUES :

——————— > 0.65

———————— $0.33 - 0.65$

– – – – < 0.35

Fig. 4.33. Pattern of interstate correlation bonds in the north-central United States. Source: Hagood and Price, 1952, p. 545.

Kentucky, had a coefficient of only $+0.01$, suggesting that the line between north and south remains strong in the United States). Part of Hagood's map is reproduced in Fig. 4.33. Here for an area of the northern United States the values of the coefficients have been replaced by lines of varying width. The result shows the scaffolding of 'regional bonds' between the thirteen states; it emphasizes the strong north–south links between Montana and Wyoming, between the two Dakotas, Nebraska and Kansas, and between Minnesota and Wisconsin. Likewise we see the rather weak links east–west across the grain of the country.

In practice, Hagood used these correlation bonds to supplement a single comparative index, the 'composite agriculture–population index', which was calculated by factor analysis in much the same way as Steiner (1965) calculated major dimensions of climate from a range of climatic parameters (Chap. 4.II(2b)). Delineation began with the easily recognizable regional nuclei which formed the centres of homogeneous regions. Once these had been established, the marginal states were allocated to one of these nuclei on the basis of both their composite index and their intercorrelations with neighbours.

This distinction between 'nuclear' and 'marginal' states is brought out clearly in Table 4.8, which shows that on average Alabama was nearly three times more like its adjoining neighbour states than was Missouri. Alabama clearly lies deep within the heart of the south with strong similarities to its neighbours, while Missouri lies on the border of four of the six major regions recognized by Hagood. As its highest link lay with the state of Illinois, Missouri was finally assigned to a Great Lakes region.

We find then that statistical methods do not solve regional problems in the sense of creating homogeneity where none exists. What they appear

---

Table 4.8. Coefficients of similarity for agriculture-population profiles*

State:	Nuclear type (Alabama)	Marginal type (Missouri)
Number of neighbouring states	4	8
Coefficients of similarity:		
Maximum similarity	0·92	0·45
Minimum similarity	0·44	0·08
Mean similarity	0·75	0·29

Source: Hagood and Price, 1952, p. 545.
    * For states adjacent to Alabama and Missouri, 1950.

---

to do is to help legislate in difficult cases and to make the reasons for the choice, however marginal that choice may be, clear to the observer. Zobler (1958) used an alternative method, variance analysis, for regional allocation in a partition of the eastern United States. Here the problem was to decide whether, in terms of its industrial population (number of workers in manufacturing and primary industries in 1950), the state of West Virginia should be grouped with one of three state regions: (1) Mid-Atlantic; (2) South Atlantic; (3) East South Central. The state regions are shown in Fig. 4.34-A with West Virginia shaded. Inspection of the figures for the states making up the three regions and for West Virginia give no decisive indication to which existing region the problem state should be added.

Zobler argued that when regions are being constructed from smaller units there are two sources of variation: one among the states within a region (within-region variation) and the other among the regions (between-region variation). Variance analysis was used to measure this variation within and between regions under three conditions, with West Virginia assigned to each of the three regions in turn (Table 4.9). Variance analysis measures the sums of squares of deviations from the arithmetic mean and separates these into components associated with specific sources of variation (Gregory, 1963, pp. 133–50). The between-region variance,

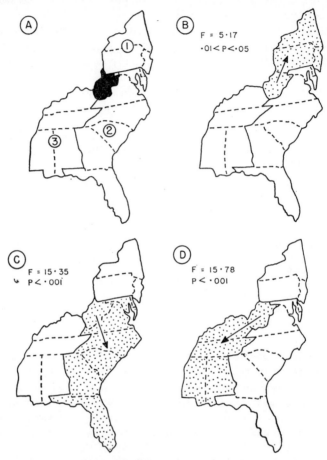

Fig. 4.34. Assignment of West Virginia to three alternative regional groups in the south-eastern United States with significance levels derived from variance analysis. Source: Zobler, 1958, p. 146.

showing the variation of the regions around the mean of all the regions, was divided by the within-region variance, showing the variation of the states around their respective regional means, to give the variance ratio or *F*-ratio. If the two variances are equal, the value of this ratio is one and the more *F* rises above one the greater the interregional differentials. In broad terms the variance ratio describes how successful the grouping

Table 4.9. Regional assignment using variance analysis*

Variance ratios:	Between-region variance (a)	Within-region variance (b)	Variance ratios (a/b)
Alternative assignment of West Virginia:			
To Mid-Atlantic region	46·09	8·91	5·17†
To South Atlantic region	71·55	4·66	15·23‡
To East South Central region	72·23	4·57	15·78‡

Source: Zobler, 1958, p. 146.
  * Eastern United States, 1950.
  † Significant at the 95 per cent confidence level.
  ‡ Significant at the 99·9 per cent confidence level.

procedure has been in keeping like states together and keeping unlike states apart.

Results in Table 4.9 show that although West Virginia could be assigned to any of the three regions the optimum allocation was to join it to Alabama, Mississippi, Kentucky and Tennessee in the East South Central division (Fig. 4.34-D). Conversely, the worst classification on this analysis would be to place it with New Jersey, New York and Pennsylvania in the Mid-Atlantic division (Fig. 4.34-B) (Haggett, 1965, pp. 259–61).

2. Non-planar graphs and non-contiguous regions

In this section we describe a series of techniques in which areal units can be grouped into regions on the basis of their like characteristics rather than their spatial contiguity. In graph-theoretic terms this is equivalent to building nodes into graphs without links having to pass through intermediate links; i.e. the graphs are non-planar rather than planar. It is of course true that the grouping methods described here can be modified to incorporate a contiguity constraint (e.g. Spence, 1968), but the attempt to match attribute space and geographic space into single regional networks runs into a number of serious difficulties (Haggett and Cliff, *in preparation*).

(a) Regional taxonomy

We may regard the general objective of an efficient regional taxonomy as to draw boundary lines between groups so as to minimize internal (within-group) variation and to maximize external (between-group) variation. Regional definitions have clear analogies with the sociologists, problem in clique definition, viz. 'allocating individuals to groups such that the connections individuals have within groups do not vary nearly to the same extent as the connections they have across groups' (Cox, 1968, p. 5). Taxonomic classification, a major concern of the natural sciences from Linnaeus and Montelius onwards, has come back into prominence within the last decade with the advent of high-speed digital computers. Reviews of recent developments are given in Sokal and Sneath (1963) and Williams and Dale (1965).

In network terms we can regard individuals as nodes within a completely connected non-planar graph in which weights (i.e. distances) are assigned to each link. Berry (1967) has reviewed four main methods by which clusters are inductively determined within the sets of distances:

(i) *Linkage analysis* (McQuitty, 1959) links each station to its nearest neighbour and defines groups in terms of the pattern of linkages. A group is characterized by the fact that 'every member of the group is more like some other member of the group than any observation not in the group' (Berry, 1967, p. 236). (ii) *Cluster analysis* (Cattell, 1944; Rao, 1948) links stations to the closest pair in order of their proximity to that pair. Groups are defined in terms of changes in the B-ratio (ratio of average within-cluster distance to average distance between points within the cluster and outside the cluster). (iii) *Metric profiling* uses profiles for each station in which the distances to all other stations are plotted. Clusters are developed on the basis of the similarity of profiles. (iv) *Bonded groups* (Olson and Miller, 1958; Haggett, 1965, Chap. 10) specify *a priori* 'acceptance levels' for group measure membership and graphs linkages between stations separated by distances less than the acceptance level. Changes in the acceptance level yields further information about the cluster structure of the population being studied.

Hierarchic grouping (Ward, 1963; King, 1967) is a modification of cluster analysis. The procedure is described by Berry (1967, p. 238): 'Start with n points, link the closest pair, then calculate their centroid and the average intergroup distance. The next step begins with n-2 points plus one centroid; these are treated as n-1 points, the closest pair is again linked, and the centroid of that pair is calculated. If the previous centroid is linked, it will have a weight of two when average intragroup distance is calculated this second time. The increment to average intragroup distance by performing this step is calculated. The process is repeated step by step and culminating in a final step where a pair of presumably complex centroids is linked and the single centroid of the n points results.' Table 4.10 shows an example of this hierarchic grouping process with respect to a very simple network of four individuals (Cox, 1968, p. 23). This grouping procedure may be simply illustrated by a hypothetical matrix of distances for four stations: A to D. Searching the first matrix shows the lowest distance value is 3·00 at the BC intersection. Since these are the cells most alike they are combined into a single region and the distance between the new region and the remaining two stations com-

Table 4.10. Stages in the grouping of a 4 × 4 distance matrix

(A)

	A	B	C	D
A		6	20	15
B	6		3	9
C	20	3		12
D	15	9	12	

(B)

	BC	A	D
BC		13	10·5
A	13		15
D	10·5	15	

(C)

	BCD	A
BCD		14
A	14	

(D)

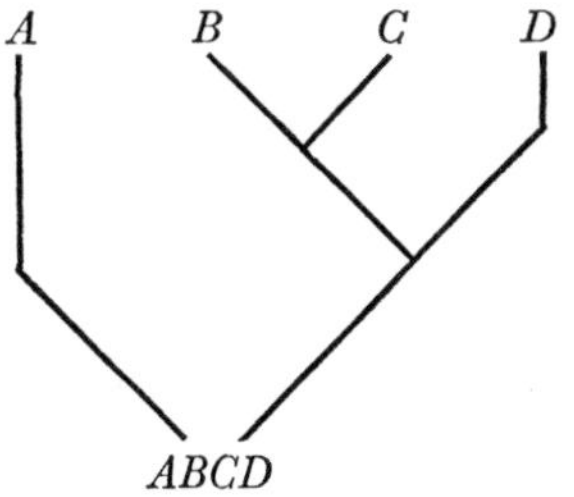

ABCD

Source: Cox, 1968, pp. 24–5.

puted. The new N-1 matrix contains the same distances between A and D, but new distance values are inserted for BC equivalent to the means of their distance values in the 4 × 4 matrix. The lowest value in the new matrix is identified and the grouping cycle repeated. Three regions are replaced by two, and two by one in successive cycles. The linkage tree formed by this grouping process is shown in Table 4.10-D.

Berry (1961) has illustrated this technique very clearly with reference to the service-industry characteristics of nine census divisions of the United States in 1954 (Fig. 4.35-B). Here distance between points (D^2) showed that of the nine census divisions, New England (α) and East North Central (β) had most in common (a D^2 value of only 0·69), while East South Central (θ), the heart of the 'South', and Pacific (ζ) were most dissimilar (with a D^2 value of nearly 35). By placing the two nearest divisions together the nine units were reduced to eight. Distances between the remaining eight regions were derived and the Middle Atlantic district added to the New England–East North Central region. By repeating this process the regions were progressively diminished till finally the whole of the United States formed a single region. These successive stages are

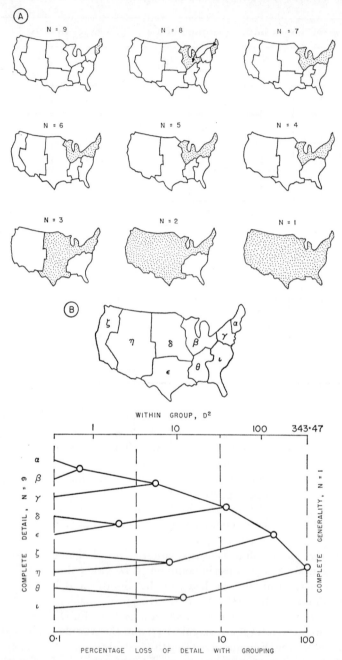

Fig. 4.35. Stages in the regional grouping of census divisions in the United States.
Source: Berry, 1961, pp. 272, 273.

shown in the nine maps of Fig. 4.35-A. We can then have nine different levels of regional breakdown of the United States, each one efficient at its particular level. Berry (1961, p. 273) has shown that in this breakdown process (in which we proceed from many to few regions) we are progressively gaining in generality and progressively losing in definition. Perfect detail is available only with all nine original regions; perfect

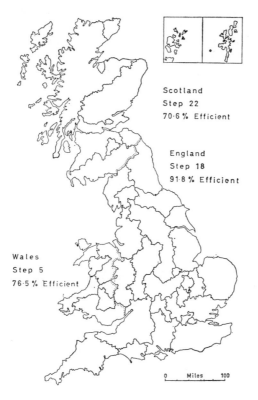

Scotland
Step 22
70·6 % Efficient

England
Step 18
91·8 % Efficient

Wales
Step 5
76·5 % Efficient

Fig. 4.36. Optimum employment regionalizations of England, Wales, and Scotland. Source: Spence, 1968, p. 101.

generality is available only if we regard the whole United States as one unit. This loss of detail can be calibrated using the distance measured previously (within-group D^2) and ranges from zero with all nine divisions to 343·47 with only one region. Fig. 4.35 plots this progressive loss of detail by the use of a 'linkage tree', which shows the progressive combination of regions. We should note in particular that the loss of detail (as measured by the within-group D^2) is plotted on a logarithmic scale, which emphasizes the very small loss of detail in the first five grouping steps. Specifically only 3·5 per cent of the detail was lost in these steps. This means that we can learn almost as much by regarding the United States as four large regions as nine smaller census districts. The implication of Berry's findings is that while we must choose the regional breakdown that serves our particular research purposes, we need to be aware of their relative efficiency. If it matters little whether we need two, three, or four service regions in the United States, then the loss of detail analysis suggests it is worth adopting four regions (only 3·5 per cent loss) or three regions (ten per cent loss) rather than two regions (forty-eight per cent loss). Similarly Spence (1968) has been able to suggest a number of rather efficient employment regions for England, Wales and Scotland on the basis of taxonomic groupings of counties (Fig. 4.36).

(b) Distance scaling of links

In the preceding discussion the values of 'distance' along each link were either assumed (as in Table 4.10) or taken as given (as in Fig. 4.35). In practice the determination and measurement of the distance values

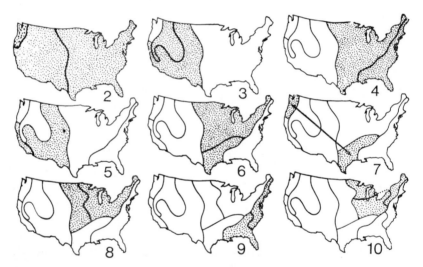

Fig. 4.37. Stages in the regional grouping of U.S. climatic stations based on factor analysis, distance grouping and discriminant analysis. Source: Steiner, 1965, p. 340.

separating each node in the non-planar graph is of critical importance. The problems encountered in establishing distance values are illustrated here through the medium of a single case study: Steiner's (1965) analysis of the climatic regions of the United States (Fig. 4.37). Climatic classification is one of the classic problems in geographic taxonomy; in this study, furthermore, it is possible to follow through the three major stages in distance scaling: selection of parameters, identification of major dimensions, and finally, distance scaling itself.

Stage I: Selection of Parameters. The selection of suitable parameters for a classification poses critical problems which can be solved only by reference to the relevant field of study. For climate, Steiner selected four main variables—mean temperature ($°F$), mean precipitation (inches), mean sunshine (per cent of possible sunshine) and relative humidity (per cent at one p.m.)—which were available for either the whole or a substantial part of the 1931–60 record. For each major variable four characteristics were measured—the mean annual value, the mean January value, the mean July value and the range (July–January)—to give a total of sixteen parameters for each of the sixty-seven stations. Stations were located within the conterminous United States with a slightly denser pattern of stations in the eastern half (Fig. 4.41).

In order to allow comparability between the sixteen indices the values were (i) normalized by the use of logarithmic or square-root transformations and (ii) converted to standard scores. Broadly the effect of such transformations is to stabilize variance and to bring the distribution more closely into line with the normal distribution so as to allow statistical analysis. The original or transformed values were replaced by standard scores (Z), where:

$$Z_i = \frac{X_i - \bar{X}_i}{S}$$

$\bar{X}_i$ is the arithmetic mean and S the standard deviation of the distribution. The effect of standardization is to reduce the sixteen distributions to common scale with a range of around $+3$ to -3.

Stage II: Identification of major dimensions. The climatic record for the stations analysed by Steiner may be conceptualized as a swarm of sixty-seven points located in terms of sixteen dimensional space. This concept of *n*-dimensional space is illustrated by the three points, α, β and γ, in Fig. 4.38. In the simple case of one dimension represented by a single vector, V_1 (Fig. 4.38-A), α and β are close together. They remain close when a second dimension, vector V_2, is added (Fig. 4.38-B). When a third dimension is added, vector V_3, we can see the position is changed and that β is now much nearer to γ (Fig. 4.38-C). Although we cannot show graphically a fourth dimension, vector V_4, there is no mathematical limit to adding this and further vectors to give a theoretical multidimensional space, i.e. *n*-dimensional space.

The problem in measuring the 'distance' between the climatic stations in Steiner's example is that we have no reason to assume that the sixteen axes are orthogonal to each other. Indeed we know that some axes are oriented in the same direction at low angles (e.g. annual temperature and July temperature with a correlation coefficient of $+0.951$), while others are oriented in different directions at higher angles (e.g. July precipitation

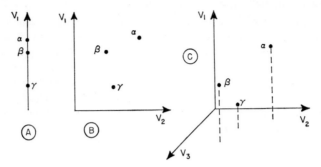

Fig. 4.38. Plotting of three values (α, β and ψ) in one-, two-, and three-dimensional space. Source: Haggett, 1965, p. 254.

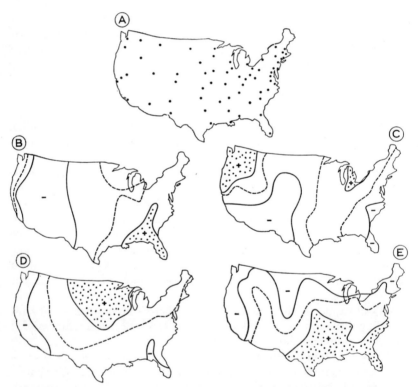

Fig. 4.39. Factor analysis of U.S. climatic data. (*A*) location of stations. (*B–E*) Factor score for first four components. Source: Steiner, 1965, pp. 332–9.

and July humidity with a correlation coefficient of −0·259). In order to replace the swarm of sixteen axes with a few orthogonal axes in which distance measurements between stations may take place, Steiner employs factor analysis.

The sixteen axes are replaced by the four major axes which together account for 88·6 per cent of the total variance. The first axis (twenty-nine per cent of variance) is strongly associated with the annual rainfall axis (correlation coefficient of +0·886) and inversely with annual sunshine (−0·922), and is characterized as a *humidity axis*. It shows (Fig. 4.39-B) a general west–east progression of values with anomalies on the west coast. The second *atmospheric turbidity axis* (twenty-six per cent) is strongly associated with sunshine ratios (+0·912) and differentiates stable sunny stations from zones of rapid air mass change (Fig. 4.39-C). The third axis, *continentality* (twenty per cent) shows highest scores near the Canadian border in the middle of the continent. The fourth axis, *thermality* (fourteen per cent) displays a pattern of general latitudinal increase in temperature from north to south (Fig. 4.39-D and -E).

Stage III: Distance scaling and grouping. Where the axes are orthogonal, distance between points in *n*-dimensional space follows from the well-known rule of the 'square of the hypotenuse'. It can be briefly written as:

$$D = \sqrt{\sum_{i=1}^{n} (x_i - y_i)^2}$$

where $i = 1, 2 \ldots n$; where D is the distance (similarity) between points x and y; and where x_i and y_i are the values of characteristic i (Berry, 1958, p. 301). More complex measures of generalized distance have been derived by Mahalanobis and others (Mahalanobis, Rao and Majumdar, 1949), where both the average value of a region and the dispersal of the units within it have to be taken into account. Distance in Euclidean space is the normal measurement system adopted, but it clearly forms one point on a spectrum of geometries that ranges from the hyperbolic space of Lobachevsky through to the hyperbolic space of Minkowski. The general form of the distance relationship may be written as:

$$D(n) = \sqrt[k]{\sum_{i=1}^{n} (x_i - y_i)^k}$$

where k may take on various values (e.g. $k = 2$ for Euclidean space and $k = 1$ for Manhattan space).

For Steiner's stations distance was measured in Euclidean terms for the four-dimensional climatic space to give a symmetric matrix of values for the distance between each station and every other. The object of classification is simply then to place in one group stations which are near together (homogeneous) in n-dimensional space, and to separate groups which are far apart in the same space field. Methods for achieving this aim are set out in the preceding section (Chap. 4.II(2a)).

(c) Discrimination between groups

Classifications drawn up in the hierarchic method described above (Chap. 4.II.(1a)) contain optimality criteria at each successive stage in the grouping procedure. There is, however, no guarantee that a succession of local optima will lead to the global optima; a classification system which produces regions by optimal grouping of its members gives slightly different sets of regions than one which proceeds by optimal splitting of the total population. This difference between the results produced by 'clumping'

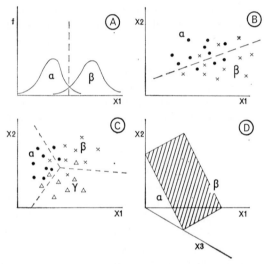

Fig. 4.40. Geometric representation of discriminant functions.

and 'splitting' procedures necessitates the use of statistical procedures to check the validity of the groupings produced.

Discriminant analysis is one of the most commonly used validation procedures. Its purpose is to allocate individuals to a correct population with a minimum of error, usually on the basis of single or multiple measurements of the individual and a prior set of similar measurements on individuals whose origin is known (Kendall and Buckland, 1957, p. 84). One central problem is the construction of functions of the known population to act as discriminator (discriminant function). Fig. 4.40-A shows a very simple case of discriminant functions where the distribution of two groups α and β is shown as a function of measurement X_1. In this highly idealized case the curves are bell-shaped distributions, and the midpoint between the means of the two distributions would provide a useful discriminant function. Thus any unknown individual with a value of X_1 greater than a given critical level would be assigned to group β with a minimum of error. The application of simple discriminant functions to the prediction of network growth is discussed in Chap. 5.I (see Fig. 5.4).

A more complex case with two characteristics (X_1 and X_2) might lead to a more complex discriminant function in which classes are separated in terms of both characteristics (Fig. 4.40-B). The problem is of course readily extendable to a larger number of populations (e.g. α, β, and γ in Fig. 4.40-C) and to a larger number of characteristics (e.g. X_1, X_2, and X_3 in Fig. 4.40-D).

One example of the application of discriminant analysis to regional boundary making has been analysed by Casetti (1964) in a study of North American climate. Discriminant analysis was performed on three arrangements of seventy climatic stations: (1) a modified Koppen classification of the climatic stations; (2) a 'disordered' classification obtained by removing and re-allocating at random the cards separating the Koppen climatic regions; and (3) a shuffled classification obtained from the climatic classification by removing the cards separating the sets, shuffling the decks and re-allocating at random the sub-set partitions. The climatic, disordered and shuffled classifications on which the analysis was performed, provide a useful test for the techniques proposed, because they are arranged so as to contain an increasing amount of noise from the first classification (1) to the last (3). In the programme used by Casetti, multiple-discriminant functions are used to determine the scores of objects in discriminant space. Each station was initially associated with twenty-four parameters of precipitation and temperature. Component analysis was used to break down these initial parameters into six major elements. The first component accounted for sixty per cent of the total variability and broke down the stations according to precipitation and temperature levels. A second component (twenty-two per cent of the variability) broke out a continuum with arid climates at one extreme and climates with low temperature and low precipitation at the other. The third component (eleven per cent of the variability) differentiated west and east coast climates. Euclidean distances between each object and its class centre were calculated in discriminant space, the nearest class determined, and objects allocated to their nearest classes. This re-ordering (i) allows the investigator to assess the quality of a given initial classification relative to the original set of data, and (ii) improves the initial classification by generating a limit classification that preserves the original rationale but is in optimum agreement with the data and with the discriminant procedure used.

Fig. 4.41 shows the application of the method to climatic classification of the seventy north American stations. The left-hand sets of maps show the 'core' areas of the classification for the climatic, disordered and shuffled arrays. The boundaries surround stations which the discriminant iteration did not remove from their initial subsets (Casetti, 1964, p. 47), and, predictably, the size of these core areas is inversely proportional to the weight of random classificatory elements. The right-hand set of maps shows the boundaries generated by discriminant analysis by the final

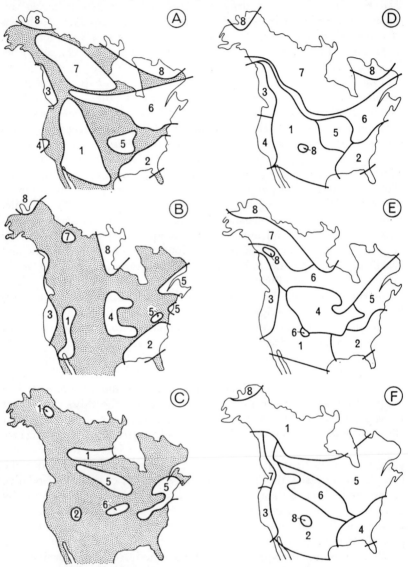

Fig. 4.41. Regionalization by discriminant iterations: North American climatic data. Source: Casetti, 1964, pp. 71–9.

(seventh) iteration. Despite the obliteration of the normal climatic classification in the second two cases, the final maps of both the disordered (Fig. 4.41-D) and shuffled (Fig. 4.41-F) card-decks show the re-establishment of systematic classificatory elements. The boundaries drawn in these two latter maps contain stations which belong in the same classes, which are not widely scattered, and which do not intermingle with stations of other classes as in the original arrays. The regions demarcated may be described by relatively simple low-degree curves. Comparison of the two sets of maps shows that although the cores in the two lower pairs of maps (Figs. 4.41-C and -E) are smaller than the original Koppen cores (Fig. 4.41-A), they are no less homogeneous. Discriminant analysis picks out the sound elements in the classification even when hidden by prevalent noise (Casetti, 1964, p. 49).

(d) Regionalization of dyadic data

The taxonomic procedures described above have assumed sets of data available for a series of single geographical locations (e.g. climatic stations, counties, etc.). Where data is available for pairs of regions (*dyads*) a wide

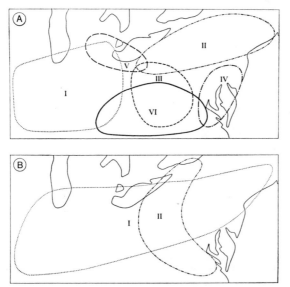

Fig. 4.42. Regionalization of connectivity matrices for (*A*) inter-state highway network (*B*) inter-city jet flight network, north-eastern United States. Source: Hebert, 1966.

range of procedures for grouping into regional networks exists. In Chapter 1 transport networks were reduced to a dyadic form in which pairs of nodes were described in a binary matrix. Fig. 1.21 shows the results of Garrison's and Marble's (1962) study of a binary matrix of airline connections for Venezuela and Fig. 4.42 shows the regions developed by

Hebert (1966) for binary matrices of transport connections in the United States manufacturing belt. The contrast between nodal regions for the interstate highway network (regions I to V in Fig. 4.42-A) and the inter-city jet flight links (regions I and II in Fig. 4.42-B) is well marked.

Where the dyadic matrix contains flow data (e.g. number of telephone

Fig. 4.43. Factor analysis of 63 commodities moved between 36 trade block areas of India: the first four dyadic factors are mapped. Source: Berry, 1966, pp. 240–7.

calls, volume of trade, etc.) for pairs of places factor analytic techniques may be preceded by elementary linkage analysis. Nystuen and Dacey (1961, pp. 38–42) describe methods of reducing a flow matrix to a series of graphs. Here order is established in terms of the aggregate values of incoming or outgoing flows, and hierarchic relations between nodes are determined by the maximum flow to a higher-order node. A more complete digest of the information within the matrix is given by the full factor-analytic procedures discussed above. Using procedures parallel to those discussed for Steiner's study, flows can be broken down into major components and the components used as axes for the mapping of dyads in factor space. Grouping procedures within that space lead to regional aggregations (*nodal* regions) in the same way that *formal* regions were derived (Chap. 4.II(2a)). Thus Goddard (1968) was able to develop a nodal-regional structure for central London from factor analysis of dyadic data on taxi flows. The fullest account of dyadic factor analysis is given by Berry (1966, pp. 189–237) as part of a massive analysis of commodity flows for sixty-three commodities moving between thirty-six trade blocks within India (Fig. 4.43). The 1260 × 63 matrix was reduced to a twelve-factor orthogonal structure, and a methodology for linking this simplified flow structure to non-flow changes was developed. The combination of regional taxonomies for both locational and dyadic data suggests a direct link between the dual problems of describing and interpreting regional networks.

It is clear from this chapter that geographers studying route-location and geographers studying regionalization have been tilling directly adjoining research fields. While the discovery of common elements may serve only the academic purpose of tidying some loose threads in our work, it is hopefully possible that useful borrowings of concepts may occur in some future work.

Part Three: Structural Change

Our knowledge of the whole complex phenomena of growth is so scanty that it may seem rash to advance even these tentative suggestions.
(D'ARCY WENTWORTH THOMPSON,
On growth and form, *1917, p. 282.*)

Chapter five   Growth and Transformation

The growth and decay of transport networks over time has been implicit
in much of the four previous chapters and a recurrent theme in geographic research (e.g. Godlund, 1952). The direct adjustments between
changes in flow and channel patterns have already been noted (Chap.
3.II(4)), while the cross-sectional relationships between indices of network
structure and measures of network environment (e.g. Chap. 2.II) have
clear implications for their relations over time. In this chapter the sequential changes through which networks appear to evolve are traced in more
detail and the models put forward to explain or replicate this growth are
examined. It will be clear that the models currently available are somewhat fragmentary and that any general theory of network growth lies
in future research. However, a final section looks at the possibilities of
transformation between major topologic classes and their implications for
a more integrated approach.

I. PATTERNS OF SPATIAL EVOLUTION

The three main categories of spatial sequence identified in network growth
are here identified as *node-connecting* sequences, *space-filling* sequences,
and *space-partitioning* sequences. In the first categories the vertices of
the network to be connected (in growth models) or disconnected (in decay
models) are identifiable as discrete nodes; in the second and third categories the nodes are samples drawn from a continuous spatial environment.

1. Node-connecting sequences
A number of attempts have been made to place the observed sequence
of transport network changes into some kind of discrete 'stage' model.
Although all stage models run the risk of partitioning what may be a
continuous process into separate segments, the discrete nature of transport
investment makes this objection less serious. As Lachene (1965, p. 184)

points out, transport links are built with average and minimum capacities which may be in excess of potential traffic at the time of building. The low marginal costs of long-term transport once the link is built and its long expected life give an accumulator effect to the network. Transport links therefore not only serve the *short-term* function of providing necessary flows between nodes, but a *long-term* function of shaping the relative growth of the nodes themselves. The close interactions between transport links and urban growth have been traced in detail by Gauthier (1967) for the São Paulo network and Kissling (1967) for the Canadian Maritime Provinces. A more general model has been proposed by Lachene (1965)

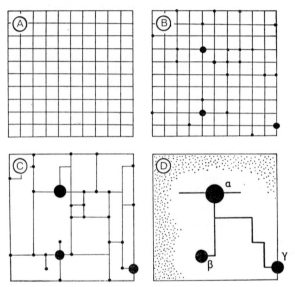

Fig. 5.1. Stages in the growth of hypothetical regional transport networks. Source: Lachene, 1965.

and by Taaffe, Morrill and Gould (1963), and their stages are used as the basis for a combined four-stage model (Fig. 5.1; Fig. 5.2).

(*i*) *Initial stage.* The basic difference between the Lachene and Taaffe models lies in the environment postulated rather than in the mechanisms of network growth. Lachene conceives a 'mid-continental' situation with a sparsely populated territory and fairly uniform activity, while Taaffe is concerned with a coastal situation in which colonial exploitation proceeds inland from the coastal baseline. Gould's study of the growth of transport in Ghana (Gould, 1960) with parallel studies in Nigeria, East Africa, Brazil and Malaya formed an empirical base for the stages proposed.

The first stages of the two models are shown in Figs. 5.1-A and 5.2-A.

Since the Lachene territory is rather uniformly developed at a low level a great number of low-investment links (e.g. dirt roads) are built and little difference in potential is envisaged; the square grid represents this rather uniform network, although there are grounds for considering a triangular or hexagonal grid more representative (see Chap. 3.I). The first phase of the Taaffe model consists of a scatter of small ports and trading posts along the coast of the hypothetical region which is being developed

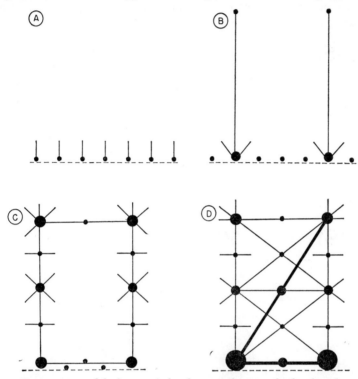

Fig. 5.2. Four-stage model of network development in an underdeveloped country. Source: Taaffe, Morrill, and Gould, 1963, p. 504.

(Fig. 5.2-A). Each small port has a small inland trading field, but there is little contact along the coast except through occasional fishing boats and irregular traders. This phase is identified in Nigeria and Ghana as running from the fifteenth century to the end of the nineteenth century with groups of indigenous peoples clustered around a European trading station.

(*ii*) *Local node differentiation.* The second stage is shown in the Lachene model (Fig. 5.1-B) by the emergence of urban centres at intersection points, and in the Taaffe model (Fig. 5.2-B) by emergence of a few major ports with inland lines of penetration and the growth of inland trading

centres at the terminals. With the growth of coastal ports the local hinterland also expands and diagonal routes begin to focus on the growing ports.
Again the second phase in the Taaffe model is identified in Ghana and
Nigeria with the growth of inland trunk routes. These appear to have been
built inland for three major reasons (1) to connect a coastal administrative
centre politically and militarily with its sphere of authority up-country;
for example in Ghana there was the desire to reach Kumasi, capital of
the rebellious Ashanti; (2) to tap exploitable mineral resources, such as

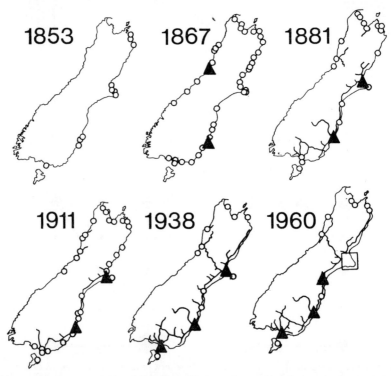

Fig. 5.3. Rail network development and changing status of seaports in the South
Island, New Zealand, 1853–1960. Source: Rimmer, 1967, pp. 21–7.

the Enugu coalfields in Nigeria; and (3) to tap areas of potential agricultural export production, such as the cocoa areas north of Accra. Although
each motive has played its part, the role of mineral exploitation has been
a critical one in African railway building, and examples from Uganda
(Kasese copper line), the Cameroons (Garoua manganese line) and
Mauritania (Fort Gourard iron-ore line) suggest this phase has not yet
ended. The process of port selection and differential growth has been
traced in detail by Pred (1966) for United States atlantic seaboard and
by Rimmer (1967) for New Zealand ports (Fig. 5.3).

(*iii*) *Node interconnection*. The growth of urban centres in the Lachene model and the evolution of transport technology make possible the creation of transport networks of a new type (e.g. railways) (Fig. 5.1-c). Since however the capacity of the new mode is greater and the capital expenditure higher than in the original network (Fig. 5.1-A), it is set up only between a limited number of nodes. The selection of these nodes has been reviewed by Black (1967) in terms of a discriminant analysis. His general problem was to disentangle the complex of factors that controlled the extension of the railway network into southern Maine, in the north-eastern United States, during the 1840–50 decade. The problem was translated into an operational form as: Why were some potential links between centres built while others were not? Centres were defined from the listing of places in the 1840 census, and all centres above a threshold figure of around two thousand inhabitants were included as potential nodes. With seventy-two centres defined in this manner (see Fig. 5.4-A) the number of potential railway links was clearly $\dfrac{72\,(72-1)}{2}$ or 2,556. Since the complete range of links was assumed to be non-planar (Chap. 1.I) this number was unrealistically high and the number of potential links was re-defined in planar terms. Using a template with 60° sectors a pattern of 196 nearest-neighbour links was built up around the seventy-two nodes. Of these potential links only twenty-seven had actually been constructed by 1850.

Seven major hypotheses were put forward to account for the selection of links constructed. It was argued that the probability of connection increases with: (I) nearness to the point at which the network began (i.e. the city of Portsmouth (*a*) in nearby New Hampshire); (II) shortness of the link; (III) importance of the centres connected (defined on a P_1P_2 basis using 1840 population census data); (IV) potential local interaction (given from a combination of the previous two hypotheses into a gravity-model formulation); (V) absence of intervening opportunity; (VI) Potential regional interaction (combining hypotheses I and IV); and (VII) closeness of link orientation with regional orientation. The regional orientation defined by least-squares procedures is shown in Fig. 5.4-A. All operational definitions involved simplifying assumptions about southern Maine as an isotropic plain.

Discriminant analysis (see Chap. 4.II(2c)) was used as a sorting procedure to separate the characteristics of the built-links (1) from the unbuilt links (0). Table 5.1 shows the values of the single discriminant functions (median point between the means) for each of the seven hypotheses. For example under hypotheses I the discriminant function in terms of distance from network origin was computed as 76·7 miles. The use of this function resulted in a misclassification of twenty-eight per cent of the links; i.e. links were 'built' under this hypothesis that were not in fact constructed, and vice versa. Note that six of the seven hypotheses were statistically significant at the ninety-nine per cent confidence level, but that the level of misclassification ranged from sixteen to thirty-six per cent. The three best hypotheses were combined into a multiple-regression equation:

$$Y = 0{\cdot}66 + 0{\cdot}0036\ X_1 + 0{\cdot}10\ X_{IV} - 0{\cdot}0043\ X_{VII}$$

involving the three leading hypotheses (I, IV and VII). Values of Y were computed for both the actual links ($Y = 0{\cdot}611$) and potential links

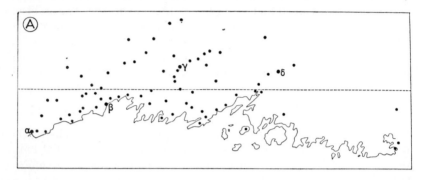

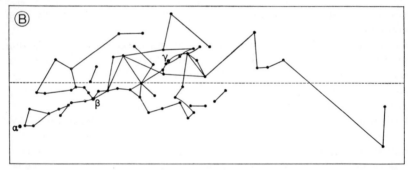

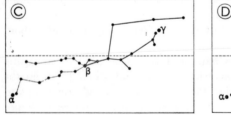
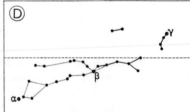

Fig. 5.4. Discriminant analysis of the growth of the railway network in southern Maine, U.S.A. 1840–50. (*A*) Distribution of potential nodes. Dashed line shows regional orientation. (*B*) Sample of potential links. (*C*) Links actually built. (*D*) Links predicted by discriminant model. Source: Black, 1967.

Table 5.1. Summary statistics for the Maine railway network

Variable:	Mean value		Correlation coefficient	Discriminant score	Errors
	Links constructed	Other possible links			
Distance from Portsmouth	58·333	94·981	−0·400*	76·657	22
Link length	8·592	17·132	−0·354*	12·862	29
Product of end node populations × 1/1000	15·055	8·158	0·362*	11·606	28
Potential flow function	2·795	0·708	0·491*	1·751	21
Intervening opportunity	0·037	0·189	−0·208	0·112	—
Gravity Model function	0·053	0·013	0·442*	0·033	13
Directional deviation	27·333	43·906	−0·302*	35·619	28

Source: Black, 1967.
    * Statistically significant at the 99 per cent confidence level.

($\Upsilon = 0.198$) to give a multiple discriminant function at the mid-point of ($\Upsilon = 0.404$. Computation of Υ values for each link using the above equation resulted in a very considerable improvement (only eleven per cent error). Extension of multiple discriminant analysis suggests that considerable progress in link extension models is possible by comparing discriminant functions over both space (e.g. regional variations in functions) and time.

In the Taaffe model (Fig. 5.2-c) the third phase is marked by the growth of feeder routes and the emergence of lateral interconnections. The feeder-routes growth is accompanied by continued growth of the main sea-coast terminals in a spiral of trade-capture and expansion. Intermediate centres grow up between the coastal and interior terminals. Taaffe (Taaffe *et al.*, 1963, pp. 511–14) shows a series of maps of road development in Ghana and Nigeria in the period since 1920 in order to suggest the lateral connections of earlier disconnected lines of penetration and exploitation (Fig. 5·5).

(*iv*) *Regional node differentiation.* The creation of the improved network in the third phase of the Lachene model improves the potential of all points in the territory, but particularly that of the two major inland centres (α and β) and the major external bridgehead (γ). The dispersion in potentials leads to further rapid urban growth so that when further networks are built (Fig. 5.1-D) the only towns with sufficient exchange to justify interconnection are the three major centres. Setting up the network to connect other smaller centres is not justified by traffic volumes and the concentration of activities in a few centres continues. 'The development of the transport network accentuates concentration for it is economically possible between a decreasing number of points; in the long run the optimum and minimum capacities of transport have increased

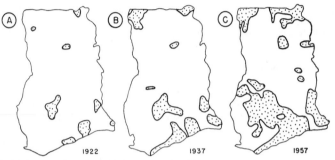

Fig. 5.5. Diffusion of zone of high-density road network in Ghana. Source: Taaffe, Morrill, and Gould, 1963, p. 512.

much more rapidly than average traffic between towns' (Lachene, 1965, p. 195). Similarly the fourth phase in the Taaffe model repeats the process of linkage and concentration and shows the emergency of 'high-priority' linkages between the most important centres (indicated by a broader line in Fig. 5.2-D). The best paved roads, the heaviest rail schedules and airline connections will follow these 'main street' links between the three major centres. The heavy traffic in the 'triangle' of southern Ghana suggests such a development here.

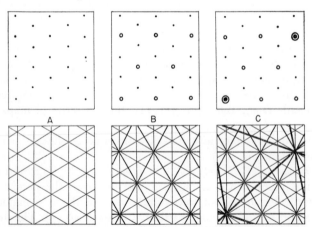

Fig. 5.6. Network development by route substitution between successively higher-order centres in a Löschian landscape. Source: Haggett, 1965, p. 82.

Haggett (1965, p. 82) has attempted to fuse this final stage of link substitution between higher-order centres with a theoretical Löschian landscape (Lösch, 1954, p. 127) which is shown in Fig. 5.6. If we begin with an idealized Löschian landscape in which desire lines connect each settlement with the next in a network of intersecting pathways, we obtain the type of pattern still discernible on maps of rural areas in tropical Africa. In the second stage (Fig. 5.6-B) the economic level has been raised to give a longer interaction-distance, to halve the number of major centres and to leave a series of by-passed smaller centres connected by smaller routes. In the third stage (Fig. 5.6-c) the interaction has been raised still

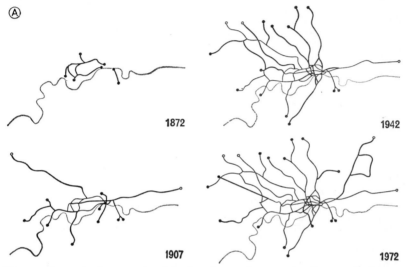

Ⓐ

1872

1942

1907

1972

Fig. 5.7. Stages in the growth of the London underground rail network. Source: Cope, 1967.

further with a new set of optimum routes, a new and smaller set of major centres, and a larger set of by-passed centres. Empirical examples of this process can be identified in the emerging pattern of both the new motorway network and the revised rail network for the United Kingdom (Ministry of Transport, 1963, pp. 71–136; British Railways Board, 1963). Although bypassing of small centres forms the most dramatic type of readjustment in this model, there is considerable evidence of a continual small-scale process of 'route straightening' between major nodes over time. Fleischer (1963) has studied this process on two major highways extending north and south from Portland, Oregon, U.S.A., over the period 1935 to 1960. The shortest-path between Portland and Seattle has been shortened from 187 to 162 miles (straight-line distance = 133 miles), and from Portland to the Grants Pass from 279 to 248 miles (straight-line distance = 212 miles) over this twenty-five year period.

Study of the historical sequence of scheduled air routes over the North
Atlantic for a similar period would reveal more dramatic realignments to
the shortest-path link.

The impact of changes in the network on its structural characteristics
has been studied by Cope (1967). The growth of the London underground

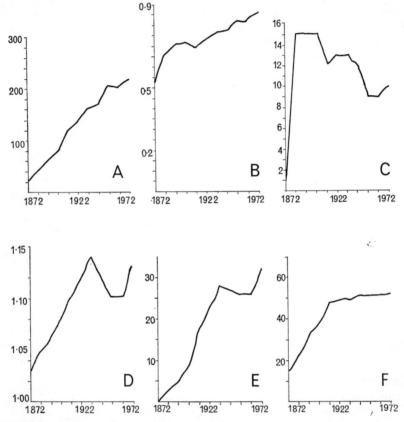

Fig. 5.8. Changing structural parameters of the London network (Fig. 5.7),
1872–1972. *A* Total network length, miles. *B* Average edge length, miles.
C Alpha index. *D* Beta index. *E* Cyclomatic number. *F* Diameter. For explanation of indices see Chap. 1.III.1. Source: Cope, 1967.

system (Fig. 5.7) over the period from its inception in 1863 to the present
was grouped into three stages: (1) initial expansion (to 1912), (2) consolidation (1912–32), and (3) rationalization (from 1932). Fig. 5.8 shows
the impact of change over the eleven decades in terms of a series of
structural parameters. Between 1872 and 1932 the total length of the
system extended from twenty to over two hundred miles (Fig. 5.8-A).
This tenfold increase was accompanied by much less spectacular changes

in its topological properties; over the same period the average edge length (Fig. 5.8-B) increased by about one half from 0·51 miles to 0·76 miles, while the connectivity ratios moved from fifteen to ten per cent redundancy as measured by the Alpha index (Fig. 5.8-C), and from 1·03 to only 1·13 on the Beta index (Fig. 5.8-D). Not all indices showed regular change over the period. Values for the cyclomatic number (Fig. 5.8-E) reflect the rationalization of the network in decades VIII and IX while the diameter (Fig. 5.8-F) shows no major changes since 1912. Cope (1967) has further distinguished the vectoral growth of a network outwards from its origin from the angular growth which represents a rotational component around the central point. By plotting the growth of the London underground network on a polar co-ordinate projection, he was able to relate the two moments to the growth of the system. The existing system

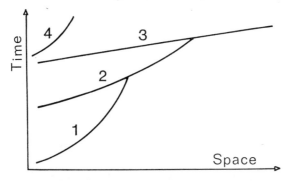

Fig. 5.9. Sequence of transport innovations with 'technological' unconformities in peripheral areas.

shows stronger rotational components near the centre with high vectoral components near the periphery; study of its growth suggested that 'high angular moments distinguished the early decades but [that] the vectoral property increases as the system becomes more mature' (Cope, 1967, p. 81).

The growth, consolidation, and selective abandonment of a given transport mode occurs within a longer term context of overall network growth. A number of workers have seen the growth of networks in terms of a succession of spatial diffusions, in which technical innovation is propagated through a regional system through a series of waves (Gould, 1960). By varying the rate of innovation and the number of innovating centres we may readily see both the succession of modal shifts near the centre of the system and the presence of modal discontinuities near the periphery (e.g. the 'mule/hovercraft' discontinuity) (Fig. 5.9). Smith's (1964) study of the changes in the road/railway network in southern New South Wales provides a clear description of this process of peripheral growth and central up-grading in a rapidly evolving economy.

With transport networks the pattern of growth is closely related to the political realities at the time of construction. Meinig (1962) has examined the historical geography of two rail nets: (i) a wholly state-directed enterprise in South Australia and (ii) one built and operated by several private companies in the north-western United States. They were chosen for comparative study because they were built and developed at roughly the same time, were both in wheat-growing regions, and were both designed largely to move export grain from the farming districts to tidewater ports. Meinig finds a number of common features in the railway net developed in the two areas. Both extended at about the same pace, the one in response to political pressure and concepts of public service and the other to profit possibilities. Both were complicated by changes in the general orientation of trade to different ports, and both were affected by the influence of local communities on routing. In both too, the number of alternative possible routes always outweighed the routes that could be built and was decided in an intimate context. Differences between the state and private networks are found, however, to be the more striking. Meinig places first the contrast in (a) duplicate routes and (b) duplicate services. In the Columbia basin the links between inland exporting centres and tidewater ports are commonly duplicated, and the exporter is faced with a choice of competitive services to different tidewater ports. There is complete absence of such alternatives in South Australia. Moreover, the hinterlands of individual lines in South Australia remain stable in contrast to the constant piracy and 'invasion' of territories in the Columbia basin. Such fluidity in the privately owned network is suggested by Meinig as a cause of the rapid conversion and subsequent development of the Columbia basin system on a uniform gauge, while the South Australian system retained its relatively watertight hinterlands each served by its own gauge. With the growth of government regulation in the United States the original contrasts in the patterns are now fading slightly (Haggett, 1965, pp. 69–70).

2. Space-filling sequences

Where networks undergo rapid change, the establishment of the sequence of evolution is a rather straightforward problem of regular observation and meaningful partition of the process; where however the rates of evolution are very slow the sequence may have to be deduced from either the present stage of the pattern or from auxiliary evidence. For stream channels distinction is therefore drawn between *observed* and *inferred* sequences.

(a) Observed patterns

For stream-channel networks the direct observation of *short-term* networks is restricted to a few environments, for example, of the gullying of artificially-produced surfaces (Schumm, 1956), of lake margins suddenly exposed by earthquakes (Morisawa, 1964), and of glacier surfaces eroded by summer meltwater (Holmes, 1955). Schumm (1956, pp. 620–2) noted the headward development of a number of small basins on the Perth Amboy industrial sand and clay dump between 1948 and 1952 (Fig. 5.10), and proposed seven ways in which the drainage networks have changed as the systems cut back into the artificial terrace. Apparently the channels subject to the fastest headward cutting were those either fed by the maximum water supply, in that they headed into swales on the terrace surface, or which happened to be located on the weakest material.

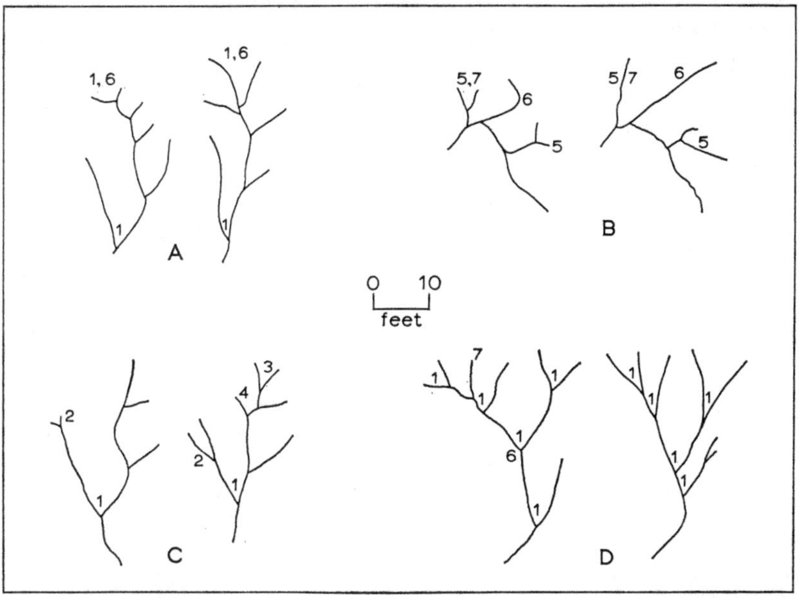

Fig. 5.10. Drainage-pattern changes in selected basins between 1948 and 1952 at Perth Amboy. Basins *A, C,* and *D* are steep gradient streams. Basin *B* is a youthful basin on the upper surface of the terrace. Drainage changes are indicated by numbers on the figures: 1. Angle of junction change. 2. Migration of junction. 3. Bifurcation. 4. Addition of tributary. 5. Angle of bifurcation change. 6. Channel straightening. 7. Elimination of tributary. Source: Schumm, 1956.

Schumm (Fig. 5.11) proposed a schematic drainage evolution for a typical Perth Amboy basin, involving progressive headward cutting, the addition of tributaries as new valley-side slopes developed where L_g exceeded X_c, and the progressive straightening of channels and decrease of junction angles through time.

   The earthquake of 1959 exposed a strip eight to twenty feet wide along the sandy and silty shores of Hebgen Lake, Montana, on which Morisawa (1964 and 1968) has observed the development of small drainage systems between 1960 and 1961. Integration of gullies into systematic networks appeared to take place very rapidly, and integration and expansion appeared to take place simultaneously. Evidence of equilibrium within the growing systems was given by their accordance with the law of stream numbers from their earliest stages, but they required somewhat longer to behave in accordance with the laws of stream lengths and gradients. This was because, apparently, of the small initial relief of the basins and of the rapid initial development of the tributaries favoured by the regional slope of the exposed surface. As downcutting progressed,

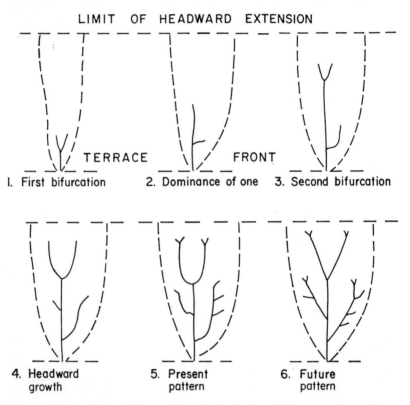

Fig. 5.11. Suggested evolution of the Perth Amboy drainage pattern. Source: Schumm, 1956.

and relief increased, the first-order tributaries extended and ramified, whereas during the later reduction of relief the total number of streams may have been decreased by integration and junction migration. Between 1960 and 1961 the number of first-order streams decreased, whereas their mean length increased (neither of these changes was statistically significant), and the junction angles did not change. Morisawa (1964, p. 352) concluded that the networks were expanding by simultaneous integration and extension, during which new tributaries were developed, old ones extended, some tributaries eliminated, some lengthened and others shortened. During this whole evolution, however, some overall equilibrium seems to have been maintained, as exemplified by the maintenance of constant bifurcation, length and slope ratios, and it is possible that headward ramification was matched by downstream integration. This

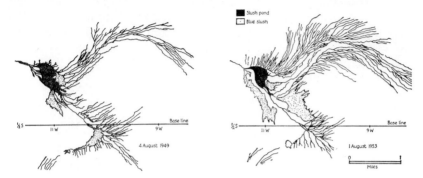

Fig. 5.12. Drainage pattern of meltwater streams on glacier, Shaw's Valley, Greenland Ice Cap. (*A*) August 4, 1949. (*B*) August 1, 1953. Source: Holmes, 1955.

conclusion is supported by Carter and Chorley (1961) who noted a similar equilibrium in a small stream system which is cutting back into a terrace of the Farmington River, Connecticut. Fig. 5.12 shows the mapped development of meltwater streams on the surface of a glacier in south-west Greenland during the period 1949–53, showing an increase in total channel length of one hundred per cent in the four years (Holmes, 1955; Leopold, Wolman, and Miller, 1964, pp. 426–7).

(b) Inferred patterns

It is natural that ideas regarding the *long-term* evolution of stream-channel networks should have been associated with those having to do with the evolution of the whole assemblage of landforms, and in particular with the cycle of erosion. A sequence of origin was implied in the early classification of streams as 'subsequent' (Jukes, 1862, p. 400) and 'consequent' (Powell, 1875, pp. 165–6), and a large amount of Davis' cyclic synthesis was concerned with the fusion and elaboration of theoretical notions relating to the initiation, evolution and equilibrium of drainage networks. Much of Davis' nomenclature implied a sequence of stream evolution on an exposed surface, together with the idea that drainage density increased rapidly to early maturity and then slowly decreased, so associating stream network evolution closely with the cycle of erosion. Even the hydraulic engineer Gravelius (1914) attempted to develop his ordering system as a reflection of a theoretical sequence of stream development. Some time later Glock (1931) postulated a 'dynamic cycle' of six stages in the development of a stream system on a new land surface, which he deduced from maps, involving: (1) initiation; (2) elongation by headward growth; (3) elaboration of the skeletal form by the addition of fingertip tributaries; (4) maximum extension; (5) early integration, decreasing the drainage density by lateral abstraction and encroachment of one stream on another, by absorption due to the loss of surface runoff as the water table falls with increasing landscape dissection, and by adjustments in the course of the main stream as it tends towards the

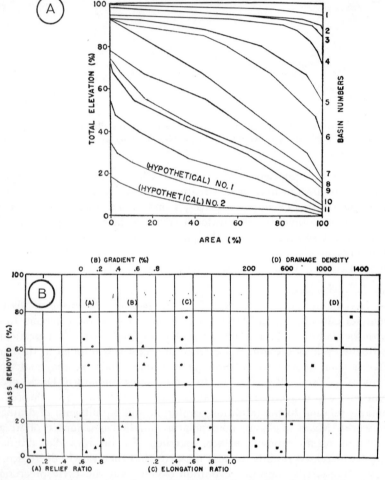

Fig. 5.13. (*A*) Sequence of second-order Perth Amboy hypsometric curves. (*B*) Per cent mass removed plotted against drainage density and three other morphometric parameters. Numbers increase from youthful to mature basins.
Source: Schumm, 1956.

shortest distance to baselevel; and (6) later integration. Johnson (1933) criticized this scheme, firstly because of the inadequacy of Glock's maps in representing the true drainage network, and secondly because drainage density is not a simple function of time and differs under different climatic and lithologic conditions. Thus the placing of a number of maps into an assumed time sequence could be no guarantee that a correct inference would be reached as to the way in which a drainage net develops through time.

In an attempt to place different networks in an assumed time sequence, Schumm (1956, pp. 615–17) assumed that the percentage of total mass removed below the original Perth Amboy terrace surface could be employed as a relative indicator of the passage of time. He therefore selected eleven second-order basins of approximately the same area and placed them in order of their eroded mass, covering a total lowering of some forty feet (i.e. one hundred per cent elevation)|(Fig. 5.13-A). The percentage of mass removed was then plotted against drainage density (Fig. 5.13-B). Schumm deduced from this that the rapid increase of relative relief until about twenty-five per cent of the basin mass has been removed is possibly associated with a rapid initial increase of drainage density (as the lengthening valley-side slopes allow the initiation of new channels), after which there is a less rapid increase (when the relative relief remains almost constant), partly accomplished by headward expansion which continues late into the erosion cycle.

Another inferential study by Ruhe (1952) examined the drainage networks of differing ages which have developed on the glacial till sheets in Iowa (Leopold, Wolman and Miller, 1964, pp. 423–5) (Fig. 5.14). The difference was particularly well marked between the older Iowan and Tazewell, on the one hand, and the younger Cary and Mankato, on the other; but it was proposed that the drainage density had increased through time (as did the number of tributaries), rapidly for about the first 20,000 years and then more slowly. One of the shortcomings of this study is that it failed to take fully into account the effect of other factors, notably differences in drift composition, which might affect drainage density independent of time.

The most elaborate inferential investigation into the possible manner of evolution of stream networks was made by Melton (1958A). Taking the 156 basins which formed the basis of his earlier study (Melton, 1957), he assumed that they must have evolved in some way during time and therefore must represent, at this time instant, a considerable range of ages, so that his large basin sample must have included representatives of many stages of development. However, there was generally apparent none of the systematic variations which one might expect if absolute age was associated with variations in drainage density (D) (Leopold, Wolman and Miller, 1964, pp. 422–3), and Melton assumed that the wide range of

NAG—K

climatic soil and geological environments represented must also affect the range of network forms observed. It is apparent that drainage density ($=$ total length of streams (L)/total basin area (A)) is distinct from stream frequency ($F=$ number of streams of all Strahler orders $(\Sigma N_s)/A$). For areas which are large with respect to first-order basins, D and F showed no systematic variations with A, but data for the 156 mature basins (i.e. smooth valley-side slopes continue up to the divides) showed a systematic relationship between D and F, with a coefficient of correlation of $+0.97$, such that for all basins it was possible to propose that F/D^2 (the relative channel density) is a dimensionless constant (Melton, 1958A, pp. 35–6). The relative channel density therefore represents 'a basic law of behaviour of planimetric elements of maturely developed drainage basins' (Melton, 1958A, p. 36), and is a dimensionless measure

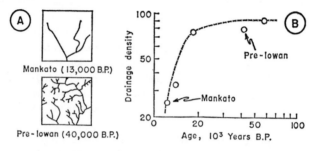

Fig. 5.14. Development of dendritic system over time: stream patterns on part of the Des Moines lobe, central U.S.A. Source: Ruhe, 1952.

of the completeness with which the channel net fills the basin outline for a given number of channel segments but for any value of A (Melton, 1958B, pp. 446–7 and 1958A, p. 38). Basins which are geometrically similar may be expected to have the same value of F/D^2 and to lie on a line parallel to that shown in Fig. 5.15. Melton (1958A, pp. 38–40) was concerned with the sources of scatter of the points and as to whether F/D^2 changes systematically with other basin parameters. Apart from basin circularity differences, which appeared to have no effect, and map errors, which were unknown, there were two basin variables which led to systematic departures from the observed relationship between F and D: (1) Departures from a steady state of complete drainage network adjustment to the basin, when all first-order streams and divides are in competition with each other (i.e. another way of defining 'maturity'). In other words, if a network is still extending headward to increase the drainage density without changing the stream frequency, the point representing the basin will move to the right, parallel to the D-axis. It is possible that all the basins employed in the analysis were not fully mature. (2) The variation of F/D^2 with valley-side slope angle (θ). The effect of these is that F/D^2 correlates significantly (negatively) with the basin ruggedness number $(H = $ total basin relief $(R)/D = $ a dimensionless number). In rugged areas F/D^2 is predictably small and the basin outline is filled with relatively longer channels than those in basins with the same value of N in less rugged areas—probably because the fingertip tributaries can extend close to the divides in rugged areas (Melton, 1958B, p. 451). However,

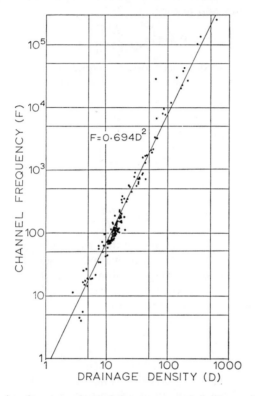

$$F = 0.694D^2$$

Fig. 5.15. Relation between channel frequency and drainage density. Source: Melton, 1958A.

ignoring for the moment these minor effects of relief and valley-side slope, the function $F = 0.694 \ D^2$ represents a drainage network growth model, which shows that as drainage density increases within a constant area, so does stream frequency (Melton, 1958A, p. 50) (Fig. 5.15).

Carrying the analysis further, Melton (1958A, pp. 40–8) used his data to analyse F/D^2 in terms of the four independent variables which (ignoring the hypsometric integral and any random effects) he assumed to determine the total geometric properties of mature drainage basins;

namely log $\sqrt{A}$, log L, log R and log P (i.e. basin perimeter). It was found that the relationship:

(1)
$$\frac{F}{D^2} = + 0\cdot8147 \; \frac{(\sqrt{A})^{0\cdot5}}{L^{0\cdot25}R^{0\cdot25}}$$

explains approximately 39·7 per cent of the observed variation in log F/D^2. From this it was possible to specify N for mature basins in terms of independent variables which are themselves functions of climate, soil conditions, geology and history:

(2)
$$N = 0\cdot8147 \; \frac{L^{1\cdot75}}{(\sqrt{A})^{1\cdot5}R^{0\cdot25}}$$

Considering a first-order basin of area equal to unity (i.e. one square mile), then from the equation relating F and D^2, and substituting for F and D^2:

(3)
$$N = 0\cdot694 \; L^2$$

Inserting A = 1 in equation (2):

(4)
$$N = 0\cdot8147 \; \frac{L^{1\cdot75}}{R^{0\cdot25}}$$

and combining (3) and (4):

(5)
$$R = \frac{1\cdot899}{L}$$

The implications of this analysis are that as R increases (through uplift or fall of baselevel), the total stream length decreases (perhaps by channel elimination due to mass wasting) (Melton, 1958A, p. 53); and that where relief is progressively reduced, the total stream length increases—although the mechanism for this remains obscure.

(c) *Theoretical patterns*

By far the most important theoretical model of drainage evolution as a space-filling process was the one in which Horton (1945) employed his infiltration theory of surface runoff. Horton assumed that surface channels are exclusively the work of surface runoff resulting from overland flow, the depth of which increases linearly away from the divide until at a critical distance (X_c) it is deep enough to entrain surface material and initiate rills which cross-grade to form streams, excavation by which produces new (valley-side) slopes on which new rills systems develop, and so on. An initial simplification was that a given drainage system could be expected to evolve within a diamond-shaped potential catchment area (Fig. 5.16). If the maximum storms produce a runoff intensity giving a value of X_c (*bg* in Fig. 5.16-A) which is less than the maximum available length of overland flow ($L_g = bo$), then rills develop in *ofgh*. Because length of flow is greatest along *go*, the maximum erosion intensity there will allow the development of a master rill which will be incised and develop two valley-side slopes each having a maximum length of overland flow of $\frac{1}{2}L_g$. Because the gradient of these slopes is greater than that of the original surface, X_c is now reduced (Fig. 5.16-B), and master tributaries develop by cross-grading and micro-piracy from the valley-side rill systems. The maximum available length of overland flow is now

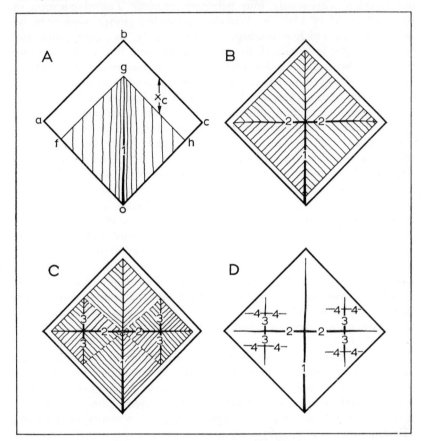

Fig. 5.16. Development of a drainage net in a stream basin. Source: Horton, 1945.

reduced to $\frac{1}{4}L_g$, but if runoff of sufficient intensity occurs to produce a value of X_c which is less than this, then rills will develop on each of the new valley-sides from which first a master rill and then a tributary will develop (Fig. 5.16-c). This addition of tributaries will continue (Fig. 5.16-D) until the drainage density is great enough to reduce the maximum available length of overland flow to a value which is so small that even the most intense runoff cannot produce a small enough value of X_c to initiate rills on the slopes. Thus the whole of Horton's concept of drainage evolution leads to a more-or-less equilibrium value of drainage density which is controlled by the intensity of overland flow.

A number of difficulties are immediately apparent with respect to such a model. Firstly, if one does not postulate a diamond-shaped catchment area which is potentially available for occupancy by the stream system, how is it possible to determine which and how many of the rills will develop sequentially into tributary streams? The elaboration of Horton's model by Leopold, Wolman and Miller (1964) encounters the same problem which is inadequately resolved by the assumption that certain rills are long enough to command sufficient flow to develop into tributaries. Secondly, it is difficult to make the necessary leap in scale from Horton's cross-grading rill systems to the large-scale drainage systems observed in nature. Thirdly, it is difficult to imagine how a pristine surface could come into existence upon which a Hortonian drainage system could develop simultaneously from scratch. This third problem was circumvented by Horton (1945, pp. 342–6) when he adapted his theoretical model to drainage development on a new land surface which was being progressively exposed by a recession of the shoreline.

Clearly the evolution of drainage networks according to the Horton model is governed by two principles (Horton, 1945, p. 345): (1) Streams develop successively at points where the length of available overland flow becomes greater than X_c. (2) Competition results in the survival of those streams which have the earliest start, or the greatest length of overland flow, or both, and which are able to absorb their competitors by cross-grading. Despite the previous criticisms of this model, Horton's ideas of network evolution seem applicable to stream channels on unvegetated surfaces with low values of infiltration capacity and little surface soil development, and are particularly so to the small-scale networks of clay and shale badlands (Schumm, 1956). Where there is a well-developed soil cover, however, and throughflow is an important hydrologic component, surface runoff is generally restricted to areas near river channels, zones of convergence and concavities, so that the depth of overland flow is not directly related to distance from the divides. This being so, the concept of the X_c distance, so central to the Horton model, breaks down (Kirkby and Chorley, 1967).

3. Space-partitioning sequences

Steady-state models of spatial partitions imply a state of equilibrium without stating the stages by which this equilibrium is achieved, i.e. they represent an instantaneously-achieved situation. It has already been noted (Chap. 1.IV) that the characteristics of the 'steady state' models of cells (i.e. the trihedal angles and the tendency to equal-area cells) have been difficult to find in administrative partitions. Some of these difficulties may be overcome, as Haggett (1965, pp. 53–5) suggests, by suitable transformtions of the base-plane (e.g. the substitution of income-space planes for distance planes in trade-cell studies), but there seem to be more fundamental problems. Steady-state models are often unhelpful since cellular networks are (i) usually formed by an iterative step-by-step process, and (ii) the early boundaries are not necessarily adjusted to late-stage conditions. Indeed the relaxation times of many geographical systems are so slow that early boundaries, once established, may themselves form an important reference line with respect to which later boundaries are formed.

Some of the most useful ideas on this iterative cell-division process can be generalized from detailed work on the physics of fracture (well summarized by Irwin, in Flugge, 1958, pp. 551–90). Here the argument rests on two points: (1) If we take a plane sheet of material then we may expect the flaws (points, lines or areas of weakness) to follow an assymetric frequency distribution, the largest flaws being least common and smaller flaws being most common. If we further assume the pattern of flaws is randomly distributed, then we may expect on average, the larger flaws to be further apart and the smaller flaws closer together. (2) Each fracture of this idealized surface will create around it a buffer zone of tensional relief, such that new fractures will be unlikely to form within a specific distance. This is because the fracture reduces the stress in its vicinity and increases the strength of the remaining, unfractured, surface by removing the points of greatest weakness.

The general relationships between fracture, stress relief and spacing may be given by the equation:

$$l = 0 \cdot 7 \left(\frac{N}{\tau_0}\right)^m, \tau_0 < N$$

where l is the average crack spacing for a given applied tension, τ_0, in terms of a flaw distribution parameter m; and N is the nominal tensile strength of a small sample (Lachenbruch, 1962, p. 43). Only where $m = \infty$ is the crack spacing determined by stress relief only, and by extension it is only in this unlikely limiting case that division of cells into a regular tessellation of hexagons is likely to occur. Smaller values of m denote more influence by flaws on crack spacing.

If we begin with a value of m greater than unity then we may assume that the earliest cracks to appear are likely to be guided by lines of weakness. The zone of stress relief caused by the early fractures is unlikely to be wide enough to achieve self-adjusting spacing and the path of such fractures may well be sinuous. However, the tension relief will give greater uniformity of strength in the remaining undivided areas and raise the value of m. The fracture of these areas under increased tension is therefore likely to give more regular cell patterns, with values of l more dependent on stress relief and less dependent on the location of flaws. Fig. 5.17 suggests a four-stage model of cell division to illustrate the process in action.

One special feature of this diagram is that cracks tend to intersect at right angles rather than in the trihedral fashion of the steady-state model. Lachenbruch (1962, pp. 45–6) points out that the zone of stress relief around a fracture line is not a uniform band but (i) dies out asymptotically with distance normal to the fracture, and (ii) is anisotropic with relief at a maximum in a direction perpendicular to the crack and at a minimum in a direction parallel with the crack. If a tensional crack extends towards an existing crack at an oblique angle it will tend to 'veer' towards the

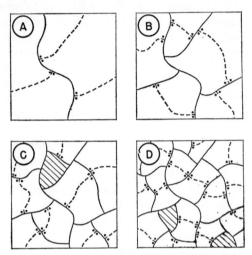

Fig. 5.17. Four-stage 'fracturing' model of the sub-division of a plane with randomly distributed zones of weakness. The shaded areas indicate areas too small to be fractured under given stress, where stress is increasingly monotonically from (*A*) to (*D*). Source: Haggett, in Chorley and Haggett, 1967, p. 653.

normal as it enters the stress-relief zone. Therefore we may argue that '. . . the intersection of a crack with a pre-existing one tends to be orthogonal . . . and, conversely, the orthogonal intersection suggests that one of the cracks involved predates the other' (Lachenbruch, 1962, p. 46).

In practice, fracture theory has been restricted in its geographical applications to the study of polygonal structures in frozen ground (Corte and Higashi, 1964). However, the relative scarcity of trihedral junctions in other cellular networks and the high number of orthogonal intersections suggests that it might be worth exploring the extension of tension-failure models to historical sequences of cell division (Haggett, 1965, p. 52). Again the analysis of the growth of administrative area boundaries is a rich area for experimentation.

II. SIMULATION MODELS

A recent approach to the explanation of certain observed features of channel networks has been through the application of simulation techniques of the Monte Carlo type to the hypothetical development of transport systems. The general idea underlying such work is that local environment seems to have been less important than chance in determining many of the properties of real networks (Smart, Surkan and Considine, 1967, p. 88). It should be noted at the outset, however, that most such studies claim to simulate certain features of completed networks, rather than giving a step-by-step picture of their evolution through time; although the final similarities which do emerge may throw considerable

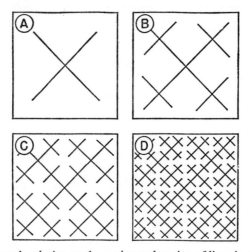

Fig. 5.18. Steinhaus's solution to the optimum location of lines in respect of areas.
Source: Bunge, 1964.

light on both the manner and sequence of evolution. Thus Cohn's (1954) work showed that it was possible to build up, from a minimum number of parameters, a satisfactory approximation to certain real working parts of the vascular system. Starting with the problem of supplying blood to a cube divided into successively smaller parts, he was able to derive expressions for (1) the number of branches, (2) the length and resistance of each branch, (3) the total resistance of the system to flow, and to match these with clinical results. A two-dimensional analogy to Cohn's three-dimensional work is provided by the mathematician Steinhaus working on the problem of ' . . . locating one finite area as near as possible to an infinite area'. The resulting sequence of graphs (Fig. 5.18) shows successive increases in the length of the tree-like structure as the perimeter of the finite area is increased further. Bunge (1964, p. 12) suggests, therefore, that the dendritic patterns represent the linear pattern which optimally satisfies the command: 'locate a set of singly connected lines of finite total length as near to an area from a given point as possible'.

1. Branching networks

The first important study of this type was carried out as recently as six years ago by Leopold and Langbein (1962). The authors began by simulating the coalescence of rills originating as equally-spaced along a divide and having, for each 'step' forward equal chances of turning to the right, left or continuing downslope. The banning of 'uphill' steps and permitting the union of any two streams stepping into the same square produced a bifurcating stream-like network (Fig. 5.19), which exhibited a logarithmic

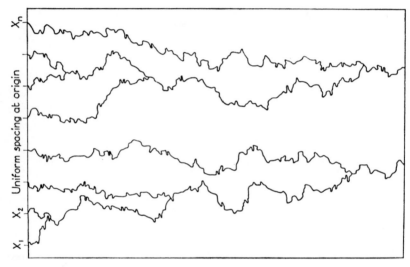

Fig. 5.19. Portion of random walk model of rill or stream network. Source: Leopold and Langbein, 1962.

relationship between stream length and Strahler order. In a more elaborate model (Leopold and Langbein, 1962) the authors began with a 60 × 120 matrix of squares from which they randomly selected one as the source of the first stream. They then extended it by steps into successive squares—allowing an equal probability of extension into any of the four adjacent squares—until it ran off the map. A second source was selected and another stream generated in a similarly random manner until it either joined the first stream or ran off the map, and so on. Subsequent streams were allowed to join existing ones at any point, including their source, and the whole process was continued until all the squares were filled with streams. A simulation of this type produced the fifth-order basin shown in Fig. 1.15-A. The important result of the simulation was that it made demonstrable the fact that such randomly-generated patterns seem to obey the relationships involving stream order observed in natural streams.

In an application of the first of the Leopold and Langbein models, Hack (1965) simulated the development of post-glacial drainage on the

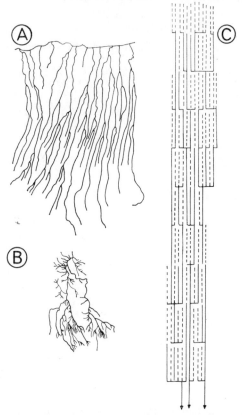

Fig. 5.20. Drainage systems on the Ontonagon Plain, Michigan, U.S.A. (*A*) Streams between Mineral and Cranberry Rivers. (*B*) Mill Creek. (*C*) Simulation model. Source: Hack, 1965.

progressively-exposed floor of Lake Duluth, Michigan, where strongly-attenuated stream systems were influenced by a series of approximately equal-spaced glacial grooves normal to the lake shore (Fig. 5.20-A). These systems exhibit the Hortonian laws and little change seems to have occurred as the result of capture, migration of divides or headward cutting, but there has been valley deepening which is roughly proportional to the contributing drainage area. Assuming that the spacing of the original channels was controlled by the glacial grooves and that, as the shoreline of the Ontonagon Plain receded, the streams of water were required to flow further and further to reach the lake, being subjected always to chance mergings with adjacent streams, Hack constructed a random-walk model. This assumed a uniform spacing of glacial grooves of 0·1 miles, a drainage density of ten and a length of first-order streams of one mile (Fig. 5.20-C), and produced the relationship:

$$L = 4A^{0.67}$$

which accorded well both with field observations and with the predictions of Leopold and Langbein (1962) that evenly-spaced streams which are

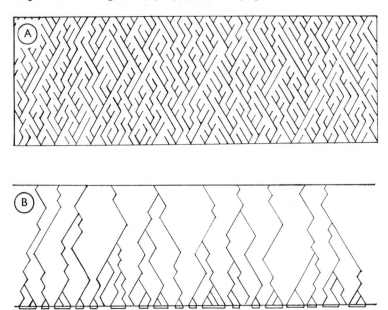

Fig. 5.21. Randomly generated drainage nets (*A*) and basins (*B*). Source: Scheidegger, 1967B.

randomly-generated will unite at distances which are proportional to order and that $L \approx A^{0.6}$. Another study by Scheidegger (1967B) simulated stream development on a steeply-sloping mountain side by a series of randomly-selected half-left or half-right steps (Fig. 5.21-A), claiming that the resulting drainage basins (Fig. 5.21-B) compare well with those draining into the Swiss Rhone.

It was clear that such laborious simulation techniques are admirably suited to generation by computer. Schenck (1963) used this method to generate a fourth-order basin on a 20 × 30 matrix, allowing the random steps of the second Leopold and Langbein model (eliminating reverse flow and closed loops). The result gave good relationships between N and L versus order, and the area/length relationship was similar to that postulated by Gray (1961) from field observations. Liao and Scheidegger (1968) generated a large number of bifurcating arborescences with twenty pendant vertices (i.e. first-order streams if the arborescence is inverted), and found that they accorded with the law of stream numbers and that the bifurcation ratio tended to decrease with stream order, as in nature.

One of the most ambitious stream network simulation studies employing a computer to generate a large number of networks was carried out by Smart, Surkan and Considine (1967) who described a programme for simulating, classifying and counting stream segments, lengths and areas. These workers modified the second Leopold and Langbein model by allowing biased networks (i.e. having greater probabilities of extending in certain directions), and by setting up a series of rules to prevent source junctions, triple junctions, looping and trapping. The programme was used to generate some 600 basins of third to fifth order (those with $N_1 \geqslant 10$ being selected for analysis), with a maximum drainage density (i.e. all matrix squares having a stream), which were divided into two classes: (1) *Random:* These were generated in a manner somewhat similar to the second model of Leopold and Langbein, using a 40 × 40 grid with equal probabilities of streams 'stepping' into each of the four adjacent squares (i.e. $P_\downarrow = P_\uparrow = P_\leftarrow = P_\rightarrow = 0.25$) (Fig. 5.22-A). (2) *Biased:* These games were modified to employ a 40 × 60 grid, with four times the probability of a stream extending in one direction than in the opposite one (i.e. $P_\downarrow = 0.4; P_\uparrow = 0.1; P_\leftarrow = P_\rightarrow = 0.25$) (Fig. 5.22-B). The parameters resulting from the biased games compared quite well with the range which Melton (1958c) suggested was exhibited by third-order natural stream networks (Table 5.2). The relationships between stream number and order also accorded quite well with those predicted by Shreve (1966), although the networks simulated by computer regularly departed from Shreve's predictions as the number of first-order streams decreased. This was probably due to the fact that Shreve's theory applies to isolated networks, whereas those generated by computer games are competing ones, as in nature (Smart, Surkan and Considine, 1967, pp. 94–5).

It is clear that the simulation models described above differ from other evolutionary approaches in that, by simulating the end result of channel network evolution, it is hoped to throw light on some general principles governing evolution, rather than to simulate the stage-by-stage development. Such studies seem to indicate that although every stream junction has its own *raison d'etre*, 'the mechanics of these small-scale effects is so complicated that the details concerning most individual . . . (events) will remain forever unknown' (Scheidegger and Langbein, 1966, p. 3). Thus the areal variability of process can never be completely explained, but the end result is as if the process was largely random (Scheidegger and Langbein, 1966, p. 1). Similarly, the comparison of Shreve's randomly-generated networks with those observed in nature 'support the conclusion that populations of natural channel networks developed in the absence of geologic controls are topologically random (which does not, however, imply that the lengths, shapes, and orientations of the links necessarily are also random) and therefore that, as proposed, the law of stream numbers is largely a consequence of random development of the topology of channel networks according to the laws of chance' (Shreve, 1966,

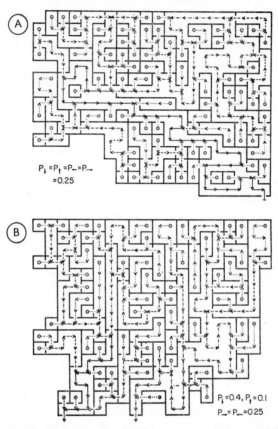

$P_↓ = P_↑ = P_← = P_→$
$= 0.25$

$P_↓ = 0.4, P_↑ = 0.1$
$P_← = P_→ = 0.25$

Fig. 5.22. Portions of channel networks generated by computer programme: (*A*) random game, (*B*) biased game. Source: Smart, Surkan, and Considine, 1967.

p. 36). Nevertheless, it must be recognized that the randomly-generated networks previously described do not simulate the actual geomorphic processes involved in the evolution of natural networks, which usually take place simultaneously over an exposed area (Schenck, 1963, p. 5743). As Leopold, Wolman and Miller (1964, p. 421) conclude: 'Although (mathematical models) provide fundamental general principles with which to begin, neither the rational nor the random walk descriptions

Table 5.2. Statistics for biased games

X	Order	$\bar{X}$	σ_x	$\sigma_x/\bar{X}$	Range of $\sigma_x/\bar{X}$ (Melton, 1958c)
N_1/N_2	3	5·015	1·30	0·258	0·15–0·35
	4	4·650	0·877	0·189	
	5	4·778	0·366	0·077	
N_2/N_3	3	3·707	1·63	0·441	0·35–0·55
	4	4·340	1·53	0·353	
	5	4·699	1·20	0·255	
$\bar{L}_1 = \bar{A}_1 = l_{ex}$	3	1·963	0·314	0·160	0·15–0·35
	4	1·989	0·213	0·107	
	5	1·975	0·114	0·058	
$\bar{l}_{in}$	3	2·075	0·493	0·238	
	4	2·098	0·246	0·117	
	5	2·077	0·109	0·053	
$\bar{L}_2$	3	5·184	2·34	0·452	0·35–0·55
	4	4·196	1·43	0·341	
	5	4·456	0·705	0·158	
$\bar{A}_2$	3	12·31	4·18	0·340	
	4	10·37	2·61	0·252	
	5	10·86	1·44	0·133	
$\bar{L}_2/\bar{L}_1$	3	2·717	1·33	0·488	0·35–0·55
	4	2·116	0·681	0·321	
	5	2·260	0·348	0·154	
$\bar{L}_3/\bar{L}_2$	3	4·366	3·55	0·813	0·55–0·75
	4	3·069	1·81	0·590	
	5	2·821	1·13	0·400	
$\bar{A}_2/\bar{A}_1$	3	6·355	2·22	0·348	
	4	5·218	1·18	0·226	
	5	5·493	0·600	0·109	
$\bar{l}_{in}/\bar{l}_{ex}$	3	1·087	0·314	0·289	
	4	1·065	0·155	0·146	
	5	1·055	0·084	0·080	

Number of basins for orders = 3, 4, and 5 are, respectively, 82, 63, and 30; mean numbers of first-order streams are, respectively, 18, 63, and 187. σ_x = standard deviation of x. $\bar{l}_{in}$ = mean interior link length. $\bar{l}_{ex}$ = mean exterior link length.

Source: Smart, Surkan and Considine, 1967.

indicate how drainage nets develop and change through time. It is seldom in nature that an area of any size is at any moment unrilled or undissected.'

In developing a different type of computer-based stream network simulation model, Howard (1968, pp. 1–3) recognized three distinct types of network models: (1) *Growth models:* which develop by headward extension and branching on an initially uneroded land surface. These have the disadvantage of predicting networks much more regular than are observed in nature, and tend to attain some static equilibrium when they have reached a given state. (2) *Random models:* which consider network processes to involve probabilistic choices leading to alternative final states. These give little insight into the processes or stages involved in network evolution, become static after generation, and produce excessive wanderings and near-loopings which are rare in nature. (3) *Capture models:* which start with an initial network and investigate its subsequent stability in the face of probabilities of capture and migration of divides. Horton's micropiracy was an example of a capture model, but his scheme of drainage system development was essentially a growth model. One of Howard's (1968) early capture models contained 450 squares in an 18 × 25 matrix and assumed that parallel streams imparted a uniform drainage density and that there was a single drainage point at (9, 25) (Fig. 5.23). Certain initial assumptions were made, namely that the relative elevations were as shown in Fig. 5.23, and that the probability of capture of a given stream segment by one in an adjacent square was a function of their upstream drainage areas (which control discharge and the amplitude of adjacent meanders), and of gradient across the site of potential capture. The operation of the capture simulation model involves the following steps: (1) Randomly select a point on the stream matrix. (2) Examine the four surrounding squares in terms of the probability of their capturing the stream selected. (3) If capture is allowed, regrade the upstream drainage area of the captured stream according to the formula empirically derived from field data: stream gradient = contributing basin area$^{-0.6}$. (4) Select another random point of possible capture, and so on, until the desired number of trials or captures have occurred. Fig. 5.23 shows one series of simulations after 4, 100 and 319 captures, respectively. Repeated runs showed, firstly, that network changes increased rapidly to a peak and then declined to a fairly constant value, and, secondly, that when the resulting networks were ordered according to the Strahler system they agreed quite well with natural networks. From the above preliminary analysis, Howard (1968, p. 27) suggested that capture may be a significant process in the development of some natural stream networks, and applied this idea specifically to stream systems observed on pediments. Work on this interesting capture model is still in progress, and it is currently under active elaboration and modification.

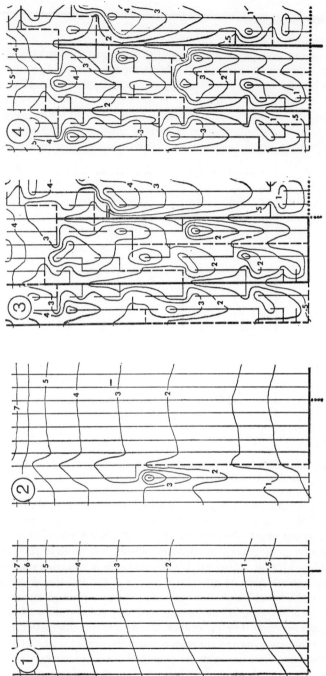

Fig. 5.23. Stages in Howard's capture model simulation, after zero (*1*), four (*2*), one hundred (*3*) and 319 captures (*4*).
Source: Howard, 1968.

2. Circuit networks

(a) Colonization models

Simulation models for the growth of transport networks have yet to reach the sophistication of stream-pattern models (Leopold and Langbein, 1962; Schenck, 1963). Nevertheless, the extension of branching models to more complex circuit networks is foreshadowed in Schenck's model where various boundary conditions may be substituted in the programme. Fig. 5.24 shows five alternative boundary conditions with the available outlets shown as heavy lines and the main drainage lines by dashed lines.

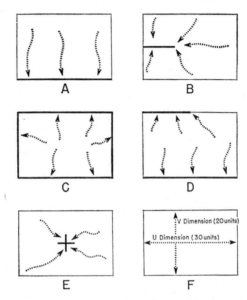

Fig. 5.24. Alternative boundary conditions (heavy lines) used in the computer simulation of dendritic networks. Source: Schenck, 1963.

Of special significance is Fig. 5.24-E where the streams drain to a central 'sink' and show patterns not unlike those derived by Haggett for road systems around small inland towns (Fig. 5.36). Routes have been included as by-products of general settlement simulation models (Hägerstrand, 1967). Thus Morrill (1965, pp. 65–82) includes a tentative road network in his eight-generation Monte Carlo model of the locational evolution of a theoretical central place hierarchy (Fig. 5.25). Although the mechanics of the migration-assignment process and the recomputing of the probability field for each time generation lie outside our discussion it is worth noting (i) that the pace of extension around the centre is uneven over both space and time, and (ii) that the extent of the road system is greater in the direction of early start. Spatial unevenness is related to the random element built in to the Monte Carlo model (Haggett, 1965, pp. 305–9), and temporal unevenness to the extending 'reach' of the migration process with improving technology; the migration probability field was successively expanded from k/D^2 in the first generation to k/D^1 in the eighth generation, where k is a constant and D the diameter of the field.

Morrill's model highlights the indirect orientation of routes with respect to the originating centre (i.e. the pattern is not rigorously stellate), and associates this with the iterative process by which roads grow by the addition of new links. When, however, the blotchy and uneven pattern of settlement is filled in, then new and more direct links can be formed. Comparison of the route north from the centre in the fifth and sixth generations (E, F) shows this realignment process in operation (Fig. 5.25).

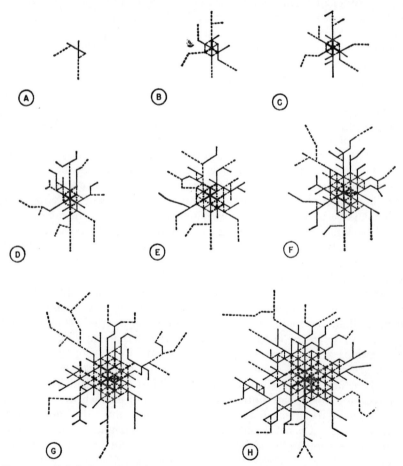

Fig. 5.25. Simulation of road network around a central node by Monte Carlo procedures. Source: Morrill, 1965.

Morrill points out (1965, p. 109) that the road net in his model is passive rather than active: that is, the road net is a product of the migration process; existing roads are *not* taken into account in the assignment of new migration routes and central places. Other artificial restrictions in the model are clear in the fixed mesh of the roads in the 'settled' area near the centre. Both the size of mesh and the triangular form are directly related to the size of the hexagonal cells in the original probability field. Where square rather than hexagonal cells are used (e.g. Morrill, 1962, p. 117), the resulting road network will also be basically rectilinear—indeed not unlike the road system over the greater part of the western United States (Thrower, 1966).

An exploratory behavioural model of transport development in an underdeveloped country has been developed by Gould (1966). The model is based on search theory and is basically a two person game played between a 'developer' and an 'environment'. The playing area is a conventional square area (representing a block of virgin territory) with a sea coast along one edge and the site for a main port in the centre of this edge. The objective of the developer is to tap the resources of the area through a programme of road building and resource development, but since these resources are in the first instance unknown, this can be achieved only through spatial exploration. The environment reveals the 'resources' and 'hazards' of each area (these may be in terms of mineral or agricultural potential) at each stage of the network expansion (Fig. 5.26).

In a typical cycle of the game the developer explores the space by investing capital in his roads. The network expands outwards from the port in a series of straight lines segments between pairs of xy coordinates. After the first period of road construction (say, one year) the information is fed back from the environment on the resources of the area transected by the roads (see t_1, t_2, t_3). The developer may then either begin developing the resources he has already tapped (by investing in all-weather roads), or may continue to search by more exploratory road building in the second period. The game is made more realistic by rules governing interest rates, capital budgets, repayment time-limits road depreciation, and more hazardous by introducing internal environmental 'shocks' in terms of floods or droughts or external variations such as fluctuations in world market prices for commodities. With the development process, capital slowly accumulates and a taxation system is introduced to fund further road development. Thus the network expands at an ever-developing rate until a density threshold is reached.

Although the game is elementary, it shares with other operational games (e.g. the Cornell Land Use Game) great advantages in educating users in the complexity of the network development process. From preliminary runs at Pennsylvania State University, Gould suggests three tentative conclusions about the process: (1) Network development appears to be a two-stage process in which a 'space-covering' search is succeeded by a 'space-organizing' search once information about economic potential is available. This two-fold distinction is also reflected in the space-searching behaviour of animals. For example Fig. 5.27-A shows the search paths of the larvae of an insect predator (*Stethorus*) for red mite prey in the presence of uneven topography. The random search paths of the predator over the flat tray were replaced by regular searching of the cone on which the prey were located (Fleschner, 1950; Gould, 1966, p. 23). (2) The utility function of players is unstable and changes over time as information

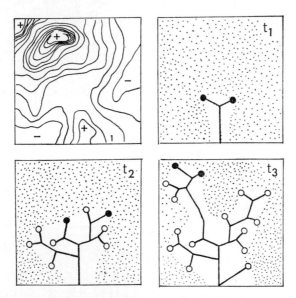

Fig. 5.26. Stages (t_1, t_2, t_3) in the development of an exploratory transport network in an unknown environment. Environmental 'quality' mapped as isarithms. Solid circles indicate terminals from which future link extensions are most likely.
Source: Gould, 1966.

about space grows. Developers are initially cautious, and even at later stages of the game may be content to aim at satisfactory rather than optimum levels of return on their transport investment. (3) The effect of 'surprise' concentrates investment in a small area. After several fruitless cycles of road expansion in which no worthwhile resources are revealed, the 'spatial success experience' (Gould, 1966, p. 31) has a very definite effect on the development of the road pattern. African road-building experience in the 1920's and 1930's tends to confirm this pattern of sudden explosions of road-building activity following resource potential (Gould, 1960, pp. 100–10). A fuller behavioural psychological interpretation of network evolution would demand consideration of a full range of stress conditions, and Gould argues that details of bad route planning and mis-alignment are often traceable to hurried decisions based on insufficient surveys. This behavioural approach to network growth is paralleled by mechanistic computer models of line systems based on probability distributions (Figs. 5.27-B and -C).

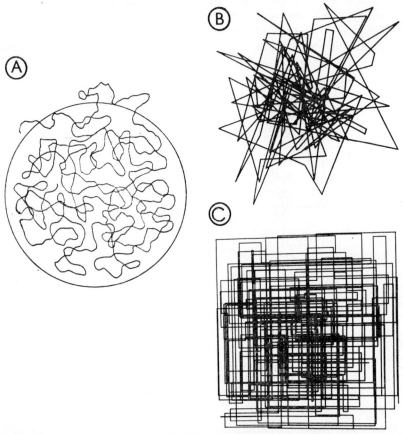

Fig. 5.27. (*A*) Search path of insect (*Stethorus*) within circular illuminated area. (*B*) Random search path generated by Gaussian bivariate distribution. (*C*) Random search path constrained by sequence of vertical and horizontal movements with uniform probability density. Source: Gould, 1966, p. 23; Noll, 1966, pp. 67, 68.

(*b*) Interconnection models

The second group of models which attempt to simulate the growth of transport networks are mainly related to a consolidation phase when (a) the number of nodes to be incorporated into the graph is known, and (b) the spatial problem is one of optimum arrangement under changes in the sources and directions of flows. Random plane models form the most primitive node-connecting models. A random plane network is one constructed by choosing nodes at random over a plane and linking together those within specific distances. Although the general properties of the model are of interest in terms of the probabilities of occurrence of graphs of given sizes, most of the work reported to date is based on simulation techniques. Gilbert (1941) investigated the effect of increasing the contact distances (R) on graph formation for 1,000 randomly located

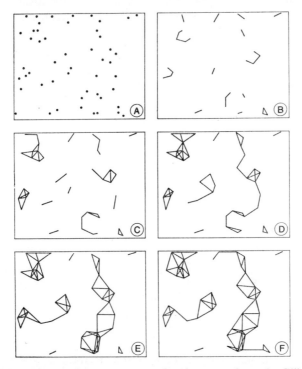

Fig. 5.28. Network growth as a sequence of node connections: the Gilbert–Brown random graph model. Source: Gilbert, 1941; Brown, 1965.

points. Brown (1965) repeated these experiments on a randomly located set of fifty nodes over a rectangular plane of 50 × 40 units (Fig. 5.28-A). The nodes were located using pairs of coordinates from random number tables. Each of five networks (Figs. 5.28-B through 5.28-F) is based on the same fifty nodes, but the critical distance over which links are permitted is progressively raised over one, two, three, four and five units respectively. As the value for the critical linkage level R goes up, the number of graphs decreases and the probable number of nodes in each graph increases. It is evident that critical values of R may be determined—a lower value below which no connections are made, and an upper value above which all connections are made between all points. The work to date is too limited to suggest general characteristics of random plane models.

Brown (1965, pp. 50–2) suggests applications for the model in epidemiology. Here the interests centres on the probability of sets of nodes becoming 'infected', given the 'infection' of a single node or sets of nodes. Certainly, by introducing variable weights into the links linked to a deterrence function (which need not be directly related to crude distance on the plane), interesting possibilities for replicating epidemic outbreaks like the foot-and-mouth diffusion from Shropshire over much of central

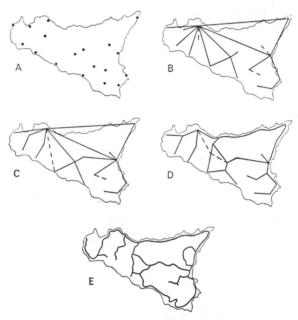

Fig. 5.29. Stages in link allocation for simulating the Sicilian 1908 railroad network. (*A*) Selected vertices. (*B–C*) Stages in link allocation. (*D*) Postdicted 1908 network. (*E*) Actual 1908 network. Source: Kansky, 1963, pp. 139–46.

England in the autumn of 1967. From the viewpoint of road networks the Gilbert–Brown models would need to be modified to take account of service ceilings. Links are unlikely to be made once the route density has risen to a level appropriate to the transport demands of the plane. Simple modification in which links are not added in areas within a specified distance of existing links would show contrasts between the original model with its overlapping links and the modified density-dependant model. The latter throws considerable weight on the *order* in which links are constructed, with earlier links 'neutralizing' a dependant supply area (compare the Lachenbruch space-partitioning model, Chap. 5.I(3)). In practice, the model would need further modification to allow feedbacks between connections and demands and variations in the plane.

Where nodes vary greatly in value the connection sequence will normally reflect these differences. Kansky (1963, pp. 132–47) developed a simulation model for connecting the population nodes of Sicily by a railroad network. The basis of the simulation was to predict the probable localization of railway routes from the geographic characteristics of the area; the length, number, and rate of extension of the routes was either

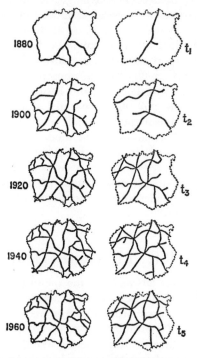

Fig. 5.30. Comparison of the (*A*) actual growth of railway network in central Sweden with (*B*) the simulated growth. Source: Morrill, 1965, pp. 130–70.

derived directly from the historical evidence or predicted from comparative studies of railway parameters in terms of an area's general level of economic development. Thus in the case of the Sicilian railway network for 1908, the number of vertices was predicted as 16·48, the connectivity as 1·13 (β-index) and the mean edge length as 17·62 miles.

The procedure for the stochastic model may be generalized as a four-stage procedure: (1) List and weigh all major settlements in the area in terms of a 'population income' score. (2) Select probable vertices from this list of settlements by a randomization process. (3) Connect the two largest vertices by a rail link. (4) Add the other edges in sequence such that ' . . . the next largest centre joins the largest and closest centre which is already located on the network' (Kansky, 1963, p. 138). If, after all the vertices have been connected to the railroad system (Fig. 5.29-B), unallocated edges remain, then (5) add the edges in such a way that the circuit between the first, second, and third largest centres is completed, and continue to complete the circuits bringing in the fourth, fifth and lower-order centres in turn (Fig. 5.29-C). The map of routes may be finally adjusted by (6) the adoption of delta-wye transformations (Akers, 1960) to simplify the links between triangles of three points, and (7) local adjustment of the routes to physical variations in the relief. Figs. 5.29-D and -E compare the 'postdicted' and real Sicilian networks.

A number of limited attempts have also been made to simulate the growth of actual transport networks using Monte Carlo methods. Garrison and Marble (1962, pp. 73–88) describe attempts to simulate the growth of the railroad system of Northern Ireland between 1830 and 1930, while Morrill (1965, pp. 130–70) reports parallel studies on the rail nets of central Sweden. All the areas are small and their networks relatively simple in structure. The success of the models varies considerably. The stochastic model of Northern Ireland yielded less clear results than a 'deterministic model' based on arbitrary rules of neighbourhood, regional and field effects. Reasonably close matches between the simulated network and historical sequences of route development are shown by Morrill's (1965) work (Fig. 5.30).

III. NETWORK TRANSFORMATION

Much of the significant advances in the quantitative analysis of simple branching systems over the last twenty years stem from Horton's (1945, p. 281) recognition of stream order, and from its modification to a simpler combinatorial system by Strahler (1952, p. 1120). Indeed, Bowden and Wallis (1964, p. 767) describe this ordering system as ' . . . the touchstone by which drainage net characteristics could be related to each other and to hydrologic and erosional processes.' The success of the Horton–Strahler approach in the recognition of fundamental regularities in drainage systems suggests that the search for comparable ordering-systems for more complex networks—notably transport circuits—would be worth pursuing. Haggett (1966, 1967) has explored some elementary problems in transforming complex to simple transport networks.

1. Order in complex transport systems

Attempts to order or categorize transport networks containing circuits have not, so far, been conspicuously successful. To be sure, useful functional classifications of railroad networks in terms of their flow characteristics have been attempted by Wallace (1958), but the recognition of the classes—'internal', 'originating', 'terminating', 'bridge', and 'balanced traffic' lines—demands data that often are just not available (Fig. 5.31). Very simple transport systems may consist merely of single lines or sets of lines arranged in a tree-like fashion. Thus Rasmusson (1962, p. 80) has mapped the intricate tracery of the lanes connecting the small granite pits on the island of Malmön with the bigger roads and railways leading

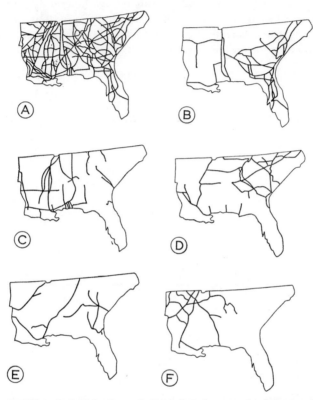

Fig. 5.31. Functional classification of the railroad network of the south-eastern United States. The network (A) is dissected into lines dominated by the following traffic regimens:—(B) Internal. (C) Originating. (D) Terminating. (E) Bridge. (F) Balanced. Source: Wallace, 1958.

to the exporting ports. Other Scandinavian geographers have mapped the seasonal flow of logs down the river systems of Sweden and Finland to the Baltic ports (e.g. Hultland, 1962). In these cases the transport net is simple enough for Strahler-type ordering systems to be applied directly, and no simplification problem need be encountered (Fig. 5.32).

These simple networks are, however, in the minority. Road systems and rail systems in almost all developed countries consist of a highly complex interlacing of circuits with trees or isolated lines as appendages.

Thus Garrison and Marble (1962, p. 35) found the average cyclomatic number for a group of fifteen countries (including units as undeveloped as the Sudan and Angola) as high as 200: by comparison, the cyclomatic number for trees is zero. The problem to be tackled is thus a formidable one as the inspection of the road trace on any single Ordnance Survey one-inch sheet will confirm.

Direct ordering of 'natural' circuits is thus inherently more difficult than that of 'natural' trees studied by Horton, in that circuits have alternative paths connecting vertices on the network. It seems, therefore, more profitable to consider ways of simplifying circuit networks into trees,

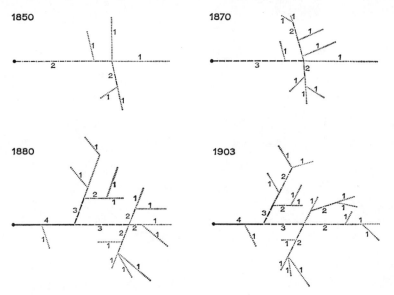

Fig. 5.32. Strahler-type ordering applied to the expansion of the railway network in north-east London. Source: Werrity, 1967.

allowing thereby the direct application of the Horton–Strahler approach to the 'artificial' trees thus formed. The fact that circuits (a higher-order topologic form) are theoretically decomposable into trees (a lower order topologic form) is clear by inspection. Interest in the number of possible trees in a fully-connected graph has been stimulated by work on maser analysis, and Ku and Bedrosian (1965) have derived general solutions for partitioning full graphs into complementary trees.

(a) Order in fixed-path networks

The notion of combinatorial order in dendritic stream systems has been discussed at length in Chap. 1.II. Following Melton (1959) we can show that the Horton–Strahler ordering system is basically a simple mathematical concept derived from the combinatorial analysis of a finite rooted tree. Fig. 5.33-A shows such a tree in the familiar but idealized form of a channel network. The network consists of three kinds of nodes:

Set A: Root nodes (marked by ⊙),
Set B: Outer nodes (open circles) with an array of one line from each node, ○
Set C: Inner nodes (closed circles) with an array of three lines from each node, ●.

For the network in question there are one node in Set A, eight nodes in Set B, and seven nodes in Set C. Lines joining nodes represent stream channels (see Fig. 1.5).

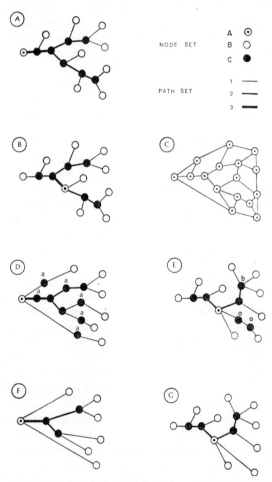

Fig. 5.33. Combinatorial order in fixed-path and variable-path networks. Source: Haggett, 1967, p. 282.

If any path from an outer node (Set B) is defined as 'downstream', then for any inner node (Set C) two lines are always 'upstream' and only one line 'downstream'. By nominating all downstream paths from Set B as first-order paths, a simple combinatorial system of ordering may be initiated. Second-order paths are defined as downstream lines from inner nodes at the junction of first-order paths, third-order paths as downstream lines from inner nodes at the junction of second-order paths, and thus nth-order paths as downstream lines from inner nodes at the junction of $(n-1)$-order paths (*cf.* Fig. 1.5).

Where stream channels are being investigated, the combinatorial definition of downstream used by Melton and hydraulic reality coincide: the root of the finite tree (Fig. 5.33-A) and the outfall of the stream system at its lowest point are identical. It is clear, however, that Melton's re-definition of the Horton–Strahler ordering system makes it a mathematical system, wholly independent of the geomorphic characteristics of height, slope, and direction of flow. Thus it is equally relevant, from the mathematical (though not the hydrological) definition, to reassign node membership of Sets A, B and C, and to derive from this reassigned network a logical path-ordering system (see Fig. 5.33-B) (Haggett, 1967).

(b) Extension to variable-path networks

While in stream network studies the graph of the system is topologically a tree (i.e. paths between any two points are unique) the graphs of most highway systems are more complex. They contain parallel lines, loops and islands, so that the paths between two points are commonly not unique: variable paths predominate. This situation is suggested in Fig. 5.33-C where twelve new lines have been added to the original fifteen, linking the nodes in Fig. 5.33-A and 5.33-B to simulate a model highway net.

Assignment of node membership in this more connected graph poses more problems. Set A membership is arbitrary since any node may be defined as a root (see Fig. 5.33-D and 5.33-E). Inner and outer nodes may be defined only in terms of tree-structures, and hinge on the construction of such trees about the original roots of the graph. Construction of minimum-path trees within a complex network are an essential part of traffic-assignment procedures, and a number of computer programmes have been devised to build such trees from an input of link data. It is clear that these minimum-path trees (unlike their hydrologic counterparts in stream-channel networks) are not unique: variable definitions in terms of travel length, travel time, or travel cost may yield different trees. Moreover, any trees constructed will be unstable over time as new links are built or the parameters of links are changed. Arbitrarily-defined trees can be constructed, none the less, and Fig. 5.33-D and 5.33-E show two such trees generated by a simple minimum-distance criterion.

From such trees, assignment of nodes to Sets B and C is straight-forward. However, the set of inner nodes differs from those shown in Fig. 5.33-A and 5.33-B, in that all inner nodes do not have an array of exactly three lines from each: some have only two lines (see nodes labelled *a* in Fig. 5.33-D and 5.33-E), while others have more than three lines (see node labelled *b* in Fig. 5.33-E). The two-line nodes are topologically redundant so that graphs *D* and *E* can be redrawn as *F* and *G* without loss of information, since the metric distance along the path is unchanged by the excision of two-line nodes. The nodes with four (or more) lines are retained since they do not disturb the combinatorial ordering system, i.e. there is no confusion over the one downstream line from each inner

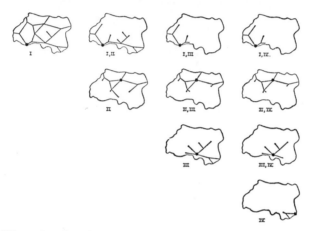

Fig. 5.34. Hierarchy of road networks based on centres I, II, III and IV. Source: Haggett, 1966.

node, regardless of the number of lines arrayed from it. The Horton–Strahler path-ordering system can therefore be applied to variable-path networks through the adoption of minimum-path trees about one or a number of arbitrarily defined roots, and by the 'pruning' of redundant two-line inner nodes.

Since any one of the nodes in the variable-path network may be designated a root, a large number of potential networks may be derived. The general rule for the maximum number of sub-networks is:

$$\max N_n = \tfrac{1}{2} \, (P_n{}^2 + P_n)$$

where N_n are networks and P_n the nodes. We should note, however, that this is a maximum, and that in practice the introduction of new lower-order centres will 'disturb' only those networks immediately about it. Fig. 5.34 shows a hypothetical network with four major centres ranked in size, I, II, III, IV; and the hierarchy of ten sub-networks generated from them. Note that the introduction of centre IV disturbs the networks

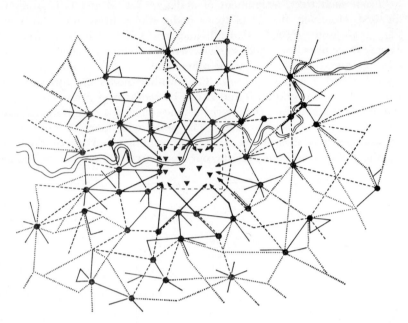

Fig. 5.35. Hierarchic structure in an urban transport network. Projected form of the London bus system.

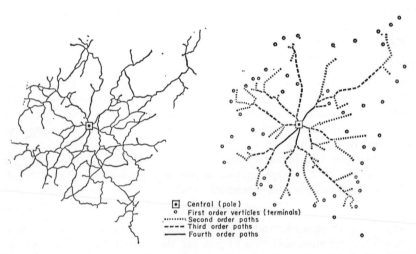

☐ Central (pole)
o First order verticles (terminals)
·········· Second order paths
---- Third order paths
—— Fourth order paths

Fig. 5.36. Decomposition of the road network around the town of Viseu, Portugal (*left*) into a dendritic form (*right*). First-order paths have been omitted for clarity.
Source: Haggett, 1966.

centred on II and III but not that centred on I; consequently the networks N (I, III) and N (I, IV) are identical. Since the smallest number
of networks we can create is only twice the number of roots less one
(i.e. assuming lower-order centres are introduced sequentially along a
straight line at increasing distances from the highest-order centre) we
know that:

$$2(P_n) - 1 > N_n > \tfrac{1}{2}(P_n{}^2 + P_n)$$

Thus with ten centres, we know the number of sub-networks will be
greater than nineteen but less than fifty-five. The idea of a nested hierarchy of centres accords well with classic central-place theory (Lösch,
1954) and with the actual organization of transport networks. Fig. 5.35
shows plans for the reorganization of the London transport system on the
basis of a hierarchy of transport services based on central and suburban
nodes (cf. also Fig. 3.11).

A practical application of this conversion procedure is shown in Fig.
5.36. Here the central root is the town of Viseu (1950, population 13,000)
in the Beira Alta region of north-central Portugal: the network (main
and secondary roads) is shown for the region around the town in Fig. 5.36.
The hierarchical position of Viseu in respect to other nodes is determined
from population data. From this general network the network centred
on Viseu rather than 'competing' centres is isolated (the dotted line in
Fig. 5.36-A), and dissected to give the ordered pattern of paths shown
in Fig. 5.36-B. For simplicity the first-order paths have not been shown,
but the dendritic structure of fourth-order paths centring on Viseu is
striking. Other applications of the ordering algorithm have been made for
major motor routes in England and Wales (Haggett, in Chorley and
Haggett, 1967, p. 661).

2. Some implications of spatial order

(a) Regularities in fixed-path networks

Horton was able to show that the adoption of the channel-ordering
systems had important implications for the spatial ordering of both the
stream segments and the associated watershed characteristics. Horton's
'laws' outlined in Chap. 2 have been subjected to a wide range of tests in
different geomorphic environments, and have been placed by Strahler
(in Chow, 1964), Leopold and Langbein (1962), Scheidegger (1966) and
by others within a general framework of hydrodynamics and systems
analysis.

Since the hydrologic and geomorphic implications of these findings
lie outside this paper, the Horton regularities may be simply summarized
as a statement of definitions, identities and structural equations. The list
given is restricted to planar features of possible relevance to highway
systems and ignores equations involving height and slope characteristics.

Definitions: U Order number

        N_u Number of paths of order U

        L_u Length of one path of order U

        A_u Catchment area served by any one path of order U

        λ_u Angle of junction between any one path of order U and a second path of order $U + 1$

        Q_u Flow along any one path of order U

Identities: Branching ratio (R_b)

$$R_b = N_{u+1}/N_u$$

Structural Equations:

$$\log_{10} N_u = a_2 - b_2 U$$
$$\log_{10} L_u = a_3 + b_3 U$$
$$\log_{10} \bar{A}_u = a_4 + b_4 U$$
$$\lambda_{u,u+1} = a_5 - b_5 U$$
$$\log_{10} \bar{Q}_u = a_6 + b_6 U$$

The structural equations relate, through the intercept a and the regression coefficient b, the combinatorial order of stream channels (U) to a wide range of characteristics of varying geometric dimensions. Experimental evidence confirms the general form of the structural equations in a wide range of environments; the standard errors for the last two equations are greater than for the other three.

(b) Extension to variable-path networks

Extension of the path-ordering system from fixed to variable-path networks allows testing of the hypothesis that regularities discovered in fixed-path networks may also exist in variable-path networks. Results of some preliminary experimental work with highway networks are reported here. Fig. 5.37 shows one of the sample of highway networks examined: a 2,750 square-mile tract of northern Portugal. Here a close network of about 1,600 miles of metalled roads served a population of 1·88 million people in 1950. The major city of the region (Oporto, 0·28 million) was designated as the root node and a minimum-distance tree constructed on the basis of mileage from this root. Flows were based on the 1949–50 traffic census.

Using the same definitions as for the fixed-path model the following structural equations were computed:

$$\log_{10} N_u = 2\cdot86 - 0\cdot552\ U$$
$$\log_{10} L_u = 0\cdot983 - 0\cdot426\ U \text{ (miles)}$$
$$\log_{10} \bar{A}_u = 0\cdot992 + 0\cdot651\ U \text{ (square miles)}$$
$$\lambda_{u,\ u+1} = 98\cdot4 - 19\cdot6\ U \text{ (degrees)}$$
$$\log_{10} \bar{Q}_u = 1\cdot99 + 0\cdot147\ U^{1\cdot56} \text{ (metric tons per day — motor traffic)}$$
$$\log_{10} \bar{Q}_u = 1\cdot14 + 0\cdot187\ U \text{ (metric tons per day — animal traffic)}$$

The first three equations may be regarded as direct consequences of the combinatorial ordering system employed: the only significant new findings of the experiment lay in the very small standard errors obtained. Relationships between order, path number, path mean length, and path mean area appear to be as close for variable-path as for fixed-path networks on the evidence of these limited results.

Results on the fourth were somewhat ambiguous as only junctions with the fourth-order paths were measured. As an alternative line of evidence, the hypothesis that the branching angle was related to relative flows along paths was examined. A sample of 243 line junctions on the

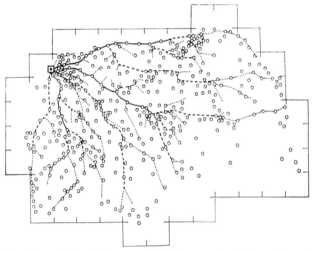

Fig 5.37. Combinatorial order in a sample highway network in north-west Portugal centered on Oporto. Path orders for second, third and fourth order routes are shown by dotted, broken and solid lines respectively; first-order paths have been omitted for clarity. Open circles show the location of flow sampling points. Source: Haggett, 1967, p. 285.

Portuguese major road network were examined; the line with the highest mean daily flow was designated the main stem, and the other two lines branches. The azimuths of all paths were measured on a vertex-to-vertex basis and the direction of the 'branching' paths expressed as angles of departure from the line of the main stem. Fig. 5.38 shows the overall form of the results: as the flow on the branch diminishes (flow expressed as percentage of main stem flow) the departure angle increases (*cf.* Roux, 1895; Murray, 1927). The equations derived from the results were:

$$\lambda = 83 \cdot 2 - 0 \cdot 721 Q_p \text{ (degrees)}$$
$$\lambda = 81 \cdot 6 - 0 \cdot 529 Q_p \text{ (degrees)}$$

where λ is the mean departure angle and Q_p the percentage of mainstem flow. The difference between the results for motor traffic and animal traffic underlines a basic difference in the flow-order equations computed in the original Oporto study. The greater correspondence between the fixed-path and variable-path results for flow as a function of order for the more primitive of the two transport media is striking. Whether we can hypothesize from this that the ancient road network in this part of Europe is more closely adjusted to the more primitive transport media, and whether we can derive any predictions about the topology of the network

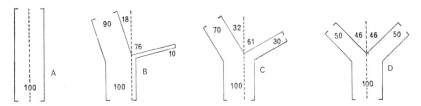

Fig. 5.38. Average relations between mean boundary angle and percentage of main stem flow: sample of 243 triple junction on Portuguese highway network. Source: Haggett, 1967, p. 286.

that would be most closely adjusted to current motor traffic, are matters for present speculation and future research.

Empirical experiments have been further extended to sub-networks within the Oporto system, to other networks within Portugal, and to limited British highway networks (Haggett, 1967A). The general form of the results suggests that regularities revealed by combinatorial ordering of fixed-path networks are also discernible in certain variable-path networks, although standard errors for flow and angular relationships may prove to be large and the forms of the functions somewhat revised.

(c) Implications for other systems

Strahler's extension of the Horton findings into a general system of watershed budget equations suggests that similar extensions of our own findings might be worth pursuing. Since there already exists a sophisticated body of central-place theory, the extension is limited to a rather tentative linking of certain structural equations to a similar set of structural equations developed by Berry (1964) for systems of cities.

For any regional network, values for the area and length of any path may be plotted to show variations in network density. Thus in Fig. 5.34-A the logarithm of path length:

$$\left[\log_{10}\sum_{i=1}^{n}\sum_{j=1}^{k}(L_u)_{ij}\right]$$

on the x-axis is plotted against the logarithm of path area:

$$\left[\log_{10}\sum_{i=1}^{n}\sum_{j=1}^{k}(A_u)_{ij}\right]$$

for the Oporto network. The resulting points for U_1, U_2, U_3 and U_4 lie close to a diagonal line of $45°$ slope. Extension of this line cuts the intercept where the logarithm of path length is at zero. The logarithm of area at this point (A_x) has the following property:

$$\log_{10}A_x = \log_{10}D_k{}^{-1}$$

where D_k is the density for a graph of the order U_k. This property may be simply checked by dividing the area of the network by the total length of the minimum-distance tree. Fig. 5.39-A may therefore be extended by

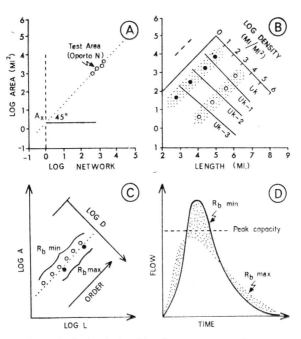

Fig. 5.39. Some hypothetical relationships between network structure and path order. Source: Strahler, 1956; Haggett, 1967, p. 288.

adding a series of diagonal lines at $45°$ to the original axes, each diagonal representing lines of constant network density. A logarithmic density scale may be added orthogonal to these diagonals (Fig. 5.39-B).

Density values for sample networks from different socio-economic environments will generally fall along the appropriate diagonals in much the same way as in Berry's (1964, p. 153) analysis—the trade areas from different environments (e.g. Chicago urban centres, Chicago suburban centres, south-western Iowan centres, and South Dakota centres) clustered along appropriate population density diagonals. There is no direct equivalent in the network case to Berry's inequalities representing the upper size limits of centres of each class as determined by factor analysis, but examination of networks at lower-orders within each network may show up irregularities at some levels. Two hypothetical networks are plotted on Fig. 5.39-B with suggested limits for the networks of order U_{k-1} to U_{k-n} where U_k represents the highest order of the complete network: this order classification forms an irregular scale orthogonal to the density diagonals.

The spacing of the points for sub-networks of lower orders along the density diagonal for the whole network U_k is a measure of the branching ratio, R_b. Closely-spaced values indicate a low R_b value, and vice versa. Extreme variations in branching ratios have been shown by Strahler (1964, in Chow, pp. 4–11) to have direct relevance to flow-concentration characteristics in stream systems (see Chap. 3.III.3b), but there is some question about extension of these findings to highway systems. First, the flow-concentration in hydrologic systems (the unit hydrograph) refers to downstream flow past the outfall (= root) after a storm over the whole basin; traffic problems include the inverse of this—i.e., peak flows on paths after a large input at the root. Second, the range of R_b values in variable-path networks may be somewhat more restricted. The minimum value is of course two, while the maximum value in a regular system of nodes (a Christaller lattice) would be five. For a random distribution of nodes the expected branching ratio is probably 4·789 (Dacey, 1963; Dacey and Tung, 1962) (Chap. 1.III). Experimental results for road systems examined in Portugal give R_b values for whole networks ranging from 3·481 to 4·793.

Parallels between the network ordering system and the central-place ordering system of Christaller with its k-value system have already been suggested by Nordbeck (1965) and by Woldenberg (1966), in terms of the concept of allometric growth. This states that 'the specific growth rate of an organ is a constant fraction of the specific growth rate of the whole organism' (Woldenberg, 1968A; p. 122). When two parts of a system are each related allometrically to the whole system, they are related to each other by a power function. Woldenberg (1966; 1967; 1968A) has thus proposed that the logarithmic relationships between aspects of stream systems are merely those which would be expected to exist between parts of a system subject to allometric growth. To him even the law of stream numbers is a logarithmic relationship, in that an integer increase in stream order increases its absolute magnitude (i.e. discharge) logarithmically (Woldenberg, 1966; 1968A, p. 122). A more realistic approach, therefore, would view stream network growth as a probabilistic version of allometry (Woldenberg and Berry, 1967, p. 131). Gibrat's law of proportionate effect introduces such a random element by stating that 'a variate subject to a process of change is said to obey the law of proportionate effect if the change in the variate at any step in the process is a random proportion of the previous value of the variate' (Woldenberg, 1967, p. 95), i.e.:

$$Y = a X^{b + E}$$

where Y is size at t_1; X is size at t_0; a and b are dimensionless constants; and E is a random element. An added complication is that stream channels of given order seem to be able to develop only above certain threshold sizes of catchment (Schumm, 1956); this implies that there is a hierarchy of basins and channels of different order, masked by a certain randomness, which tends to grow allometrically in a series of quantum jumps to produce a 'multimodal continuum' (Woldenberg, 1967, p. 101). The application of the concept of allometric growth to stream systems has been

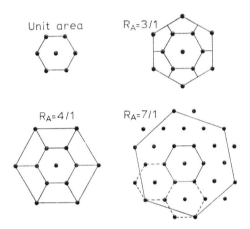

Fig. 5.40. Three basic hexagonal hierarchies of differing area ratios. Source: Woldenberg, 1968A, 1968B.

criticized by Scheidegger (1968B; 1968c), however, in that natural networks are not structurally Hortonian throughout (i.e. do not have a stable value of R_b at all areal scales within a region: Chap. 1.II(4)), which they might be expected to be if they had grown allometrically.

The foregoing considerations have led Woldenberg (1968A; 1968B; 1968c; 1969) to extend Nordbeck's (1965) idea that river networks represent an example of hierarchical allometric growth, by considering them as merely one type of hierarchical space-filling phenomenon, analogous to nested urban market areas. Woldenberg (1968A, p. 31) developed Christaller's concept of mixed hierarchies of hexagonal space-filling (i.e. the simultaneous spatial operation of different space-partitioning mechanisms) by taking the three primary geometric progressions of area ratios ($R_A = 3$, 4 and 7) (Fig. 5.40) and combining their first sixteen ranks (Table 5.3). When rank (i.e. area index—which is inversely related to areal magnitude or order) is plotted against number of areas, the

Table 5.3. Space-filling hierarchies

Rank: (Area index)	$R_A = 3$	$R_A = 4$	$R_A = 7$
1	1		
2		1	
3			1
4	3		
5		4	
6			7
7	9		
8		16	
9	27		
10			49
11		64	
12	81		
13	243		
14		256	
15			343
16	729		

Source: Woldenberg, 1968A.

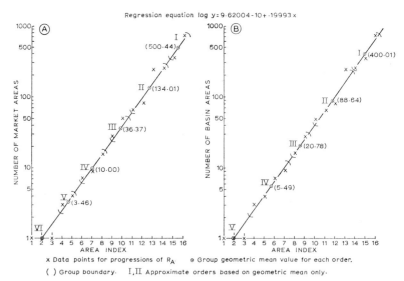

Fig. 5.41. Plots of (*A*) number of market areas, South Bothnia, Finland; (*B*) Number of Strahler stream segments in Wolfskill Canyon, California versus area index showing hierarchical groupings. Source: Woldenberg, 1968A, pp. 32, 72.

Table 5.4. Comparison of ordering systems for market areas and stream segments

	Order	Groups $(R_A = 3, 4 \& 7)$	Geometrical mean of group	Empirical data	Convergent mean of group	Christaller's model
Market Areas of	6	1, 1, 1	1	1	1	1
South Bothnia,	5	3, 4	3·46	3	3·48	3·3
Finland	4	7, 9, 16	10·00	9	10·33	10
(see Fig. 5.41-A)	3	27, 49	36·37	36	37·18	33
	2	64, 81, 243, 256	134·01	151	147·19	100
	1	343, 729	500·44	516	517·86	333
Number of Strahler	5	1, 1, 1	1	1	1	—
Stream Segments in	4	3, 4, 7, 9	5·24	5	5·49	—
Wolfskill Canyon,	3	16, 27	20·78	21	21·14	—
California	2	49, 64, 81, 243	88·64	96	97·91	—
(Maxwell, 1960)	1	256, 343, 729	400·01	409	420·25	—
(see Fig. 5.41-B)						

Source: Woldenberg, 1968A.

discrete progression can be approximated by a regression line (Fig. 5.41). It is then possible to group consecutive numbers together in a variety of ways so that the geometric means of each group form progressions which satisfy both the regression line and observed hierarchical sequences as nearly as possible. These means express the number of areas of different order of magnitude which represent optimum space-filling sequences in mixed hierarchies, and provide quite good fits to observed data of both number of markets (Fig. 5.41-A) and number of streams (Fig. 5.41-B) in a homogeneous region.

A still better fit is given by the convergent mean (a compromise between the group geometric and arithmetic means) (Table 5.4), and Woldenberg (1968A, pp. 39–41) associates this model with a balance between least overland work and economies of scale, such that the mixed hierarchical space filling represented by sequences derived from the group convergent means is held to be a response to a tendency towards an equilibrium state of least work. Nested mixed hierarchies of drainage basins are therefore viewed as equilibrium systems involving both least work (itself a compromise between the small values of length of overland flow needed to minimize theoretically the inefficient overland removal of water and debris, and the need to have large channels of high order to remove the channel runoff most efficiently) and maximum entropy, with basin areas of similar order tending to be reasonably equal in size as well as in channel and valley-side slope (i.e. giving a uniform regional distribution of available free energy) (Woldenberg, 1968A, p. 65; Leopold and Langbein, 1962; Langbein and Leopold, 1964). This refined method is further based on a grouping which begins by approximating the number of first-order basins as closely as possible, on the assumption that because their areas are determined and constrained by basic physical processes they are the fundamental building blocks of the spatial hierarchy.

The fact that this convergent mean model of mixed hierarchies is so powerful in predicting, among other things, number of basin areas of given order in a homogeneous region is indicative to Woldenberg (1968A, p. 68) that nested systems of drainage basins behave as though an hexagonal determinism were at work, even though natural basins are not hexagonal in terms of their conventional geometry. It is suggested that an appropriate spatial transformation proper to the areal energy expenditure of the system might reveal this underlying hexagonal behaviour. Further, Warntz (1968) has suggested that the hexagonal nesting might be better exemplified by considering the stream channels themselves as the edges of the hexagons, presumably containing overland flows from the hilltops. It is clear that Woldenberg's deterministic network model is in conflict with the probability models of Shreve (1966; 1967) and Scheidegger (1968B; 1968C), and it is claimed that the former has much greater flexibility in accommodating the observed natural variations of the bifurcation ratio (Woldenberg, 1968A, pp. 74–5; 1969, p. 104).

It is clear from Woldenberg's work that it is possible to move from the study of networks to that of other combinatorial structures. One of the dual strengths of network analysis is that it not only builds some links between the physical and behavioural sides of geography—a strength foreseen by Packe (1743) over two centuries ago—but also enables transformations to be made between different spatial structures within geography. The increased flexibility that stems from this approach has yet to be fully exploited in contemporary research.

References

AKERS, S. B., JR. (1960). The use of the Wye-Delta transformations in network simplification. *Operations Research,* **8,** 311–23.
ALEXANDER, C. (1964). *Notes on the synthesis of form.* Cambridge, Mass.
APPLETON, J. H. (1962). *The geography of communications in Great Britain.* London.
ASHLEY, R. H. and W. F. BERARD (1965). Interchange development along 180 miles of I-94. *Highway Research Record,* **96,** 46–58.
AVONDO-BODINO, G. (1962). *Economic applications of the theory of graphs.* New York.
BATES, C. C. (1953). Rational theory of delta formation. *Bulletin of the Association of American Petroleum Geologists,* **37,** 2119–62.
BATTERSBY, A. (1964). *Network analysis for planning and scheduling.* London.
BEATY, C. B. (1963). Origin of alluvial fans, White Mountains, California and Nevada. *Annals of the Association of American Geographers,* **53,** 516–35.
BECKENBACH, E. F. (1964). *Applied combinatorial mathematics.* New York.
BECKMANN, M. J. (1952). A continuous model of transportation. *Econometrica,* **20,** 643–60.
BECKMANN, M. J. (1967). Principles of optimum location for transportation networks. *Northwestern University, Studies in Geography,* **13,** 95–119.
BECKMANN, M. J. and T. MARSCHAK (1955). An activity approach to location theory. *Kyklos,* **8,** 125–43.
BELL TELEPHONE LABORATORIES (1966). A rapid route to the shortest path. *Science,* **154,** 295. (Paper by A. M. Noll.)
BENES, V. E. (1965). *Mathematical theory of connecting networks and telephone traffic.* New York.
BENSON, M. A. (1959). Channel-slope factor in flood-frequency analysis. *Proceedings of the American Society of Civil Engineers, Journal of the Hydraulics Division.* HY **4,** 1994, 1–9.
BERGE, C. (1962). *Theory of graphs and its application.* London.
BERRY, B. J. L. (1960). An inductive approach to the regionalization of economic development. *University of Chicago, Department of Geography, Research Paper,* **62,** 78–107.
BERRY, B. J. L. (1961). A method for deriving multifactor uniform regions. *Przeglad Geograficzny,* **33,** 263–82.
BERRY, B. J. L. (1964). Cities as systems within systems of cities. *Papers, Regional Science Association,* **13,** 147–63.
BERRY, B. J. L. (1966). Essays on commodity flows and the spatial structure of the Indian economy. *University of Chicago, Department of Geography, Research Papers,* **111.**
BERRY, B. J. L. (1967). Grouping and regionalization: an approach to the problem using multivariate analysis. *Northwestern University, Studies in Geography,* **13,** 219–51.
BLACK, W. R. (1967). Growth of the railway network of Maine: a multivariate approach. *University of Iowa, Department of Geography, Discussion Papers,* **5.**
BLACKMAN, R. B. and J. W. TUKEY (1959). *The measurement of power spectral.* New York.
BLAIR, D. J. and T. H. BISS (1967). The measurement of shape in geography. *Nottingham University, Department of Geography, Quantitative Bulletin,* **11.**
BLAISDELL, R. F. (1964). *Sources of information in transportation.* Evanston, Illinois.
BORCHERT, J. R. (1961). The twin cities urbanized area: past, present, and future. *Geographical Review,* **51,** 47–70.

BOS, H. C. and L. M. KOYCK (1961). The appraisal of road construction projects: a practical example. *Review of Economics and Statistics*, **43**, 13–20.

BOWDEN, K. L. and J. R. WALLIS (1964). Effect of stream-ordering technique on Horton's laws of drainage composition. *Bulletin of the Geological Society of America*, **75**, 767–74.

BOYCE, R. B. and W. A. V. CLARK (1964). The concept of shape in geography. *Geographical Review*, **54**, 561–72.

BRICE, J. C. (1960). Index for description of channel braiding (Abst.). *Bulletin of the Geological Society of America*, **71**, 1833.

BRICE, J. C. (1964). Channel patterns and terraces of the Loup Rivers in Nebraska. *U.S. Geological Survey Professional Paper*, **422-D**.

BRITISH RAILWAYS BOARD, (1963). *The reshaping of British railways*. London.

BROSCOE, A. J. (1959). Quantitative analysis of longitudinal stream profiles of small watersheds. *Office of Naval Research, Geography Branch, Project NR 389-42, Technical Report*, **18**.

BROWN, L. A. (1965). Models for spatial diffusion research. *Office of Naval Research, Geography Branch, Task 389-140, Technical Report*, **3**.

BRUSH, L. M. (1961). Drainage basins, channels, and flow characteristics of selected streams in central Pennsylvania. *United States Geological Survey, Professional Paper*, **282-F**.

BUNGE, W. (1962). Theoretical geography. *Lund Studies in Geography, Series C, General and Mathematical Geography*, **1**.

BUNGE, W. (1964). Patterns of location. *Michigan Inter-University Community of Mathematical Geographers, Discussion Paper*, **3**.

BUNGE, W. (1966). Gerrymandering, geography and grouping. *Geographical Review*, **55**, 256–63.

BUNTON, R. B. and W. R. BLUNDEN (1962). An analysis of route factors for one-way and two-way street systems. *Australian Road Research Board, Proceedings, First Conference*.

BURSTALL, R. M. (1966). Computer design of electricity supply networks by a heuristic method. *The Computer Journal*, **9**.

BURTON, I. (1963). *Accessibility in northern Ontario: an application of graph theory to a regional highway network*. Ontario.

BUSACKER, R. G. and T. L. SAATY (1965). *Finite graphs and networks: an introduction with applications*. New York.

CARLSTON, C. W. (1963). Drainage density and streamflow. *United States Geological Survey, Professional Paper*, **422-C**.

CARLSTON, C. W. (1965). The relation of free meander geometry to stream discharge and its geomorphic implications. *American Journal of Science*, **263**, 864–85.

CARLSTON, C. W. (1966). The effect of climate on drainage density and streamflow. *Bulletin of the International Association of Scientific Hydrology*, **11**(3), 62–9.

CARTER, C. S. and R. J. CHORLEY (1961). Early slope development in an expanding stream system. *Geological Magazine*, **98**, 117–30.

CASETTI, E. (1964). Classificatory and regional analysis by discriminant iterations. *Office of Naval Research, Geography Branch, Contract 1228(26), Task 389-135, Technical Report*, **12**.

CATTELL, R. B. (1944). A note on correlation clusters and cluster search methods. *Psychometrika*, **9**, 169–84.

CAYLEY, G. (1879). On the colouring of maps. *Royal Geographical Society, Proceedings*, **1**, 259–61.

CHISHOLM, M. D. I. (1962). *Rural settlement and land use: an essay in location*. London.

CHORLEY, R. J. (1957A). Illustrating the laws of morphometry. *Geological Magazine*, **94**, 140–50.

CHORLEY, R. J. (1957B). Climate and morphometry. *Journal of Geology*, **65**, 628–68.

CHORLEY, R. J. (1958). Group operator variance in morphometric work with maps. *American Journal of Science*, **256**, 208–18.

CHORLEY, R. J. (1962). Geomorphology and general systems theory. *United States, Geological Survey, Professional Paper*, **500–B.**

CHORLEY, R. J. and P. HAGGETT, Editors (1965A). *Frontiers in geographical teaching*. London.

CHORLEY, R. J. and P. HAGGETT (1965B). Trend-surface mapping in geographical research. *Institute of British Geographers, Publications*, **37**, 47–67.

CHORLEY, R. J. and P. HAGGETT, Editors (1967). *Models in geography*. London.

CHORLEY, R. J., Editor (1969). *Water, earth and man*. London.

CHORLEY, R. J. and M. A. MORGAN (1962). Comparison of morphometric features, Unaka Mountains, Tennessee and North Carolina, and Dartmoor, England. *Bulletin of the Geological Society of America*, **73**, 17–34.

CHOW, VEN TE, Editor (1964). *Handbook of applied hydrology*. New York.

CHRISTALLER, W. (1933). *Die Zentralen Orte in Süddeutschland*. Jena.

CLAPHAM, J. C. (1964). Adjustment of regional boundaries to keep the contained workloads within a prescribed tolerance. *Canadian Operations Research Society Journal*, **2**, 71–82.

CLARK, C. (1967). *Population growth and land use*. London.

CLARK, C. O. (1945). Storage and the unit hydrograph. *Transactions of the American Society of Civil Engineers*, **110**, 1419–88

CLARK, P. J. (1956). Grouping in spatial distributions. *Science*, **123**, 373–4.

CLOZIER, R. (1963). *Géographie de la circulation*. Paris.

COATES, D. R. (1958). Quantitative geomorphology of small drainage basins of southern Indiana. *Office of Naval Research, Project NR 389-042, Technical Report*, **10.**

COBURN, T. M., M. E. BEESLEY, and D. J. REYNOLDS (1960). The London–Birmingham motorway: traffic and economics. *Department of Industrial and Scientific Research, Road Research Technical Paper*, **46.**

COHN, D. L. (1954). Optimal systems: I. The vascular system. *Bulletin of Mathematical Biophysics*, **16**, 59–74.

COOKENBOO, L., JR. (1955). *Crude oil pipe lines and competition in the oil industry*. Cambridge, Mass.

COOPER, L. (1967). Solutions of generalized locational equilibrium models. *Journal of Regional Science*, **7**, 1–18.

COPE, D. R. (1967). A network analysis of the growth of the London underground system and its relation to population changes, 1863–1972. University of Cambridge, B.A. Dissertation.

CORTE, A. and A. HIGASHI (1964). Experimental research on desiccation cracks in soil. *United States Army Material Command, Cold Regions Research and Engineering Laboratory, Research Report*, **66.**

COTTON, C. A. (1963). Development of fine-textured landscape relief in temperate pluvial climates. *New Zealand Journal of Geology and Geophysics*, **6(4)**, 528–33.

COURANT, R. (1937). *Differential and integral calculus*. New York.

COX, K. R. (1965). The application of linear programming to geographic problems. *Tijdschrift voor Economische en Sociale Geografie*, **56**, 228–36.

COX, K. R. (1968). On the utility and definition of regions in comparative political sociology. Ohio State University, Department of Geography, Unpublished Paper.

COXETER, H. S. M. (1961). *Introduction to geometry*. New York.

CRAWFORD, N. H. and R. K. LINSLEY (1966). Digital simulation in hydrology: Stanford Watershed Model. IV. *Stanford University, Department of Civil Engineering, Technical Report*, **39**, 210.

CREIGHTON, R. L., I. HOCH, and M. SCHNEIDER (1959). The optimum spacing of arterials and expressways. *Traffic Quarterly*, **13**, 447–94.

CUMMINGS, L. P. (1967). The structure of networks and network flows. University of Iowa, Ph.D. Dissertation.

DACEY, M. F. (1963). Certain properties of edges on a polygon in two dimensional aggregate of polygons having randomly distributed nuclei. *Report*. Philadelphia, Penn. Mimeo.

DACEY, M. F. (1965). A review of measures of contiguity for two and *k*-colour maps. *Office of Naval Research Contract Nonr 1228(33), Technical Report*, **2.**

DACEY, M. F. (1967). Description of line patterns. *Northwestern University, Studies in Geography*, **13,** 277–87.

DACEY, M. F. and T. TUNG (1962). The identification of randomness in point patterns, I. *Journal of Regional Science*, **4,** 83–96.

DANTZIG, G. B. (1963). *Linear programming and extensions*. Princeton.

DAVIS, W. M. (1899). The geographical cycle. *Geographical Journal*, **14,** 481–504.

DE WIEST, R. J. M. (1965). *Geohydrology*. New York.

DIRICHLET, G. L. (1850). Über die Reduction der positeven quadratischen. Formen mit drei unbestimmten ganzen Zahlen. *Journal für die reine und angewandte Mathematik*, **40,** 209–27.

DOMANSKI, R. (1963). Zespoy sieci komunikacyjnych. *Instytut Geografii Polskiej Akademii Nauk, Prace Geograficzne*, **41.**

DORAN, J. E. and D. MICHIE (1966). Experiments with the graph transverser program. *Proceedings of the Royal Society, Series A*, **294,** 235–95.

DURY, G. H. (1953). The shrinkage of the Warwickshire Itchen. *Proceedings of the Coventry Natural History and Scientific Society*, **2,** 208–14.

DURY, G. H. (1964A). Principles of underfit streams. *United States Geological Survey Professional Paper*, **452-A.**

DURY, G. H. (1964B). Subsurface exploration and chronology of underfit streams. *United States Geological Survey Professional Paper*, **452-B.**

ENKE, S. (1951). Equilibrium among spatially separated markets: solution by electric analogues. *Econometrica*, **19,** 40–8.

EYLES, R. J. (1966). Stream representation on Malayan maps. *Journal of Tropical Geography*, **22,** 1–9.

EYLES, R. J. (1968). Stream net ratios in west Malaysia. *Bulletin of the Geological Society of America*, **79,** 701–12.

FARBEY, B. A. and J. D. MURCHLAND (1967). Towards an evaluation of road system designs. *Regional Studies*, **1,** 27–37.

FISHER, H. J. and N. A. BOUKIDIS (1963). The consequences of obliquity in arterial systems. *Traffic Quarterly*.

FLAMENT, C. (1963). *Applications of graph theory to group structure*. Englewood Cliffs, N.J.

FLEISCHER, G. A. (1963). Effect of highway improvement on travel time. *Highway Research Record*, **12,** 19–47.

FLOOD, M. M. (1956). The travelling salesman problem. *Operations Research*, **4,** 61–75.

FLUGGE, S. (1958). *Handbuch der Physik*. Berlin.

FOGEL, R. W. (1964). *Railroads and American economic growth: essays in econometric history*. Baltimore.

FOOT, D. H. S. (1965). The shortest-route problem: Algol programs and a discussion of computational problems in large network applications. *University of Bristol, Department of Economics, Discussion Papers*, **10.**

FORD, L. R., JR. and D. R. FULKERSON (1962). *Flows in networks*. Princeton.

FOSTER, E. E. (1949). *Rainfall and runoff*. New York.

FRIEDLANDER, A. F. (1965). *The interstate highway system: a study in public investment*. Amsterdam.

GARRISON, W. L. (1959–60). Spatial structure of the economy. *Annals of the Association of American Geographers*, **49,** 232–9, 471–82; **50,** 357–73.

GARRISON, W. L. (1960). Connectivity of the interstate highway system. *Regional Science Association, Papers and Proceedings*, **6,** 121–37.

GARRISON, W. L. and D. F. MARBLE (1958). Analysis of highway networks: a linear programming formulation. *Highway Research Board, Proceedings*, **37,** 1–17.

GARRISON, W. L. and D. F. MARBLE (1962). The structure of transportation networks. *U.S. Army Transportation Command, Technical Report*, **62–11.**

GARRISON, W. L. and D. F. MARBLE (1964). Factor analytic study of the connectivity of a transportation network. *Regional Science Association, Papers*, **12,** 231–38.

GARRISON, W. L., B. J. L. BERRY, D. F. MARBLE, J. D. NYSTUEN, and R. L. MORRILL (1959). *Studies of highway development and geographic change.* Seattle.

GAUTHIER, H. L. (1966). Highway development and urban growth in Sao Paulo, Brazil: a network analysis. Northwestern University, Ph.D. Dissertation.

GAZIS, D. C. (1967). Mathematical theory of automobile traffic. *Science*, **157,** 273–81.

GETIS, A. (1963). The determination of the location of retail activities with the use of a map transformation. *Economic Geography*, **39,** 1–22.

GILBERT, E. N. (1941). Random plane networks. *Journal of the Society for Industrial and Applied Mathematics*, **9,** 533–43.

GINSBURG, N. (1961). *Atlas of economic development.* Chicago.

GIUSTI, E. V. and W. J. SCHNEIDER (1965). The distribution of branches in river networks. *United States Geological Survey Professional Paper*, **422–G.**

GLOCK, W. S. (1931). The development of drainage systems: a synoptic view. *Geographical Review*, **21,** 475–82.

GODDARD, J. (1968). Functional regions within the city centre, a factor analytic study of taxi flows in central London. *London School of Economics, Graduate School of Geography, Discussion Papers*, **17.**

GODLUND, S. (1952). Ein Innovationsverlauf in Europa. *Lund Studies in Geography, Series B, Human Geography*, **6.**

GOULD, P. R. (1960). The development of the transportation pattern in Ghana. *Northwestern University, Studies in Geography*, **5.**

GOULD, P. R. (1965). On the geographical access of points to points. University Park, Penn. MS.

GOULD, P. R. (1966). Space-searching procedures in geography and the social sciences. *University of Hawaii, Social Science Research Institute, Papers*, **1.**

GOULD, P. R. (1967). On the geographical interpretation of eigenvalues. *Institute of British Geographers, Publications.*

GOULD, P. R. and T. R. LEINBACH (1966). An approach to the geographic assignment of hospital services. *Tijdschrift voor Economische en Sociale Geografie*, **57,** 203–6.

GRAVELIUS, H. (1914). *Flusskunde.* **1.** Berlin and Leipzig.

GRAY, D. M. (1961). Interrelationships of watershed characteristics. *Journal of Geophysical Research*, **66,** 1215–23.

GREGORY, K. J. (1966). Dry valleys and the composition of the drainage net. *Journal of Hydrology*, **4,** 327–40.

GREGORY, K. J. and D. E. WALLING (1968). The variation of drainage density within a catchment. *Bulletin of the International Association of Scientific Hydrology*, **13(2),** 61–8.

GREGORY, S. (1963). *Statistical methods and the geographer.* London.

GUMBEL, E. J. (1958). Statistical theory of floods and droughts. *Journal of the Institution of Water Engineers*, **12,** 157–84.

HACK, J. T. (1957). Studies of longitudinal stream profiles in Virginia and Maryland. *United States Geological Survey Professional Paper*, **294–B.**

HACK, J. T. (1965). Postglacial drainage evolution and stream geometry in the Ontonagon Area, Michigan. *United States Geological Survey Professional Paper*, **504-B.**

HÄGERSTRAND, T. (1967). *Innovation diffusion as a spatial process.* Chicago.

HAGGETT, P. (1965). *Locational analysis in human geography.* London.

HAGGETT, P. (1966). On certain statistical regularities in the structure of transport networks. Mimeo.

HAGGETT, P. (1967). On the extension of the Horton combinatorial algorithm to regional highway networks. *Journal of Regional Science*, **7,** 281–90.

HAGGETT, P., G. JAMES, A. D. CLIFF, and J. K. ORD (1969). Some discrete distributions for graphs. Unpublished MS.

HAGGETT, P. and A. D. CLIFF (In preparation). Contiguity constraints in regionalization programs. *S.S.R.C., Technical Report.*

HAGOOD, M. J. and D. O. PRICE (1952). *Statistics for sociologists.* New York.

HAIRE, M. (1959). *Modern organization theory.* New York.

HARARAY, F., R. Z. NORMAN, and D. CARTWRIGHT (1965). *Structural models: an introduction to the theory of directed graphs.* New York.

HARRIS, C. C. J. R. (1964). A scientific method of districting. *Behavioral Science*, **9,** 219–25.

HEBERT, B. (1966). *Use of factor analysis in graph theory to identify an underlying structure of transportation networks.* Ohio State University, Department of Geography.

HERMAN, R., Editor (1961). *Theory of traffic flow: Proceedings of the symposium on the theory of traffic flow held at the General Motors research laboratories, Warren, Michigan (U.S.A.).* Amsterdam.

HESS, S. W., J. B. WEAVER, H. J. SIEGFELT, J. N. WHELAN, and P. A. ZITLAU (1965). Nonpartisan re-districting by computer. *Operations Research*, **13,** 998–1006.

HITCHCOCK, F. L. (1941). The distribution of a product from several sources to numerous localities. *Journal of Mathematical Physics*, **20,** 224–30.

HOLMES, G. W. (1955). Morphology and hydrology of the Mint Julep area, Southwest Greenland. *Mint Julep Reports, Part II, Arctic, Desert, Topic Information Centre, United States Air University Publication*, **A-104-B.**

HOLROYD, E. M. (1966). Theoretical average journey lengths in circular towns with various routing systems. *Ministry of Transport, Road Research Laboratory, Reports*, **43.**

HORTON, F., Editor (1968). Geographic studies of urban transportation and network analysis. *Northwestern University, Studies in Geography*, **16.**

HORTON, R. E. (1932). Drainage basin characteristics. *Transactions of the American Geophysical Union*, **13,** 350–61.

HORTON, R. E. (1945). Erosional development of streams and their drainage basins: hydrophysical approach to quantitative morphology. *Bulletin of the Geological Society of America*, **56,** 275–370.

HOWARD, A. D. (1967). Drainage analysis in geologic interpretation: a summation. *Bulletin of the American Association of Petroleum Geologists*, **51,** 2246–2259.

HOWARD, A. D. (1968). Stream capture, bifurcation angle modification, and rate of work in stream systems. Department of Geography, The Johns Hopkins University, Baltimore. Mimeo.

HOWE, G. M., H. O. SLAYMAKER, and D. M. HARDING (1967). Some aspects of the flood hydrology of the upper catchments of the Severn and Wye. *Transactions of the Institute of British Geographers*, **41,** 33–58.

HULTLAND, G. (1962). Virkestransporterna i Kalix älvdal, 1951–60. *Geographica, Shrifter från Uppsala Universitets Geografiska Institution*, **27.**

ISARD, W. (1956). *Location and space-economy: a general theory relating to industrial location, market areas, land use, trade and urban structure.* New York.

ISARD, W., D. F. BRAMHALL, G. A. P. CARROTHERS, J. H. CUMBERLAND, L. N. MOSES, D. O. PRICE, and E. W. SCHOOLER (1960). *Methods of regional analysis: an introduction to regional science.* New York.

JAMES, P. E., C. F. JONES, and J. K. WRIGHT, Editors (1954). *American geography: inventory and prospect.* Syracuse.

JOHNSON, D. W. (1933). Development of drainage systems and the dynamic cycle. *Geographical Review,* **23,** 114–21.

JOHNSON, H. B. (1957). Rational and ecological aspects of the quarter section: an example from Minnesota. *Geographical Review,* **47,** 330–48.

JOHNSON, W. A. and R. F. MEHL (1939). Reaction kinetics in processes of nucleation and growth. *Transactions of the American Institute of Mining and Metalurgical Engineers,* **135,** 416–58.

JUDSON, S. and G. W. ANDREWS (1955). Pattern and form of some valleys in the Driftless Area, Wisconsin. *Journal of Geology,* **63,** 328–36.

JUKES, J. B. (1962). The mode of formation of some river valleys in the south of Ireland. *Quarterly Journal of the Geological Society of London,* **18,** 378–403.

KAISER, H. F. (1966). An objective method for establishing legislative districts. *Midwest Journal of Political Science,* **10.**

KANSKY, K. J. (1963). Structure of transport networks: relationships between network geometry and regional characteristics. *University of Chicago, Department of Geography, Research Papers,* **84.**

KENDALL, M. G. and W. R. BUCKLAND (1957). *A dictionary of statistical terms.* Edinburgh.

KING, B. (1967). Step-wise clustering procedures. *Journal of the American Statistical Association,* **62,** 200–10.

KIRKBY, M. J. and R. J. CHORLEY (1967). Throughflow, overland flow and erosion. *Bulletin of the International Association of Scientific Hydrology,* **12(3),** 5–21.

KISSLING, C. C. (1966). Transportation networks, accessibility, and urban functions: an empirical and theoretical analysis. McGill University, Department of Geography, Ph.D. Dissertation.

KLEINROCK, L. (1964). *Communication nets: stochastic message flow and delay.* New York.

KNEESE, A. V. and S. C. SMITH, Editors (1966). *Water Research.* Baltimore.

KOHL, J. G. (1850). *Der Verkehr und die Ansiedelungen der Menschen in ihrer Abhangigkeit von der Gestaltung der Erdoberflache.* Leipzig.

KONIG, D. (1950). *Theorie der endlichen und unendlichen Graphen: kombinatorische Topologie der Streckenkomplexe.* New York.

KOOPMANS, T. C. (1949). Optimum utilization of the transportation system. *Econometrica,* **17,** 136–46.

KOPEC, R. J. (1963). An alternative method for the construction of Thiessen polygons. *Professional Geographer,* **15(5),** 24–6.

KRUMBEIN, W. C. (1967). FORTRAN IV computer programs for Markov chain experiments in geology. *State Geological Survey, University of Kansas, Computer Contribution,* **13.**

KRUMBEIN, W. C. and F. A. GRAYBILL (1965). *An introduction to statistical models in geology.* New York.

KU, Y. U. and S. D. BEDROSIAN (1965). On topological approaches to network theory. *Journal of the Franklin Institute,* **279,** 11–21.

LACHENBRUCH, A. H. (1962). Mechanics of thermal contraction cracks and ice-wedge polygons in permafrost. *Geological Society of America, Special Paper,* **70.**

LACHENE, R. (1965). Networks and the location of economic activities. *Regional Science Association, Papers,* **14,** 183–96.

LALANNE, L. (1863). Essai d'une theorie des reseaux de chemin de fer, fondee sur l'observation des faits et sur les lois primordiales qui president au groupement des populations. *Comptes Rendus Hebdomadaires des Seances de l'Academie des Sciences,* **42,** 206–10.

LANGBEIN, W. B. (1947). Topographic characteristics of drainage basins. *United States Geological Survey Water Supply Paper*, **968–C**, 125–57.

LANGBEIN, W. B. and L. B. LEOPOLD (1964). Quasi-equilibrium states in channel morphology. *American Journal of Science*, **262**, 782–94.

LANGBEIN, W. B. and L. B. LEOPOLD (1966). River meanders-theory of minimum variance. *United States Geological Survey, Professional Paper*, **422–H.**

LANGBEIN, W. B. and L. B. LEOPOLD (1968). River channel bars and dunes —Theory of kinetic waves. *United States Geological Survey, Professional Paper*, **422–L.**

LANGBEIN, W. B. and S. A. SCHUMM (1958). Yield of sediment in relation to mean annual precipitation. *Transactions of the American Geophysical Union*, **39**, 1076–84.

LAURENSON, E. M. (1964). Catchment storage model for runoff routing. *Journal of Hydrology*, **2**, 141–63.

LAWLER, E. L. and D. E. WOOD (1966). Branch-and-bound methods: survey. *Operations Research*, **14**, 699–719.

LEOPOLD, L. B. (1962). Rivers. *American Scientist*, **50**, 511–37.

LEOPOLD, L. B. and W. B. LANGBEIN (1962). The concept of entropy in landscape evolution. *United States Geological Survey, Professional Paper*, **500–A.**

LEOPOLD, L. B. and J. P. MILLER (1956). Ephemeral streams: hydraulic factors and their relation to the drainage net. *United States Geological Survey, Professional Paper*, **282–A.**

LEOPOLD, L. B. and M. G. WOLMAN (1957). River channel patterns: braided, meandering and straight. *United States Geological Survey, Professional Paper*, **282–B**, 39–85.

LEOPOLD, L. B. and M. G. WOLMAN (1960). River meanders. *Bulletin of the Geological Society of America*, **71**, 769–94.

LEOPOLD, L. B., M. G. WOLMAN, and J. P. MILLER (1964). *Fluvial processes in geomorphology*. San Francisco.

LEVINSON, H. S. and K. R. ROBERTS (1965). System configuration in urban transportation planning. *Highway Research Board, Record*, **64**, 71–83.

LEWIN, K. (1936). *Principles of topological psychology*. New York.

LIAO, L. H. and A. E. SCHEIDEGGER (1968). A computer model of some branching-type phenomena in hydrology. *Bulletin of the International Association of Scientific Hydrology*, **13(1)**, 5–13.

LIGHTHILL, M. J. and G. B. WHITHAM (1955). On Kinematic waves II. A theory of traffic flow on long crowded roads. *Royal Society of London, Proceedings, Series A*, **229**, 317–45.

LIN, SHEN (1965). Computer solutions of the travelling salesman problem. *Bell System Technical Journal*, **44**, 2245–69.

LINSLEY, R. K. and J. B. FRANZINI (1964). *Water resources engineering*, New York.

LINSLEY, R. K., M. A. KOHLER, and L. H. PAULHUS (1949). *Applied hydrology*, New York.

LITTLE, J. D. C., K. G. MURTY, D. W. SWEENEY, and C. KAREL (1963). An algorithm for the travelling salesman problem. *Operations Research*, **11**, 972–89.

LÖSCH, A. (1954). *The economics of location*. New Haven, Conn.

LOWRY, I. S. (1965). A short course in model design. *Journal of the American Institute of Planners*, **31**, 158–66.

LUBOWE, J. K. (1964). Stream junction angles in the dendritic drainage pattern. *American Journal of Science*, **262**, 325–39.

LUSTIG, L. K. (1965). Clastic sedimentation in Deep Springs Valley, California. *United States Geological Survey, Professional Paper*, **352–F**, 131–92.

MACKIN, J. H. (1948). Concept of the graded river. *Bulletin of the Geological Society of America*, **59**, 463–512.

MCQUITTY, L. L. (1957). Elementary linkage analysis for isolating orthogonal and oblique types and typal relevancies. *Educational and Psychological Measurement*, **17**, 207–29.

MAHALANOBIS, P. C., C. R. RAO, and D. M. MAJUMDAR (1949). Anthropometric survey of the United Provinces, 1941: a statistical study. *Sankhya*, **9**, 89–324.

MANDELBROT, B. (1967). How long is the coast of Britain ? Statistical self-similarity and fractional dimension. *Science*, **156**, 636–8.

MANHEIM, M. L. (1964). Highway route location as a hierarchically-structured sequential decision process: an experiment in the use of Bayesian decision theory for guiding an engineering process. *Massachusetts Institute of Technology, Civil Engineering Systems Laboratory, Research Report*, **R64-15.**

MANHEIM, M. L. (1966). Problem solving processes in planning and design. *Design Quarterly*, **66–67,** 31–40.

MARANZANA, F. E. (1964). On the location of supply points to minimize transportation costs. *Operational Research Quarterly*, **15,** 261–70.

MARBLE, D. F. (1965). NODAC: a computer programme for the computation of two simple node accessibility measures in networks. *Office of Naval Research, Geography Branch, Contract 1228-33 (Task 389-135), Technical Report*, **10.**

MAXWELL, J. C. (1955). The bifurcation ratio in Horton's law of stream numbers. *Transactions of the American Geophysical Union*, **36,** 520.

MAXWELL, J. C. (1960). Quantitative geomorphology of the San Dimas Experimental forest, California. *Office of Naval Research, Geography Branch, Project NR 389-042, Technical Report*, **19.**

MEAD, W. R. and E. H. BROWN (1962). *United States and Canada*. London.

MEIJERING, J. L. (1953). Interface area, edge length and number of vertices in crystal aggregates with random nucleation. *Phillips Research Reports*, **8,** 270–90.

MEINIG, D. W. (1962). A comparative historical geography of two railnets: Columbia basin and South Australia. *Annals of the Association of American Geographers*, **52,** 394–413.

MELTON, F. A. (1959). Aerial photographs and structural geomorphology. *Journal of Geology*, **67,** 351–70.

MELTON, M. A. (1957). An analysis of the relations among elements of climate, surface properties, and geomorphology. *Office of Naval Research, Geography Branch, Project NR 389-042, Technical Report*, **11.**

MELTON, M. A. (1958A). Geometric properties of mature drainage systems and their representation in an E_4 phase space. *Journal of Geology*, **66,** 25–54.

MELTON, M. A. (1958B). Correlation structure of morphometric properties of drainage systems and their controlling agents. *Journal of Geology*, **66,** 442–460.

MELTON, M. A. (1958C). Lists of sample parameters of quantitative properties of landforms: their use in determining the size of geomorphic experiments. *Office of Naval Research, Geography Branch, Project NR 389-042. Technical Report*, **16.**

MELTON, M. A. (1959). A derivation of Strahler's channel-ordering system. *Journal of Geology*, **67,** 345–6.

MIEHLE, W. (1958A). Link-length minimization in networks. *Operations Research*, **6,** 232–43.

MIGAYI, M. (1966). Nodal regionalization of the interstate highway system. *Ohio State University, Department of Geography Report*. Mimeo.

MILLER, A. J. (1967). On spiral road networks. *Transportation Science*, **1,** 109–25.

MILLER, R. L. and J. S. KAHN (1962). *Statistical Analysis in the Geological Sciences*, New York.

MILLER, V. C. (1953). A quantitative geomorphic study of drainage basin characteristics in the Clinch Mountain area, Virginia and Tennessee. *Office of Naval Research, Geography Branch, Project NR 389-042, Technical Report*, **3.**

MILLS, G. (1966A). A decomposition algorithm for the shortest-route problem. *Operations Research*, **14,** 279–91.

MILLS, G. (1966B). The determination of electoral boundaries. *University of Bristol, Department of Economics, Discussion Papers*, **18.**

MILLS, G. (1968). A heuristic approach to some shortest route problems. *Journal of the Canadian Operational Research Society*, **6,** 20–5.

MILTON, L. E. (1965). Quantitative expression of drainage net patterns; *Australian Journal of Science*, **27(8),** 238–40.

MILTON, L. E. (1966). The geomorphic irrelevance of some drainage net laws. *Australian Geographical Studies*, **4,** 89–95.

MILTON, L. E. and C. D. OLLIER (1965). A code for labelling streams, basins, and junctions in a drainage net. *Journal of Hydrology*, **3,** 66–8.

MINGHI, J. (1963). Boundary studies in political geography. *Annals of the Association of American Geographers*, **53,** 407–28.

MINISTRY OF TRANSPORT (1953). *Traffic in towns: a study of the long term problems of traffic in urban areas.* London.

MOMSEN, R. P. (1963). Routes across the Serra do Mar: the evolution of transportation in the highlands of Rio de Janeiro and Sao Paulo. *Revista Geografica*, **32,** 5–167.

MOORE, E. F. (1959). The shortest path through a maze. *Annals of the Computation Laboratory of Harvard University*, **30.**

MORISAWA, M. E. (1957). Accuracy of determination of stream lengths from topographic maps. *Transactions of the American Geophysical Union*, **38,** 86–8.

MORISAWA, M. E. (1959). Relation of quantitative geomorphology to stream flow in representative watersheds of the Appalachian plateau province. *Office of Naval Research, Geography Branch, Project NR 389–042, Technical Report*, **20.**

MORISAWA, M. E. (1962). Quantitative geomorphology of some watersheds in the Appalachian Plateaus. *Bulletin of the Geological Society of America*, **73,** 1025–46.

MORISAWA, M. E. (1963). Distribution of stream-flow direction in drainage patterns. *Journal of Geology*, **71,** 528–9.

MORISAWA, M. E. (1964). Development of drainage systems on an upraised lake floor. *American Journal of Science*, **262,** 340–54.

MORISAWA, M. E. (1968). *Streams: Their dynamics and morphology.* New York.

MORRILL, R. L. (1962). Simulation of central place patterns over time. *Lund Studies in Geography, Series B, Human Geography*, **24,** 109–20.

MORRILL, R. L. (1965). Migration and the growth of urban settlement. *Lund Studies in Geography, Series B, Human Geography*, **26.**

MUELLER, J. E. (1968). An introduction to the hydraulic and topographic sinuosity indexes. *Annals of the Association of American Geographers*, **58,** 371–85.

MURCHLAND, J. D. (1965). A new method for finding all elementary paths in a complete directed graph. *London School of Economics, Transport Networks Theory Group, Report*, **22.**

MURCHLAND, J. D. (1966). Traffic assignment by digital computer. *London School of Economics, Transport Network Theory Group, Reports*, **32.**

MURRAY, C. D. (1927). On the branching-angles of trees. *Journal of General Physiology*, **10,** 725.

NAGEL, S. S. (1965). Simplified bipartisan computer redistricting. *Stanford Law Review*, **17,** 863–99.

NASH, J. E. (1960). A unit hydrograph study, with particular reference to British catchments. *Proceedings of the Institution of Civil Engineers*, **17,** 249–82.

NOLL, A. M. (1966). Computers and the visual arts. *Design Quarterly*, **66–67,** 65–71.

NORDBECK, S. (1964). Computing distances in road nets. *Regional Science Association, Papers*, **12,** 207–20.

NORDBECK, S. (1965). The law of allometric growth. *Michigan Inter-University Community of Mathematical Geographers, Discussion Paper*, **7.**

NYSTUEN, J. D. (1966). Effects of boundary shape and the concept of local convexity. *Michigan Inter-University Community of Mathematical Geographers, Discussion Papers*, **10A.**

NYSTUEN, J. D. and M. F. DACEY (1961). A graph theory interpretation of nodal regions. *Regional Science Association, Papers and Proceedings*, **7**, 29–42.

O'DONNELL, T. (1966). Computer evaluation of catchment behaviour and parameters significant in flood hydrology. *Symposium on River Flood Hydrology, Institution of Civil Engineers, London*, 103–13.

OLSON, E. C. and R. L. MILLER (1958). *Morphological integration*. Chicago.

OLSSON, G. (1965). Distance and human interaction: a review and bibliography. *Regional Science Research Institute, Bibliography Series*, **2**.

ONGLEY, E. D. (1968A). An analysis of the meandering tendency of Serpentine Cave, N.S.W. *Journal of Hydrology*, **6**, 15–32.

ONGLEY, E. D. (1968B). Towards a precise definition of drainage basin axis. *Australian Geographical Studies*, **6**, 84–8.

ORD, J. K. (1967). On a system of discrete distributions. *Biometrika*, **54**, 649–656.

ORDEN, A. (1956). The trans-shipment problem. *Management Science*, **2**, 276–85.

ORE, O. (1963). *Graphs and their uses*. New York.

OWENS, D. (1968). Estimates of the proportion of space occupied by roads and footpaths in towns. *Ministry of Transport, Road Research Laboratory Report*, LR–74.

PACKE, C. (1743). *A discourse upon the generation and form of the River Stour of East Kent delivered in a letter to the Royall-Society by the hands of Cromwell Mortimer, Sec. R.S. MS.*

PEDERSEN, P. O. (1967). On the geometry of administrative areas. Copenhagen, MS. Report.

PELTIER, L. C. (1962). Area sampling for terrain analysis, *Professional Geographer*, **14(2)**, 24–8.

PERKAL, J. (1966). On the length of empirical curves and an attempt at objective generalization. *Michigan Inter-University Community of Mathematical Geographers, Discussion Papers*, **10B**.

PERLE, E. D. (1964). The demand for transportation: regional and commodity studies in the United States. *University of Chicago, Department of Geography, Research Papers*, **95**.

PETERSON, J. M. (1961). Freeway spacing in an urban freeway system. *American Society of Civil Engineers, Transactions*, **126**, 385.

PHILBRICK, A. K. (1957). Principles of areal functional organization in regional human geography. *Economic Geography*, **33**, 299–336.

PITTS, F. R. (1965). A graph theoretic approach to historical geography. *Professional Geographer*, **17(5)**, 15–20.

POLLACK, M. and W. WIEBENSON (1960). Solutions of the shortest-route problem: a review. *Operations Research*, **8**, 224–30.

POLYA, G. (1954). *Induction and analogy in mathematics*, **1**. Princeton.

POTTER, W. D. (1953). Rainfall and topographic factors that affect runoff. *Transactions of the American Geophysical Union*, **34**, 67–73.

POTTER, P. E. and F. J. PETTIJOHN (1963). *Paleocurrents and basin analysis*. Berlin.

POWELL, J. W. (1875). *Exploration of the Colorado River of the West 1869–72*. Washington D.C.

PRED, E. R. (1966). *The spatial dynamics of U.S. urban-industrial growth, 1800–1914: interpretive and theoretical essays*. Cambridge, Mass.

PRIHAR, Z. (1956). Topological properties of telecommunication networks. *Proceedings of the Institution of Radio Engineers*, **44**, 929–33.

QUANDT, R. E. (1960). Models of transportation and optimal network construction. *Journal of Regional Science*, **2**, 27–45.

QUARMBY, D. A. (1967A). Choice of travel mode for the journey to work: some findings. *Journal of Transport Economics and Policy*, **1**, 1–42.

QUARMBY, D. A. (1967B). On the concept of generalized cost in traffic models. *Ministry of Transport, Mathematical Advisory Unit, Research Notes*, **92**.

RANALLI, G. and A. E. SCHEIDEGGER (1968A). A test of the topological structure of river nets. *Bulletin of the International Association of Scientific Hydrology*, **13(2)**, 142–53.

RANALLI, G. and A. E. SCHEIDEGGER (1968B). Topological significance of stream labeling methods. *Bulletin of the International Association of Scientific Hydrology*, 13(4), 77–85.

RAO, C. R. (1948). The utilization of multiple measurements in problems of biological classification. *Journal of the Royal Statistical Society, Series B*, **10**, 159–203.

RAPAPORT, H. and P. ABRAMSON (1959). An analog computer for finding an optimum route through a communications network. *Institute of Radio Engineers, Transactions on Communications Systems*, **CS–7**, 37–42.

RASMUSSON, G. (1962). Granite quarrying and the landscape: a comparative photogeographical study of the Malmon island, the Swedish west coast. *Lund Studies in Geography, Series C, General and Mathematical Geography*, **4.**

RATZEL, F. (1912). *Die Geographische Verbreitung des Menschen*, II. Stuttgart.

READ, R. C. (1962). Contributions to the cell-growth problem. *Canadian Journal of Mathematics*, **14**, 1–20.

REOCK, E. C. (1961). Measuring compactness as a requirement of legislative apportionment. *Midwest Journal of Political Science*, **5**, 70–4.

RICHARDSON, L. F. (1960). *Statistics of deadly quarrels*. Pittsburgh.

RIMMER, P. J. (1967). The changing status of New Zealand seaports, 1853–1960. *Annals of the Association of American Geographers*, **57**, 88–100.

ROAD RESEARCH LABORATORY (1965). *Research on road traffic*. London.

ROBERTS, P. O. (1966). The role of transport in developing countries: a development planning model. *Harvard Transportation and Economic Development Seminar, Discussion Paper*, **40.**

ROBERTS, P. O. and M. L. FUNK (1964). Toward optimum methods of link addition in transportation networks. *Massachusetts Institute of Technology, Department of Civil Engineering, Report*.

ROBERTS, P. O. and J. H. SUHRBIER (1963). Highway location analysis: an example problem. *Massachusetts Institute of Technology, Civil Engineering Systems Laboratory, Research Report*, **R62–40.**

ROCKWOOD, D. M. (1961). Columbia Basin streamflow routing by computer. *Transactions of the American Society of Civil Engineers*, **126**, 32–56.

ROSTOW, W. W. (1960). *The stages of economic growth*. Cambridge.

ROUX, W., (1895). *Ges. Abhandlungen über Entwicklungsmechanik der Organismen, Band I, Funktionelle Anpassung*. Leipzig.

RUHE, R. V. (1952). Topographic discontinuities of the Des Moines lobe. *American Journal of Science*, **250**, 46–56.

RZHANITSYN, N. A. (1964). *Morphological and hydrological regularities of the structure of the river net*. Leningrad.

SCHEIDEGGER, A. E. (1965). The algebra of stream-order numbers. *United States Geological Survey, Professional Paper*, **525–B**, 187–9.

SCHEIDEGGER, A. E. (1966A). Effect of map scale on stream orders. *Bulletin of the International Association of Scientific Hydrology*, **11(3)**, 56–61.

SCHEIDEGGER, A. E. (1966B). Stochastic branching processes and the law of stream orders. *Water Resources Research*, **2**, 199–203.

SCHEIDEGGER, A. E. (1967A). On the topology of river nets. *Water Resources Research*, **3**, 103–6.

SCHEIDEGGER, A. E. (1967B). A stochastic model for drainage patterns into an intramontaine trench. *Bulletin of the International Association of Scientific Hydrology*, **12(1)**, 15–20.

SCHEIDEGGER, A. E. (1967C). A thermodynamic analogy for meander systems. *Water Resources Research*, **3**, 1041–6.

SCHEIDEGGER, A. E. (1968A). Horton's law of stream order numbers and a temperature-analog in river nets. *Water Resources Research*, **4**, 167–71.

SCHEIDEGGER, A. E. (1968B). Horton's law of stream numbers. *Water Resources Research*, **4**, 655–8.

SCHEIDEGGER, A. E. (1986C). Horton's laws of stream lengths and drainage areas. *Water Resources Research*, **4**, 1015–21.

SCHEIDEGGER, A. E. and W. B. LANGBEIN (1966). Probability concepts in geomorphology. *United States Geological Survey, Professional Paper*, **500**–C.

SCHELLING, H. VON (1951). Most frequent particle paths in a plane. *Transactions of the American Geophysical Union*, **32**, 222–6.

SCHENCK, H. (1963). Simulation of the evolution of drainage basin networks with a digital computer. *Journal of Geophysical Research*, **68**, 5739–45.

SCHICK, A. P. (1965). The effects of lineative factors on stream courses in homogeneous bedrock. *Bulletin of the International Association of Scientific Hydrology*, **10(3)**, 5–11.

SCHNEIDER, W. J. (1961). A note on the accuracy of drainage densities computed from topographic maps. *Journal of Geophysical Research*, **66**, 3617–8.

SCHUMM, S. A. (1956). The evolution of drainage systems and slopes in badlands at Perth Amboy, New Jersey. *Bulletin of the Geological Society of America*, **67**, 597–646.

SCHUMM, S. A. (1960). The shape of alluvial channels in relation to sediment type. *United States Geological Survey, Professional Paper*, **352**–B, 17–30.

SCHUMM, S. A. (1961A). The effect of sediment characteristics on erosion and deposition in ephemeral stream channels. *United States Geological Survey, Professional Paper*, **352**–C, 31–70.

SCHUMM, S. A. (1961B). Dimensions of some stable alluvial channels. *United States Geological Survey, Professional Paper*, **424**–B, 26–7.

SCHUMM, S. A. (1963). Sinuosity of alluvial rivers on the Great Plains. *Bulletin of the Geological Society of America*, **74**, 1089–100.

SCHUMM, S. A. and R. F. HADLEY (1957). Arroyos and the semiarid cycle of erosion. *American Journal of Science*, **255**, 161–74.

SCHUMM, S. A. and R. W. LICHTY (1963). Channel widening and flood-plain construction along Cimarron River in Southwestern Kansas. *United States Geological Survey, Professional Paper*, **352**–D, 71–88.

SCHUMM, S. A. and R. W. LICHTY (1965). Time, space, and causality in geomorphology. *American Journal of Science*, **263**, 110–19.

SCOTT, A. J. (1967). A programming model of an integrated transportation network. *Regional Science Association, Papers*, **19**, 215–22.

SCOTT, A. J. (1968). On the optimal partitioning of spatially distributed point sets. In *Studies in Regional Science*, London.

SCOTT, A. J. (1969). *Approaches to the solution of the optimal network problem.* Unpublished MS.

SEBESTYEN, G. S. (1962). *Decision-making processes in pattern recognition.* London.

SHERMAN, L. K. (1932). Streamflow from rainfall by unit-graph method. *Engineering News-Record*, **108**, 501–5.

SHIMBEL, A. (1953). Structural properties of communication networks. *Bulletin of Mathematical Biophysics*, **15**, 501–7.

SHIMBEL, A. and W. KATZ (1953). A new status index derived from sociometric analysis. *Psychometrika*, **18**, 39–43.

SHREVE, R. L. (1963). Horton's 'law' of stream numbers for topographically random networks (Abst.). *Transactions of the American Geophysical Union*, **44**, 44–5.

SHREVE, R. L. (1964). Analysis of Horton's law of stream numbers (Abst.). *Transactions of the American Geophysical Union*, **45**, 50–1.

SHREVE, R. L. (1966). Statistical law of stream numbers. *Journal of Geology*, **74**, 17–37.

SHREVE, R. L. (1967). Infinite topologically random channel networks. *Journal of Geology*, **75**, 178–86.

SHYKIND, E. B. (1956). *Quantitative studies in geomorphology; subaerial and submarine erosional environments*. University of Chicago, Department of Geology, Ph.D. Dissertation.

SIDDALL, W. R. (1964). Transportation geography: a bibliography. *Kansas State University, Bibliography Series*, **1.**

SILK, J. A. (1965). Road network of Monmouthshire. University of Cambridge, Department of Geography, B.A. Dissertation.

SILVA, R. C. (1965). Reapportionment and re-districting. *Scientific American*, **213**(5), 20–7.

SITTER, L. U. DE (1956). *Structural Geology*. New York.

SMART, J. S. (1967). A comment on Horton's Law of Stream Numbers. *Water Resources Research*, **3**, 773–6.

SMART, J. S. (1968A). Statistical properties of stream lengths. *Water Resources Research*, **4**, 1001–14.

SMART, J. S. (1968B). The topological properties of channel networks. *IBM Research, RC 2310 (No. 11312)*.

SMART, J. S. and A. J. SURKAN (1967). The relation between mainstream length and area in drainage basins. *Water Resources Research*, **3**(4), 963–74.

SMART, J. S. and A. J. SURKAN and J. P. CONSIDINE (1967). Digital simulation of channel networks. *International Association of Scientific Hydrology, General Assembly of Berne, Sept.–Oct. 1967, Symposium on River Morphology*, 87–98.

SMEED, R. J. (1963). Road development in urban areas: the effect of some kinds of routing systems on the amount of traffic in the central areas of towns. *Journal of the Institution of Highway Engineers*, **10**(1), 5–26.

SMEED, R. J. (1968). Traffic studies and urban congestion. *Journal of Transport Economics and Policy*, **2**, 1–38.

SMITH, K. G. (1958). Erosional processes and landforms in Badlands National Monument, South Dakota. *Bulletin of the Geological Society of America*, **69**, 975–1008.

SMITH, K. G. (1950). Standards for grading texture of erosional topography. *American Journal of Science*, **248**, 655–68.

SMITH, R. H. T. (1964). Development and function of transport routes in southern New South Wales. *Australian Geographical Studies*, **2**, 47–65.

SMOCK, R. B. (1963). A comparative description of a capacity-restrained traffic assignment. *Highway Research Record*, **6**, 12–40.

SNYDER, F. F. (1938). Synthetic unit-graphs. *Transactions of the American Geophysical Union*, **19**, 447–54.

SOBERMAN, R. M. (1966). *Transport technology for developing regions; a study of road transportation in Venezuela*. Cambridge.

SOKAL, R. R. and P. H. A. SNEATH (1963), *Principles of numerical taxonomy*. San Francisco.

SPEIGHT, J. G. (1965). Meander spectra of the Angabunga river. *Journal of Hydrology*, **3**, 1–15.

SPENCE, N. A. (1968). A multifactor uniform regionalization of British counties on the basis of employment data for 1961. *Regional Studies*, **2**, 87–104.

STACKELBERG, H. VON (1938). Das Brechungsgesetz des Verkehrs. *Jahrbucher fur Nationalokonomie und Statistik*, **148**, 680–94.

STAIRS, S. (1965). Bibliography on the shortest-route problem. *London School of Economics, Transport Network Theory Group, Report*, **6.**

STAIRS, S. (1967). A review of computational problems of selecting an optimal road traffic network. *London Business School, Transport Network Theory Group, Report*, **51.**

STEINER, D. (1965). A multivariate statistical approach to climatic regionalization and classification. *Tijdschrift van het Koninklijk Nederlandsche Aardrijks-kundig Genootschap*, **15**, 23–35.

STODDART, D. R. (1965). The shape of atolls. *Marine Geology*, **3**, 369–83.

STODDART, D. R. (In Press.) The measurement of drainage density.

STRAHLER, A N. (1946). Elongate entrenched meanders of Conodoguinet Creek, Pa. *American Journal of Science*, **244**, 31–40.

STRAHLER, A. N. (1952). Hypsometric (area-altitude) analysis of erosional topography. *Bulletin of the Geological Society of America*, **63**, 1117–42.

STRAHLER, A. N. (1953). Revisions of Horton's quantitative factors in erosional terrain (Abst.). *Transactions of the American Geophysical Union*, **34.**

STRAHLER, A. N. (1954). Quantitative geomorphology of erosional landscapes. *C.R. of the 19th International Geological Congress, Algiers 1952*, **XV**, 341–54.

STRAHLER, A. N. (1957). Quantitative analysis of watershed geomorphology. *Transactions of the American Geophysical Union*, **38**, 913–20.

STRAHLER, A. N. (1958). Dimensional analysis applied to fluvially eroded land-forms. *Bulletin of the Geological Society of America*, **69**, 279–300.

STRAHLER, A. N. (1965). *Introduction to physical geography.* New York.

TAAFFE, E. J. and H. L. GAUTHIER (1969) *Geography of transportation.* Englewood Cliffs, N.J.

TAAFFE, E. J., R. L. MORRILL, and P. R. GOULD (1963). Transport expansion in underdeveloped countries: a comparative analysis. *Geographical Review*, **53**, 503–29.

TANNER, J. C. (1961). Factors affecting the amount of travel. *Department of Scientific and Industrial Research, Road Research Technical Paper*, **51.**

TANNER, J. C. (1966). A theoretical model for the design of a motorway system. *Ministry of Transport, Road Research Laboratory Report*, **23.**

TANNER, J. C. (1967A). Hexagonal motorway networks. *Ministry of Transport, Road Research Laboratory, Reports*, **LR78.**

TANNER, J. C. (1967B). Layout of road systems on plantations. *Ministry of Transport, Road Research Laboratory Report*, **LR68.**

TANNER, J. C., H. D. JOHNSON, and J. R. SCOTT (1962). Sample survey of the roads and traffic of Great Britain. *Road Research Laboratory, Technical Papers*, **62.**

TAYLOR, A. B. and H. E. SCHWARZ (1952). Unit-hydrograph lag and peak flow related to basin characteristics. *Transactions of the American Geophysical Union*, **33**, 235–46.

THAKER, T. R. and A. E. SCHEIDEGGER (1968). A test of the statistical theory of meander formation. *Water Resources Research*, **4**, 317–29.

THOMPSON, D'ARCY W. (1917, 1942). *On growth and form.* Cambridge.

THROWER, N. J. (1966). *Original survey and land subdivision.* Chicago.

TIMBERS, J. A. (1967). Route factors in road networks. *Traffic Engineering and Control*, **9**, 392–4, 401.

TINBERGEN, J. (1957). The appraisal of road construction: two calculation schemes. *Review of Economics and Statistics*, **39.**

TÖPFER, F. (1967). Die Ausnetzung des Wurzelgesetzes bei der Darstellung und Generalisierung von Wasserlaufen. *Petermanns Geographische Mitteilungen*, **111**, 242–54.

VANCE, J. E., JR. (1961). The Oregon Trail and the Union Pacific Railroad: a contrast in purpose. *Annals of the Association of American Geographers*, **51**, 357–79.

WALLACE, W. H. (1958). Railroad traffic densities and patterns. *Annals of the Association of American Geographers*, **48**, 352–74.

WARD, J. H. (1963). Hierarchical grouping to optimize an objective function. *Journal of the American Statistical Association*, **58**, 236–44.

WARNTZ, W. (1957). Transportation, social physics, and the law of refraction. *Professional Geographer*, **9**, 2–7.

WARNTZ, W. (1961). Transatlantic flights and pressure patterns. *Geographical Review*, **51**, 187–212.

WARNTZ, W. (1965). A note on surfaces and paths and applications to geographical problems. *Michigan Inter-University Community of Mathematical Geographers, Discussion Papers*, **6**.

WARNTZ, W. (1966). The topology of a socio-economic terrain and spatial flows. *Regional Science Association, Papers*, **17**, 47–61.

WARNTZ, W. (1968). A note on stream ordering and contour mapping. *Harvard Papers in Theoretical Geography*, **18**, 1–30.

WASIUTYNSKI, Z. (1959). *O ksztaltowaniu ukladow komunikacyjnych*. Warsaw.

WEAVER, J. B. and S. W. HESS (1963). A procedure for non-partisan districting: development of computer techniques. *Yale Law Journal*, **73**, 288–308.

WEBB, W. A. (1955). Analysis of the Martian canal network. *Astronomical Society of the Pacific, Proceedings*, **67**, 283–92.

WELLINGTON, A. M. (1887). *The economic theory of the location of railways*. New York.

WERNER, C. (1966). Zur Geometrie von Verkehrsnetzen: die Beziehung zwischen raumlicher Netzgestaltung und Wirtschaftlichkeit. *Abhandlungen des 1. Geographischen Institut der Freien Universitat Berlin*, **10**.

WERNER, C. (1967). *The role of topology and geometry in optimal network design*. Unpublished MS.

WERNER, C. (1968A). The law of refraction in transportation geography: its Multivariate extension. *Canadian Geographer*, **12**, 28–40.

WERNER, C. (1968B). Research seminar in theoretical transportation geography. *Northwestern University, Studies in Geography*, **16**, 128–70.

WERNER, C. (1969). Networks of minimum length. *Canadian Geographer*, **13**, 47–69.

WERRITY, A. (1967). The expansion of the railway network in north-east London: 1831–1967. University of Cambridge, B.A. Dissertation.

WILLIAMS, W. T. and M. B. DALE (1965). Fundamental problems in numerical taxonomy. *Advances in botanical research*, **2**.

WILSON, A. G. (1968). Models in urban planning, a synoptic review of recent literature. *Centre for Environmental Studies, Working Papers*, **3**.

WILSON, G. W., B. R. BERGMANN, L. V. HIRSCH, and M. S. KLEIN (1966). *The impact of highway investment on development*. Washington.

WISLER, C. O. and E. F. BRATER (1959). *Hydrology*. New York.

WITHEFORD, D. K. (1963). Traffic assignment analysis and evaluation. *Highway Research Board, Highway Research Record*, **6**, 1–11.

WOHL, M. and B. V. MARTIN (1967). *Traffic systems analysis*. New York.

WOLDENBERG, M. J. (1966). Horton's laws justified in terms of allometric growth and steady state in open systems. *Bulletin of the Geological Society of America*, **77**, 431–4.

WOLDENBERG, M. J. (1967). Geography and properties of surfaces. *Harvard Papers in Theoretical Geography*, **1**, 95–189.

WOLDENBERG, M. J. (1968A). Hierarchical systems: cities, rivers, alpine glaciers, bovine livers and trees. Columbia University, Ph.D. Dissertation.

WOLDENBERG, M. J. (1968B). Energy flow and spatial order: Mixed hexagonal hierarchies of central places. *Geographical Review*, **58**, 552–74.

WOLDENBERG, M. J. (1968C). Spatial order in fluvial systems: Horton's laws derived from mixed hexagonal hierarchies of drainage basin areas. *Harvard Papers in Theoretical Geography*, **13**, 1–36.

WOLDENBERG, M. J. (1969). Spatial order in fluvial systems: Horton's laws derived from mixed hexagonal hierarchies of drainage basin areas. *Bulletin of the Geological Society of America*, **80**, 97–112.

WOLDENBERG, M. J. and B. J. L. BERRY (1967). Rivers and central places: analogous systems? *Journal of Regional Science*, **7(2)**, 129–39.

WOLFE, R. I. (1961). An annotated bibliography of the geography of trans-portation. *University of California, Institute of Transportation and Traffic Engineering, Information Circular,* **29.**

WOLLMER, R. D. (1963). Some methods for determining the most vital link in a railway network. *Rand Corporation Memorandum,* **RM-3321.**

WOLMAN, M. G. (1955). The natural channel of Brandywine Creek, Pennsylvania, *United States Geological Survey, Professional Paper,* **271.**

WOLMAN, M. G. and J. P. MILLER (1960). Magnitude and frequency of forces in geomorphic processes. *Journal of Geology,* **68,** 54–74.

WONG, S. T. (1963). A multivariate statistical model for predicting mean annual flood in New England. *Annals of the Association of American Geographers,* **53,** 298–311.

WOODFORD, A. O. (1951). Stream gradients and Monterey Sea Valley. *Bulletin of the Geological Society of America,* **62,** 799–852.

YATSU, E. (1955). On the longitudinal profile of a graded river. *Transactions of the American Geophysical Union,* **36,** 655–63.

YEATES, M. (1963). Hinterland delimitation: a distance minimizing approach. *Professional Geographer,* **15(6),** 7–10.

YUILL, R. S. (1965). A simulation study of barrier effects in spatial diffusion problems. *Michigan Inter-University Community of Mathematical Geographers, Discussion Papers,* **5.**

ZERNITZ, E. R. (1932). Drainage patterns and their significance. *Journal of Geology,* **40,** 498–521.

ZOBLER, L. (1958). Decision making in regional construction. *Annals of the Association of American Geographers,* **48,** 140–8.

Further Reading

For students approaching the study of network structures within geography the extended list of references (given on pages 319–35) may prove too long to be useful. A useful starting point for physical geographers is given by LEOPOLD, WOLMAN & MILLER (1964)* or by Strahler's chapter in CHOW (1964), and for human geographers by TAAFFE & GAUTHIER (1969) or by Ullman's chapter in JAMES, JONES & WRIGHT (1954). A summary of the interactions between physical and human geography is contained in the papers edited by CHORLEY (1969).

An elementary introduction to the use of graph theory in network analysis is contained in KANSKY (1963) and ORE (1963) while more advanced treatments are available in FLAMENT (1963), BUSCAKER & SAATY (1965), and BERGE (1962). The applications of linear programming and associated optimizing techniques to networks is given in DANTZIG (1963) and FORD & FULKERSON (1962). For geographers concerned with the search for information on network flows, BLAISDELL (1964) is a mine of information with sources catalogued under modes.

The rapid rate of research makes the use of journals and reports of current research of prime importance. A review of major geographic series was made in an earlier volume (HAGGETT, 1965, pp. 327–8) and the leading journals listed there carry an increasing number of significant papers; geographic series that have risen in importance since that list was drawn up are the CANADIAN GEOGRAPHER (Quarterly), GEOGRAPHICAL ANALYSIS (Quarterly), REGIONAL STUDIES (Quarterly) and the NORTHWESTERN UNIVERSITY STUDIES IN GEOGRAPHY (Occasional). Papers in network analysis are however published in such a wide range of journals that it is necessary to move beyond this list. Physical geographers will wish to keep in contact with WATER RESOURCES RESEARCH (Quarterly) and the BULLETIN OF THE INTERNATIONAL ASSOCIATION OF SCIENTIFIC HYDROLOGY (Quarterly), while human geographers should have access to the published reports of the ROAD RESEARCH LABORATORY (Occasional) and of the United States's HIGHWAY RESEARCH BOARD (Occasional). The JOURNAL OF TRANSPORT ECONOMICS AND POLICY (Quarterly) is one of a number of recently established journals in the transport field which carries papers of interest; while algorithms for the solution of network problems are published in OPERATIONS RESEARCH (Quarterly).

* For full titles see list of references preceding this section

Locational Index

References to places mentioned in the text are listed under their specific location. Countries or equivalent units are given in small capitals (e.g. UNITED STATES) and stream systems, regardless of size, are given in italics (e.g. *Brandywine Creek*).

General Index

In view of the detailed lists of contents and the extensive use of cross-references the index to the subject matter of the text is abbreviated with major topics in small capitals (e.g. SIMULATION). References to authors are given in italics (e.g. *P. Abramson*); the index lists the names of all authors in multiple-author works and extends to cover citations in the reference list (pp. 319–35).